The Leishmaniases: Old Neglected Tropical Diseases

Fabrizio Bruschi • Luigi Gradoni
Editors

The Leishmaniases: Old Neglected Tropical Diseases

Springer

Editors
Fabrizio Bruschi
Department of Translational Research
University di Pisa
Pisa
Italy

Luigi Gradoni
Department of Infectious Diseases
Istituto Superiore di Sanità
Roma
Italy

ISBN 978-3-319-72385-3 ISBN 978-3-319-72386-0 (eBook)
https://doi.org/10.1007/978-3-319-72386-0

Library of Congress Control Number: 2017964487

Printed on acid-free paper

This Springer imprint is published by Springer Nature
The registered company is Springer International Publishing AG
The registered company address is: Gewerbestrasse 11, 6330 Cham, Switzerland

Preface

Leishmaniasis is an ancient disease, and indeed some lesions suggesting Old World cutaneous leishmaniasis were already described in tablets of the library of King Ashurbanipal from seventh century BC, while others were probably derived from earlier texts dating 1500 to 2500 BC.

Leishmaniasis is a worldwide, high-burden vector-borne disease with diverse clinical manifestations caused by protozoa belonging to the *Leishmania* genus. Different species of leishmaniasis have re-emerged in recent years with increasing global prevalence over a wide geographic range. A number of factors, including environmental, demographic and human behaviour, have contributed to the changing epidemiology of the disease and to its recent spread throughout the world.

The aim of this book is to update the readers interested in the different aspects of this ancient disease, included among the neglected tropical diseases by the World Health Organization (WHO) in 2010. Three years later, the 66th WHO General Assembly approved a resolution that represented an important milestone for the prevention, control, elimination and eradication of the disease.

In this book, various aspects are considered such as taxonomy of the different *Leishmania* species, vector biology, host immune response, immunopathological processes led by the parasite, diagnosis, clinical picture in both immunocompetent and immunocompromised patients, treatment of tegumentary and visceral forms and, finally, control perspectives.

Several eminent scientists in the field of parasitology have collaborated in the volume by providing an overview of topical issues on the different aspects of leishmaniasis.

In Chap. 1 Gradoni introduces the epidemiology of leishmaniasis, in order to help the reader to better understand the analyses of the more specialised chapters that will follow.

The chapter by Maurício is devoted to the rather complex taxonomy of the *Leishmania* genus, which includes at least 39 species that may differ highly not only in the ability to infect vertebrate hosts and vectors but also in the resulting clinical picture in humans. According to the author, as a result of recent molecular data and phylogenetic analyses, a simplification of *Leishmania* taxonomy could be possible by reducing the number of human-pathogenic species to six: *L. donovani*, *L. major*, *L. tropica* and *L. mexicana* within the sub-genus *L.* (*Leishmania*) and *L. braziliensis* and *L. guyanensis* within *L.* (*Viannia*). A consensus on *Leishmania* taxonomy should be reached among different scientists, especially clinicians and researchers, so as to avoid confusion and facilitate ease of interpretation.

In the chapter on vectors, Dvorak, Shaw and Volf describe finely the species of sandflies (Diptera: Psychodidae, Phlebotominae) and focus on their geographical distribution in both the Old and New World. Particular attention is given to the various mechanisms adopted by the parasite in order to survive in the vector's gut, and in the immunomodulatory molecules of sandfly saliva, which play a crucial role for successful transmission and infection in the vertebrate host. Both the classical phlebotomine vectors and the non-phlebotomines such as midges (recently suspected to transmit infection in some geographical regions like Australia, Martinica or Thailand) may fulfil a significant role.

In the chapter on the reservoirs of the parasite, Maia, Dantas-Torres and Campino consider different aspects of zoonotic and anthroponotic transmission, focusing in particular on the large number of host species that may be responsible for zoonotic transmission in the different geographical areas. Future research should be concentrated on food sources, breeding season, movement and migration activities and longevity of the potential reservoir host(s).

The immune response against *Leishmania* is a multifactorial process comprehensive of several components having different roles in the transmission chain since the bite of the sandfly in its very early phase of the infection. The molecules present in the vector's saliva trigger the initial inflammatory response up to the onset of the disease, where immunopathological phenomena such as autoimmune reactions occur.

Lauthier and Korenaga describe the most important mechanisms of the host immune response to the parasite, which differ between cutaneous, mucocutaneous and visceral forms of leishmaniasis. A first line of intervention is represented by neutrophils, which can kill the parasite by means of different mechanisms including NETosis (death of the cell by the release of nuclear extracellular traps). Particular focus is placed on the regulation of immune response, involving both CD4+ and CD8+Th1 cells. On the other hand, Th2 polarisation is responsible for the exacerbation of the disease.

In Chap. 6 Rojelio considers the clinical aspects of leishmaniasis, which encompass a spectrum of signs and symptoms, from nodular or ulcerative lesions occurring in cutaneous leishmaniasis (in some cases evolving to mucocutaneous leishmaniasis) to disseminated syndromes known as diffuse cutaneous leishmaniasis to visceral leishmaniasis. These different evolutions depend mostly on the infecting *Leishmania* species but also on the immunological status of the host.

A diagnosis of leishmaniasis is arduous, as it reflects the complexity of the disease. However, Gramiccia and Di Muccio have been able to elucidate the most important laboratory tools that make the parasitological diagnosis essential, in order to correctly identify the causes of such an extremely multifaceted clinical picture. In addition to the classic microscopical, immunoparasitological and more advanced molecular analyses, the *in vitro* study of cellular immunity is promising, not only in visceral but also in cutaneous leishmaniasis. However, further studies in different geographical regions are needed to achieve more accurate results and to confirm these data.

Begoña Monge-Maillo and Rogelio López-Vélez discuss the most recent acquisitions concerning the treatment of visceral leishmaniasis, by differentiating the different choices, which depend on the geographical region of the infection and on the immunological status of the host. In consideration of the development of resistance to the traditional antimonials, amphotericin B deoxycholate or lipid formulations of amphotericin B are indicated as first-line treatment for their efficacy and lower toxicity in different world areas, in particular in the Mediterranean region. However, in low-income countries, the possibility of using parenteral paromomycin should be encouraged. The promising orally administered miltefosine drug, used mostly in India, Pakistan and Bangladesh, has shown very good cure rates, but clinical failures are still relevant and probably depend on the increased development of parasite resistance.

Blum, Neumayr and Lockwood have reviewed the criteria for selecting a reliable method of treatment for tegumentary leishmaniasis, mainly determined by the infecting *Leishmania* species. Treatment options include systemic treatment with antileishmanial drugs, local topical treatment with antileishmanial ointments/creams, local intralesional injection of antileishmanial drugs and local physical treatment (cryotherapy, thermotherapy). The choice of systemic or local treatment depends on the species, size, number and location of the lesions but also changes in reason of possible comorbidities. Treatment suggestions for cutaneous leishmaniasis are presented for each species. The treatment of mucosal leishmaniasis and leishmaniasis in pregnant women, in children and in patients with immunosuppression is discussed separately.

According to Boelaert, Burza and Romero, the control of leishmaniasis depends on a limited number of key control strategies, the most important of which are early diagnosis and case management, and limit of transmission of the disease, at least in its anthroponotic form. However, sandfly control measures (insecticide spraying, insecticide-treated materials, environmental management and personal protection) are also useful to reduce transmission, especially when the procedures are conducted near human dwellings or are well addressed to vector exposure. In case of zoonotic transmission of the disease, reservoir hosts need to be controlled. Unfortunately, no human vaccine is yet available and those commercialised against canine leishmaniasis have not shown good efficacy in reducing the transmission to humans.

Pisa, Italy — Fabrizio Bruschi

Contents

A Brief Introduction to Leishmaniasis Epidemiology

1

Luigi Gradoni

1.1 Introduction

The leishmaniases are a group of human diseases caused by kinetoplastid protozoa of the genus *Leishmania*. Despite having a long history, dating back from more than 4500 years according to ancient descriptions (Akhoundi et al. 2016), leishmaniasis still ranks in the top three of the neglected tropical diseases caused by protozoa (Fenwick 2012). Why is that? Certainly not because of a negligible disease burden. A total of 98 countries and three "territories"—a United Nations definition for lands that do not possess full political sovereignty—in tropical, subtropical and temperate areas have reported endemic transmission (Alvar et al. 2012). With 350 million people considered at risk, some 1.3 million new cases of leishmaniasis are estimated to occur every year that means the ninth largest disease burden among infectious diseases and the second and fourth most common cause of death and disease, respectively, among tropical infections (Bern et al. 2008). Over the past 20 years, significant progress has been made for the improvement of diagnosis and treatment of the leishmaniases; these developments should have facilitated the implementation of sustainable national and regional control programs. However, we are a long way from it, and mortality and morbidity from leishmaniasis still show a worrying increasing trend worldwide.

There is not a single explanation of why the leishmaniases are largely ignored globally. Apart from the fact that they affect "the poorest of the poor" in most of their wide geographical range—a common denominator of neglected tropical diseases—the most probable reason for neglecting leishmaniasis, in the author's view, is because of its epidemiological and medical complexity. Crucial factors include (a) the wide genetic diversity of *Leishmania* parasites; (b) complex interactions

L. Gradoni
Unit of Vector-Borne Diseases, Department of Infectious Diseases,
Istituto Superiore di Sanità, Roma, Italy
e-mail: luigi.gradoni@iss.it

F. Bruschi, L. Gradoni (eds.), *The Leishmaniases: Old Neglected Tropical Diseases*, https://doi.org/10.1007/978-3-319-72386-0_1

among vertebrate and invertebrate natural hosts and their environment; (c) marked regional differences in clinical features; (d) and widely varying human-associated risk factors.

1.2 Epidemiological Complexity and Neglect of Leishmaniasis

At least 20 recognized *Leishmania* species are pathogenic to humans (Akhoundi et al. 2017), most of which can also be transmitted to and cause established infections in mammal species from at least eight orders. The increasing use of diagnostic molecular tools has considerably extended the list of mammals found positive for leishmanial genetic material, thereby making harder a clear distinction between dead-end and genuine reservoir hosts—the latter being mammal species capable of ensuring long-term maintenance of a *Leishmania* population and its taking up by a vector (Quinnell and Courtenay 2009).

All *Leishmania* members pathogenic to humans are transmitted by the bite of phlebotomine sand flies. Among over 900 species of sand fly recorded, about 100 are proven or suspected vectors of human leishmaniases; these include *Phlebotomus* species in the Old World and *Lutzomyia* species (*sensu* Young and Duncan 1994) in the New World (Maroli et al. 2013). According to the taxonomical revision proposed by Galati (2015), phlebotomines potentially implicated in Neotropical *Leishmania* transmission belong to 13 genera at least. The list of sand fly species found to harbour leishmanial DNA in nature has also increased considerably in the past few years. Assessing vectorial importance of a sand fly species in a specific endemic scenario can be a complex task as several biological, behavioural and ecological variables need to be considered (Ready 2013).

Clinical manifestations of human leishmaniasis are also largely diverse, although two main clinical forms are prevalent worldwide: visceral leishmaniasis (VL), a life-threatening condition that results from the dissemination of *Leishmania* in macrophage-rich tissues, and cutaneous leishmaniases (CL), a benign but often disfiguring skin condition which has a tendency towards spontaneous resolution. Other acute or chronic clinical forms may be less common or simply get unrecognized. They consist of primary conditions or sequelae of the main clinical forms and include localized leishmanial lymphadenopathy, localized mucosal leishmaniasis (such as laryngeal or lingual leishmaniasis), mucocutaneous leishmaniasis (MCL), diffuse and disseminated CL and post-kala-azar dermal leishmaniasis (PKDL). Asymptomatic human infections are increasingly recorded by means of modern diagnostics; however, their epidemiological role in different endemic settings has not been elucidated (Michel et al. 2011; Hirve et al. 2016).

As a consequence of such medical complexity, universal tools and guidelines for leishmaniasis case management—from clinical suspicion to post-therapy follow-up—are lacking or incomplete, and where available, they can only be applicable to regional contexts (see, for example, Gradoni et al. 2017). Case reporting to current surveillance systems, therefore, is inadequate so that accurate information on leishmaniasis extent and distribution is incomplete or missing. Hence, leishmaniasis is not recognized and prioritized at the public health policy level, and its visibility is eventually not proportionate to its true burden.

Similarly, the complexity of eco-epidemiological interactions among reservoir hosts, phlebotomine vectors and their environment makes it difficult to implement national programs addressed to *Leishmania* transmission control. A schematic approach aimed at targeting epidemiological cycles with appropriate control measures was proposed by the World Health Organization (WHO) in 1990 and reiterated in 2010 (WHO 1990, 2010). Each pathogenic *Leishmania* agent associated with specific vectors and reservoir hosts (if known) and with predominant clinical manifestations in humans was mapped geographically and grouped into the so-called "nosogeographical entities." Eleven entities were listed in 1990, which increased to 15 in 2010. Some of them are labelled as "anthroponotic," thus providing an informative epidemiological tag within the entity definition, for example, "anthroponotic VL (AVL) caused by *L. donovani* in the Indian subcontinent" or "anthroponotic CL (ACL) caused by *L. tropica* in Old World countries." However the latter entity had to be limited to "densely populated settlements," since a zoonotic cycle of *L. tropica* was later demonstrated in rural/peri-urban settings of Middle East. Some entities are traditionally labelled as "zoonotic," such as "zoonotic CL (ZCL) caused by *L. major* in the Old World" or "zoonotic VL (ZVL) caused by *L. infantum* in the Old and New Worlds"; others are not, despite having an obvious zoonotic nature such as all the entities of tegumentary leishmaniasis in Latin America. Two or more nosogeographical entities can be endemic in one country (e.g. four entities are found in parts of the Maghreb area, North Africa), thus making the implementation of national control programs even more complex.

1.3 Determinants of Leishmaniasis Epidemiology

In 2010, the members of the WHO Expert Committee on the Control of Leishmaniases attempted to provide a consolidated view of the epidemiological determinants impacting on patterns and trends of human leishmaniasis worldwide (WHO 2010). A systematic list of such determinants is difficult to make, as several of them are deeply interlinked, for example, "poverty-migration-state of immunity." This section makes large use of the epidemiological definitions reported in the above WHO document.

First of all, it is important to note that unlike other neglected vector-borne diseases (e.g. American trypanosomiasis), leishmaniasis transmission occurs almost exclusively via vector. Even though congenital, blood transfusion or mechanical modes of transmission (e.g. syringe) have been reported, they are definitely rare events. Intrinsic tropism and virulence of *Leishmania* species, ecological characteristics of the transmission sites and widely varying human-associated risk factors (Desjeux 2001) are the main epidemiological determinants.

1.3.1 Age, Sex and Acquired Immunity

With some exceptions, parasites of the *L. donovani*/*L. infantum* complex exhibit predominant tropism to internal tissues and cause VL, whereas all other *Leishmania* species are predominantly dermotropic and cause tegumentary clinical manifestations in humans. In VL, the susceptibility of naive individuals varies greatly as not everyone exposed to transmission develops a typical fatal disease. In endemic areas, asymptomatic or subclinical cases outnumber far in excess clinical cases. The age range affected by clinical VL depends on the parasite species and the history of population exposure. Where the causative parasite is *L. infantum* and the disease incidence is elevated, VL patients tend to be younger (the most affected age group being 0–5 years) than in endemic foci of *L. donovani* in various sites from Asia and Africa (median age, 13–23 years).

By contrast, it seems that naive individuals are universally susceptible to clinical disease caused by the commonest agents of CL. The affected age range depends on the intensity of transmission ("force of infection") to which populations are exposed (Lysenko and Beljaev 1987).

The main entities of leishmaniasis are reported to affect more males than females. This may have different explanations, such as higher exposure of males to sand fly bites because of work or social activities; gender disparity in access to health care, especially in less developed countries; and sex-linked biological factors associated with natural immune responses to parasites. At least for VL, the latter could be the most probable, as a similar M:F ratio of about 1.5:1 was reported by active-case detection studies performed in quite different endemic settings of VL and social contexts, namely, two southern European countries endemic for *L. infantum* and a group of villages in Bihar state, India, endemic for *L. donovani* (Gradoni et al. 1996; Arce et al. 2013; Siddiqui et al. 2016).

In populations exposed to high levels of transmission over long periods, a large proportion of adults will have acquired immunity. In VL foci, this is indicated by the prevalence of positive leishmanian skin-test results, which typically rises with increasing age. Immunosuppressive conditions, either due to comorbidities (e.g. HIV infection) or therapies (e.g. organ transplantation or treatment of immunological disorders) may result in increased clinical susceptibility to primary infections or in the reactivation of latent infections.

In CL foci, the presence of old scars in exposed individuals is usually associated with the protection from newly acquired lesions. Actually, the deliberate syringe infection with *L. major* in covered parts of the body ("leishmanization") has been used for years as a vaccination against infections causing potential disfiguring lesions, and still it represents the gold standard for the development of modern *Leishmania* vaccines (Khamesipour et al. 2005). In general, acquired immunosuppressive conditions have a lower clinical impact in tegumentary forms of leishmaniases.

1.3.2 Human Behaviour

In AVL foci of Bangladesh, India and Nepal, the disease occurs in agricultural villages where houses are frequently built with mud walls and earthen floors. Cattle is kept close to human dwellings, and the sand fly vector *P. argentipes* feeds on both bovines and humans; consequently, sleeping outside or on the ground increases VL risk (Ranjan et al. 2005).

ZVL foci in the Mediterranean littoral and in Latin America consist of rural areas, villages and urban districts with vegetation, where many people have dogs, the domestic reservoir host for *L. infantum*. In both endemic scenarios, the location of other domestic animals close to the main habitation can be an additional risk factor because they attract sand flies and permit their resting. Transmission takes place micro-focally due to the opportunistic behaviour of *Phlebotomus* (*Larroussius*) species and *Lutzomyia longipalpis* in the respective endemic settings (Velo et al. 2017; de Araújo et al. 2013).

In Old World ZCL (*L. major*), the risk for infection is increased when agricultural projects are launched and irrigation systems are extended into arid or semi-desert areas. Usually, these man-made ecological changes are accompanied by the intrusion of large numbers of non-immune immigrants into an existing sylvatic cycle sustained by rodents and the vector *P. papatasi*. Transmission to humans is favoured by the practice of sleeping outside during the hot season.

Geographical range and risk factors for New World tegumentary diseases have expanded during the last decades. Traditionally, leishmaniases were predominantly occupational diseases related to specific activities such as rubber tapping, military operations, road construction and new agricultural development in the forest and other enzootic areas. Recent massive deforestation has led to a rapid increase in CL cases due to peridomestic, peri-urban and even urban transmission. Skin lesions caused by different *Leishmania* species are increasingly recorded in ecotourists who spend their holidays in nature reserves without appropriate protections against sand fly bites.

1.3.3 Poverty, Malnutrition and Migration

Poverty increases the risk for leishmaniasis by many mechanisms and can potentiate morbidity and mortality of the disease (Alvar et al. 2006). In endemic areas, poor housing and bad peridomestic sanitary conditions (cattle manure, garbage collection or open sewerage) are ecological factors favouring sand fly breeding and resting, resulting in high vector densities and easier access to human blood. Small rooms crowded with a large number of people may attract anthropophilic sand flies by providing a large biomass for blood meals and facilitate inter-human transmission in anthroponotic entities of leishmaniasis. Sylvatic cycles may approach poor dwellings at the periphery of new settlements in deforested areas or irrigated arid zones.

Poverty is often accompanied by poor nutritional status, including deficiency in protein energy, iron, vitamin A and zinc, which was shown to increase the likelihood that VL infections rapidly progress to fatal disease in animal models (Anstead et al. 2001). Protein-energy malnutrition has also been associated with an increased risk for MCL. Aggravation of clinical leishmaniasis is not only a consequence of immunological deficits and concurrent infections frequently associated with poverty but also of the lack of resources to pay for care or have access to remote health-care facilities.

Poverty-driven migration is a main cause of epidemics of both VL and CL worldwide. In East Africa, AVL outbreaks associated with repeated seasonal labour movements have occurred in Ethiopia, in a territory where the disease was previously absent or reported sporadically (Herrero et al. 2009). Drought and poverty have forced people to move from Darfur into VL-endemic areas of the Gedaref State in Sudan, resulting in a high number of fatal cases. Agricultural development and construction of infrastructures in North Africa and the Arabian peninsula have caused ZCL epidemics among non-immune young migrants. In the Andean countries, the massive migration of non-immune populations from the highland to the Amazonian basin resulted in one-fourth of the population with CL lesions within a few years (Desjeux 2001).

1.3.4 Military Operations and Population Displacement

Thousands of ZCL and ACL cases have occurred in soldiers from the United Kingdom, USA and other NATO military forces serving in Iraq and Afghanistan wars since the 2000s. In Colombia, military operations in the forest resulted in more than 45,000 cases of CL in soldiers between 2005 and 2010. In both military operations and civil war scenarios, however, civilians are those most affected by leishmaniasis because of the destruction of health-care infrastructures (lack of diagnosis, shortage of drugs and interruption of vector control activities) and forced displacement of populations. The latter caused one of the biggest epidemic of VL ever reported, with an impressive fatality rate (100,000 deaths) when a civil war in southern Sudan forced people to move into VL-endemic territories in the western Upper Nile. Originally, the disease was probably zoonotic and caused sporadic disease in that area, but the massive displacement of non-immune people gave origin to active anthroponotic transmission of *L. donovani* among 280,000 individuals in the 1980s and 1990s (Seaman et al. 1996). The migration of Afghan refugees into southern Pakistan is thought to have introduced ACL into areas where it had not previously been known. An investigation, however, found similar age-related pattern of ACL endemicity in both Afghan refugee camps and neighbouring Pakistani villages (Brooker et al. 2004). The ongoing Syrian war is having a dramatic impact on the incidence of CL among the population displaced internally (about 6.5 million) and in countries of Middle East and Africa (Turkey, Jordan, Lebanon, Iraq and Egypt) where some four million Syrians were forced to seek refuge. Before the war, the annual incidence of CL (including anthroponotic and zoonotic entities) within Syria was estimated to be 23,000 cases (2004–2008). In 2012, 53,000 cases were reported, and in 2013, approximately 100,000 cases were estimated. In southeast Anatolia, Turkey, which is endemic for ACL, there is an indication of increased incidence of the disease among the Turkish population in relation with the entry of refugees from Syria. CL has emerged in Lebanon as well, although outbreaks have been largely contained to refugee populations (Du et al. 2016).

1.3.5 Human-Made Environmental Changes

Examples of environmental changes that can affect the incidence of leishmaniasis have been mentioned in previous sections. Most significant changes include urbanization and the incursion of agricultural farms and settlements into arid or forested natural areas, which may cause domestication and expansion of the natural transmission cycles. Latin American countries, particularly Brazil, have experienced epidemics of ZVL in the fast-growing peripheries of large- and medium-sized cities. In some epidemiological contexts, however, the destruction of natural habitats can reduce or even eliminate transmission of leishmaniasis, and such approach is commonly listed among the available tools for ZCL control. This is the case of the destruction of burrows and vegetation habitats of rodent population's natural reservoir of *L. major*, i.e. *Rhombomys opimus* in countries of Central Asia and *Psammomys obesus* in parts of northern Africa and Middle East. Rapid recolonization by rodents and vegetation is a common occurrence when environmental changes are not permanent because unsustainable economically; furthermore, a shift from sylvatic to synanthropic rodents acting as primary *L. major* reservoir (e.g. *Meriones*) may occur. In parts of Latin America, forest-free zones created for cattle breeding resulted in a buffer zone between the sylvatic cycle and human housing and reduced the transmission of CL. In other parts, however, deforestation appears to have led to an increase in the incidence of CL through a shift to a peridomestic transmission cycle.

A recent example of small-scale environmental change which caused to the largest community outbreak of leishmaniasis in Europe is that of Fuenlabrada, a city at the south of Madrid, in 2009–2012 (Arce et al. 2013). The area had long been known to be endemic for *L. infantum*, with sporadic human cases and a low prevalence in dogs. The outbreak peaked in 2011 and involved some 400 cases, of which one-third were VL and two-thirds CL. An investigation showed that most patients lived in proximity of a new green park afforested since 2005, enclosed by roads and other infrastructures recently built. A population of hares grew abnormally by colonizing the park in the absence of natural predators and was unable to migrate. A new sylvatic/peridomestic cycle of leishmaniasis was thus established, having hares as a major blood source for the local competent vector *P. perniciosus* and becoming an important reservoir of *L. infantum* (Molina et al. 2012). Interestingly, the susceptibility of wild hares to natural *L. infantum* infections was reported from Tarim Basin, China, in the same year the Fuenlabrada outbreak began (Liao et al. 2009).

1.3.6 Climate Change

As with other vector-borne infections, leishmaniasis is considered a climate-sensitive disease being potentially affected by changes in rainfall, temperature and humidity. Current evidence and predicted scenarios of global warming were recently made available at regional scale by the Intergovernmental Panel on Climate Change (IPCC 2014). A mere cross-check of prediction temperature maps with current distribution maps of the leishmaniases strongly suggests that climate changes are expected to affect greatly the epidemiology of the disease by a number of mechanisms. Changes in average annual temperature and humidity can have direct effects on the ecology of sand flies (poikilothermic invertebrates) and, to a lesser extent, of mammal reservoirs (homeothermic vertebrates) by altering their geographical distribution and influencing their survival, generation time and population sizes. Temperature changes may have a direct effect on the developmental cycle of *Leishmania* in sand flies in some specific parasite/vector associations; however, recent experimental evidence suggests that parasite metacyclogenesis in *L. braziliensis* and *L. infantum* is not inhibited when the respective vectors are exposed to a broad range of temperatures compatible with sand fly survival (Hlavacova et al. 2013). Indirect impacts such as drought, famine and flood resulting from changes in climate conditions could lead to massive displacement and migration of people to areas with transmission of leishmaniasis, and poor nutrition could compromise their immune resistance to parasites.

The close relation among climate conditions and sand fly seasonality is well documented (Alten et al. 2016); however, only a few studies have attempted to link interannual fluctuations in the incidence of leishmaniasis to climate cycles. In Colombia, Panamá and Brazil, sand fly densities and CL incidences were found associated with El Niño Southern Oscillation indices (Chaves et al. 2014; Ferreira de Souza et al. 2015). Rainfall and humidity variations were found positively correlated with the rise of ZCL incidence in Tunisia, with a likely effect on the increase in vegetation density followed by the expansion of the rodent reservoir populations (Toumi et al. 2012).

An effect of long-term climate change on the geographical expansion of leishmaniasis was shown in a northern Mediterranean region. Starting from 1990, areas from northern continental Italy traditionally considered free from *Leishmania* transmission had become focally endemic for *L. infantum*. This was shown through active investigations performed in 2003–2006 involving canine serosurveys and phlebotomine sand fly monitoring, which demonstrated the de novo colonization by competent vector species of territories at the foot of the Alps which had been found sand fly negative some 30 years earlier (Maroli et al. 2008).

1.3.7 Quantitative Epidemiology and Mathematical Models

Over the past 25 years, quite a bit of new information on leishmaniasis epidemiology, including the identification of a wide range of risk factors and ecological predictors, has been obtained by computer-assisted mathematical models and statistics. Multivariate logistic regression models have been extensively used to identify environmental, social or clinical parameters associated with significant leishmaniasis risk. They include, for example, (1) likelihood of sand fly and *Leishmania* transmission occurrence in territories (Alcover et al. 2014; Chalghaf et al. 2016); (2) biological factors associated with clinical outcome of VL, e.g. severe presentation vs asymptomatic condition (Dos Santos Marques et al. 2017); (3) various parameters related with likelihood of occurrence of sequelae, namely, PKDL after cured VL (Uranw et al. 2011) or MCL after primary *L.* (*Viannia*) lesions (Machado-Coelho et al. 2005) or treatment failure and lethality in VL (Coura-Vital et al. 2014; Welay et al. 2016).

Potential geographical distribution of *L. major* vectors and ZCL in Tunisia was assessed by an ecological niche modelling based on georeferenced *P. papatasi*-positive trapping sites and ZCL-positive households. Several environmental layers were used as predictor variables to model both parameters, and the disease risk was predicted in territories that were not sampled (Chalghaf et al. 2016). A similar approach, which however did not include vector information but only human cases, was used to predict the potential distribution of AVL in Bangladesh (Abdullah et al. 2017).

The scarcity of quantitative data on many aspects of the natural history of *Leishmania*, especially on the parameters relating to the vector (which is in sharp contrast to the vectors of malaria), makes it impossible to develop reliable mathematical models of transmission. Development of less ambitious submodels of parts of the transmission cycle, wherever the data are available, is however possible. The pioneering studies carried out by Lysenko and Beljaev (1987) are a good example. An epidemetric model was constructed to describe the acquisition of *L. major* infection by humans and accumulation of ZCL cases in human communities, leading to the first application of the "force of infection" concept in leishmaniasis and to the prediction of non-immune population rates based on this parameter. Unique features of ZCL caused by *L. major* made it particularly easy to "feed" the model because humans are not essential to the transmission, mortality is not affected by ZCL, human susceptibility is considered universal and diagnosis is fairly easy—including retrospective diagnosis based on scars.

Much research is needed to fill the entomological gaps in mathematical models describing *Leishmania* transmission. The missing information includes the life expectancy of infected female sand flies; gonotrophic concordance versus discordance; the natural period of ovipositional cycles; the frequency of feeding on humans and on reservoir hosts, and on animals that play no part in transmission. Mathematical models of the transmission of anthroponotic forms of leishmaniasis would be simpler to construct than models of the transmission of zoonotic forms, which are complicated by the need to consider animal reservoir hosts.

Despite the above limitations, an interesting attempt was made to calculate the average probability of introduction and establishment of canine leishmaniasis (CanL) in a theoretical dog network, assuming the presence of competent vectors in CanL-free areas and the impact of mitigation measures applied to dogs in order to prevent such events (EFSA 2015). The model assessed the average probability of disease establishment, defined as the local *Leishmania* transmission from vector to host and vice versa, leading to the presence of at least one indigenous infectious host and at least one indigenous infectious vector. By "feeding" the model with incomplete parameters largely based on data from reared sand flies colonies and personal observations (which were used to estimate a vectorial capacity index), the probability of disease establishment was found very high. Results from the model indicated that the probability of leishmaniasis introduction and establishment could be reduced by mitigation measures, the most effective being the use of topically applied insecticides.

1.3.8 Geographical Information Systems

Geographical information systems (GIS) are computer-based techniques used to input, analyse, manipulate and output geographically spatial data. They are important tools for understanding the epidemiology of the leishmaniases worldwide and can be adapted to the needs of endemic countries and geographical regions. Combined with surveillance data, GIS can be used routinely to generate base maps, delineate the distribution of vectors and reservoir hosts and prepare maps based on leishmaniasis incidence. Combined with relevant environmental and social correlates available at subnational location details, such as rainfall, temperature, land cover, poverty, urbanization, etc., GIS modelling can help to infer the likelihood of leishmaniasis being present at other locations within the country or across a region (Pigott et al. 2014). Therefore, GIS can facilitate decision-making and support strategic planning for resource allocation and smart application of control measures against leishmaniasis.

1.4 Perspectives

The high morbidity and mortality from leishmaniasis in many endemic zones of the world is unacceptable, given that affordable and effective tools for diagnosis and treatment are available (WHO 2010). However the epidemiological and medical complexity of the leishmaniases is likely to increase, as suggested by the discovery of new *Leishmania* parasites and natural cycles involving humans; the dynamic evolution of transmission foci, with newly identified hosts and geographical spreading; the impact of global phenomena due to anthropogenic factors such as climate change, migrations and conflicts and the increase of susceptible population because of iatrogenic factors, comorbidities or aging. Priority efforts should be made to improve the living conditions of populations in endemic areas because severe leishmaniasis is above all a poverty-related disease. Where prosperity and wellbeing have spread among the entire endemic population (e.g. in countries of southern Europe), morbidity and lethality of VL have dropped dramatically.

Despite the advent of modern technologies for leishmaniasis diagnosis and parasite typing and for the computer-assisted analysis of risk factors and predictors, there is an evident lack of traditional competences in parasitology and entomology specific to leishmaniasis epidemiology. Meaningful interpretation of epidemiological models and life cycles require a broader view and experience in different entities of leishmaniasis, which means deep knowledge of various ecological and social realities.

References

Abdullah AYM, Dewan A, Shogib MRI, Rahman MM, Hossain MF. Environmental factors associated with the distribution of visceral leishmaniasis in endemic areas of Bangladesh: modeling the ecological niche. Trop Med Health. 2017;45:13.

Akhoundi M, Kuhls K, Cannet A, Votýpka J, Marty P, Delaunay P, Sereno D. A historical overview of the classification, evolution, and dispersion of Leishmania parasites and sandflies. PLoS Negl Trop Dis. 2016;10:e0004349.

Akhoundi M, Downing T, Votýpka J, Kuhls K, Lukeš J, Cannet A, Ravel C, Marty P, Delaunay P, Kasbari M, Granouillac B, Gradoni L, Sereno D. Leishmania infections: molecular targets and diagnosis. Mol Asp Med. 2017;57:1–29.

Alcover MM, Ballart C, Martín-Sánchez J, Serra T, Castillejo S, Portús M, Gállego M. Factors influencing the presence of sand flies in Majorca (Balearic Islands, Spain) with special reference to Phlebotomus pernicious, vector of Leishmania infantum. Parasit Vectors. 2014;7:421.

Alten B, Maia C, Afonso MO, Campino L, Jiménez M, González E, Molina R, Bañuls AL, Prudhomme J, Vergnes B, Toty C, Cassan C, Rahola N, Thierry M, Sereno D, Bongiorno G, Bianchi R, Khoury C, Tsirigotakis N, Dokianakis E, Antoniou M, Christodoulou V, Mazeris A, Karakus M, Ozbel Y, Arserim SK, Erisoz Kasap O, Gunay F, Oguz G, Kaynas S, Tsertsvadze N, Tskhvaradze L, Giorgobiani E, Gramiccia M, Volf P, Gradoni L. Seasonal dynamics of phlebotomine sand fly species proven vectors of Mediterranean leishmaniasis caused by *Leishmania infantum*. PLoS Negl Trop Dis. 2016;10:e0004458.

Alvar J, Vélez ID, Bern C, Herrero M, Desjeux P, Cano J, Jannin J, den Boer M; WHO Leishmaniasis Control Team. Leishmaniasis worldwide and global estimates of its incidence. PLoS One. 2012;7:e35671.

Alvar J, Yactayo S, Bern C. Leishmaniasis and poverty. Trends Parasitol. 2006;22:552–7.

Anstead GM, Chandrasekar B, Zhao W, Yang J, Perez LE, Melby PC. Malnutrition alters the innate immune response and increases early visceralization following *Leishmania donovani* infection. Infect Immun. 2001;69:4709–18.

Arce A, Estirado A, Ordobas M, Sevilla S, García N, Moratilla L, de la Fuente S, Martínez AM, Pérez AM, Aránguez E, Iriso A, Sevillano O, Bernal J, Vilas F. Re-emergence of leishmaniasis in Spain: community outbreak in Madrid, Spain, 2009 to 2012. Euro Surveill. 2013;18:20546.

Bern C, Maguire JH, Alvar J. Complexities of assessing the disease burden attributable to leishmaniasis. PLoS Negl Trop Dis. 2008;2:e313.

Brooker S, Mohammed N, Adil K, Agha S, Reithinger R, Rowland M, Ali I, Kolaczinski J. Leishmaniasis in refugee and local Pakistani populations. Emerg Infect Dis. 2004;10:1681–4.

Chalghaf B, Chlif S, Mayala B, Ghawar W, Bettaieb J, Harrabi M, Benie GB, Michael E, Salah AB. Ecological niche modeling for the prediction of the geographic distribution of cutaneous leishmaniasis in Tunisia. Am J Trop Med Hyg. 2016;94:844–51.

Chaves LF, Calzada JE, Valderrama A, Saldaña A. Cutaneous leishmaniasis and sand fly fluctuations are associated with El Niño in Panamá. PLoS Negl Trop Dis. 2014;8:e3210.

Coura-Vital W, Araújo VE, Reis IA, Amancio FF, Reis AB, Carneiro M. Prognostic factors and scoring system for death from visceral leishmaniasis: an historical cohort study in Brazil. PLoS Negl Trop Dis. 2014;8(12):e3374.

de Araújo VE, Pinheiro LC, Almeida MC, de Menezes FC, Morais MH, Reis IA, Assunção RM, Carneiro M. Relative risk of visceral leishmaniasis in Brazil: a spatial analysis in urban area. PLoS Negl Trop Dis. 2013;7:e2540.

Desjeux P. The increase in risk factors for leishmaniasis worldwide. Trans R Soc Trop Med Hyg. 2001;95:239–43.

Dos Santos Marques LH, DA Rocha IC, Reis IA, DA Cunha GM, Oliveira E, Pfeilsticker TR, DE Araújo VE, Morais MH, Rabello A, Carneiro M. Leishmania infantum: illness, transmission profile and risk factors for asymptomatic infection in an endemic metropolis in Brazil. Parasitology. 2017;144(4):546–56.

Du R, Hotez PJ, Al-Salem WS, Acosta-Serrano A. Old world cutaneous leishmaniasis and refugee crises in the Middle East and North Africa. PLoS Negl Trop Dis. 2016;10:e0004545.

EFSA Panel on Animal Health and Welfare. Scientific opinion on canine leishmaniosis. EFSA J. 2015;13:4075.

Fenwick A. The global burden of neglected tropical diseases. Public Health. 2012;126:233–6.

Ferreira de Souza RA, Andreoli RV, Toshie Kayano M, Lima Carvalho A. American cutaneous leishmaniasis cases in the metropolitan region of Manaus, Brazil: association with climate variables over time. Geospat Health. 2015;10:314.

Galati EAB. Phlebotominae (Diptera, Psychodidae) Classificação, morfologia, terminologia e identificação de Adultos, vol. 1; 2015. p. 136. http://www.fsp.usp.br/egalati/ApostilaPhlebotominae_2015_vol_I.pdf.

Gradoni L, Scalone A, Gramiccia M, Troiani M. Epidemiological surveillance of leishmaniasis in HIV-1-infected individuals in Italy. AIDS. 1996;10:785–9.

Gradoni L, López-Vélez R, Mokni M. Manual on case management and surveillance of the leishmaniases in the WHO European region. Copenhagen: World Health Organization Regional Office for Europe; 2017.

Herrero M, Orfanos G, Argaw D, Mulugeta A, Aparicio P, Parreño F, Bernal O, Rubens D, Pedraza J, Lima MA, Flevaud L, Palma PP, Bashaye S, Alvar J, Bern C. Natural history of a visceral leishmaniasis outbreak in highland Ethiopia. Am J Trop Med Hyg. 2009;81:373–7.

Hirve S, Boelaert M, Matlashewski G, Mondal D, Arana B, Kroeger A, Olliaro P. Transmission dynamics of visceral Leishmaniasis in the Indian subcontinent—a systematic literature review. PLoS Negl Trop Dis. 2016;10:e0004896.

Hlavacova J, Votypka J, Volf P. The effect of temperature on Leishmania (Kinetoplastida: Trypanosomatidae) development in sand flies. J Med Entomol. 2013;50:955–8.

IPCC. Climate change 2014. Impacts, adaptation and vulnerability. Part B: regional aspects. New York: Cambridge University Press; 2014.

Khamesipour A, Dowlati Y, Asilian A, Hashemi-Fesharki R, Javadi A, Noazin S, Modabber F. Leishmanization: use of an old method for evaluation of candidate vaccines against leishmaniasis. Vaccine. 2005;23:3642–8.

Liao L, Shun Sheng Y, Wuso B, Min W, Bing X, Yong Z, YanYan H, Gang L. Leishmania infantum firstly isolated from Yarkand hare (Lepus yarkandensis). Chin J Vector Biol Control. 2009;20:45–7. [in Chinese].

Lysenko AJ, Beljaev AE. Quantitative approaches to epidemiology. In: Peters W, Killick-Kendrick R, editors. The leishmaniases in biology and medicine, vol. 1. London: Academic Press; 1987. p. 263–90.

Machado-Coelho GL, Caiaffa WT, Genaro O, Magalhães PA, Mayrink W. Risk factors for mucosal manifestation of American cutaneous leishmaniasis. Trans R Soc Trop Med Hyg. 2005;99:55–61.

Maroli M, Rossi L, Baldelli R, Capelli G, Ferroglio E, Genchi C, Gramiccia M, Mortarino M, Pietrobelli M, Gradoni L. The northward spread of leishmaniasis in Italy: evidence from retrospective and ongoing studies on the canine reservoir and phlebotomine vectors. Tropical Med Int Health. 2008;13:256–64.

Maroli M, Feliciangeli MD, Bichaud L, Charrel RN, Gradoni L. Phlebotomine sandflies and the spreading of leishmaniases and other diseases of public health concern. Med Vet Entomol. 2013;27:123–47.

Michel G, Pomares C, Ferrua B, Marty P. Importance of worldwide asymptomatic carriers of *Leishmania infantum* (*L. chagasi*) in human. Acta Trop. 2011;119:69–75.

Molina R, Jiménez MI, Cruz I, Iriso A, Martín-Martín I, Sevillano O, Melero S, Bernal J. The hare (*Lepus granatensis*) as potential sylvatic reservoir of *Leishmania infantum* in Spain. Vet Parasitol. 2012;190:268–71.

Pigott DM, Bhatt S, Golding N, Duda KA, Battle KE, Brady OJ, Messina JP, Balard Y, Bastien P, Pratlong F, Brownstein JS, Freifeld CC, Mekaru SR, Gething PW, George DB, Myers MF, Reithinger R, Hay SI. Global distribution maps of the leishmaniases. eLife. 2014;3:e02851.

Quinnell RJ, Courtenay O. Transmission, reservoir hosts and control of zoonotic visceral leishmaniasis. Parasitology. 2009;136:1915–34.

Ranjan A, Sur D, Singh VP, Siddique NA, Manna B, Lal CS, Sinha PK, Kishore K, Bhattacharya SK. Risk factors for Indian kala-azar. Am J Trop Med Hyg. 2005;73:74–8.

Ready PD. Biology of phlebotomine sand flies as vectors of disease agents. Annu Rev Entomol. 2013;58:227–50.

Seaman J, Mercer AJ, Sondorp E. The epidemic of visceral leishmaniasis in western Upper Nile, southern Sudan: course and impact from 1984 to 1994. Int J Epidemiol. 1996;25:862–71.

Siddiqui NA, Rabidas VN, Sinha SK, Verma RB, Pandey K, Singh VP, Ranjan A, Topno RK, Lal CS, Kumar V, Sahoo GC, Sridhar S, Pandey A, Das P. Snowball vs. house-to-house technique for measuring annual incidence of kala-azar in the higher endemic blocks of Bihar, India: a comparison. PLoS Negl Trop Dis. 2016;10:e0004970.

Toumi A, Chlif S, Bettaieb J, Ben Alaya N, Boukthir A, Ahmadi ZE, Ben Salah A. Temporal dynamics and impact of climate factors on the incidence of zoonotic cutaneous leishmaniasis in central Tunisia. PLoS Negl Trop Dis. 2012;6:e1633.

Uranw S, Ostyn B, Rijal A, Devkota S, Khanal B, Menten J, Boelaert M, Rijal S. Post-kala-azar dermal leishmaniasis in Nepal: a retrospective cohort study (2000–2010). PLoS Negl Trop Dis. 2011;5(12):e1433.

Velo E, Bongiorno G, Kadriaj P, Myrseli T, Crilly J, Lika A, Mersini K, Di Muccio T, Bino S, Gramiccia M, Gradoni L, Maroli M. The current status of phlebotomine sand flies in Albania and incrimination of Phlebotomus neglectus (Diptera, Psychodidae) as the main vector of Leishmania infantum. PLoS One. 2017;12:e0179118.

Welay GM, Alene KA, Dachew BA. Visceral leishmaniasis treatment outcome and its determinants in northwest Ethiopia. Epidemiol Health. 2016;39:e2017001.

World Health Organization. Control of the leishmaniases. WHO Tech Rep Ser 793. Geneva: World Health Organization; 1990.

World Health Organization. Control of the leishmaniases. WHO Tech Rep Ser 949. Geneva: World Health Organization; 2010.

Young DG, Duncan MA. Guide to the identification and geographic distribution of Lutzomyia sand flies in Mexico, the West Indies, Central and South America (Diptera: Psychodidae). Mem Am Entomol Inst. 1994;54:1–881.

Leishmania Taxonomy

2

Isabel L. Maurício

2.1 Introduction

Taxonomy stems from the human need for order, as it is much easier for the human brain to deal with things that can be put away neatly in boxes. *Leishmania* taxonomists have long been known to create and use many boxes, based on different and often inconsistent criteria, in an attempt to organize the huge clinical and geographical diversity observed in this genus. In the last three decades, with the use of molecular biology and ever more reliable phylogenetic tree building methods, as well as the findings of new variants, researchers have reorganized the various *Leishmania* boxes.

This chapter will present and discuss the most current consensus classification, the identity of recently described species and the added complexity of hybrid or recombinant lineages, in the context of the methods and markers used in taxonomic studies.

2.2 The Genus *Leishmania*

The genus *Leishmania* Ross, 1903, is considered to belong to:

Empire Eukaryota Cavalier-Smith, 1998
Kingdom Protozoa Cavalier-Smith, 2002
Infrakingdom Excavata Cavalier-Smith, 2003
Phylum Euglenozoa Cavalier-Smith, 1993

I.L. Maurício
Global Health and Tropical Medicine Centre (GHTM), Unidade de Parasitologia Médica, Instituto de Higiene e Medicina Tropical (IHMT), Universidade Nova de Lisboa, Lisboa, Portugal
e-mail: isabel.mauricio@ihmt.unl.pt

F. Bruschi, L. Gradoni (eds.), *The Leishmaniases: Old Neglected Tropical Diseases*, https://doi.org/10.1007/978-3-319-72386-0_2

Class Kinetoplastida Honigberg, 1963 (or the synonymous Kinetoplastea Cavalier-Smith, 1981), although Kinetoplastida is also often used for order
Order Trypanosomatida Kent, 1880
Family Trypanosomatidae Doflein, 1901

The genus *Leishmania* has previously been proposed to be divided into section Paraleishmania and section Euleishmania (Cupolillo et al. 2000), which is comprised of three subgenera: *L.* (*Leishmania*), *L.* (*Viannia*) and *L.* (*Sauroleishmania*). At least 39 described species of *Leishmania* can be found in the literature, with additional, yet unnamed or informally named, parasites. Many of these described species have, since, been shown to be synonymous, and taxonomy simplification has been argued for (Fraga et al. 2010; Schönian et al. 2010). It is, thus, worth to briefly look into what a species is.

2.3 Concepts of Species

The biological concept of species (Mayr, 1942) is based on reproductive isolation between populations. Briefly, if individuals from different groups are not able to produce fertile offspring, then the two groups can be considered separate species. The biological concept of species has not been and cannot be applied for *Leishmania* taxonomy, because sexual reproduction is difficult to detect and its importance in nature is still controversial (Ramírez and Llewellyn 2014; Rougeron et al. 2017).

Initial definitions of species in *Leishmania* followed relatively ad hoc principles, in a mixture of ecological, morphological, phenetic and clinical concepts. Thus, many species definitions and their names reflect this heritage. One such example is *L. major*, which was once considered a sub-species of *L. tropica*—*L. tropica major*—based on morphology, and in relation to the also since abandoned subspecies *L. tropica minor*, which is now the species *L. tropica* which corresponded to smaller forms (Saf'janova and Aliev 1973). *L. infantum* was based on epidemiological data, as it is considered to be more prevalent in infants, whereas species such as *L. guyanensis*, *L. mexicana*, *L. braziliensis*, etc. were named, as well as defined, based on their geographical distribution or region of first description. Some of the current issues in *Leishmania* taxonomy and classification are heirs to such criteria.

Other concepts that have been applied to microorganisms, particularly those without recognizable sexual reproduction, such as bacteria, include the phylogenetic concept of species, which is based on common descent. Briefly, a species will be defined as a group of individuals with a common ancestor and that are closely related or share certain traits. It has been proposed that the phylogenetic criteria should form the basis of a Universal Species Concept (Staley 2009). However, species thus defined can be somewhat arbitrary, because it depends on the level of distance chosen as cut-off point.

The phylogenetic concept of species has gained prominence in *Leishmania*. However, for the species or groups to be useful they should make biological and clinical sense. In *Leishmania*, systematic revisions have re-evaluated several groups of species, based on phylogenetic analyses of a range of markers, from the most polymorphic, such as microsatellite loci, which perform well in population studies, to the most conserved, such as RNA polymerases and the ribosomal small subunit (SSU), which perform well in the study of deeper phylogenetic relationships. Unfortunately, to date, no single marker has been applied to all described taxa within the genus *Leishmania*.

2.4 Organizing *Leishmania*

Organisms in the genus *Leishmania* cause a wide range of clinical manifestations, from visceral (VL) to cutaneous leishmaniasis (CL) (see Chap. 6). They can infect and be transmitted by a wide range of hosts and vectors, respectively (see Chaps. 3 and 4).

Recent taxonomic revisions have shown that several species designations correspond to low diversity genetic groups, effectively indistinguishable from another species, which often includes other similar sized genetic groups. Examples include *L. shawi* within *L. guyanensis*, *L. peruviana* within *L. braziliensis*, *L. killicki* within *L. tropica* and even *L. infantum* within *L. donovani*. The main question is whether *Leishmania* research requires more or less boxes. Should we recognize the intricate and complex genetic diversity within a larger group and assign species names to all groups? Or should we work with a much reduced number of species? Increasing the number of species would, undoubtedly, increase confusion among researchers and would introduce greater uncertainty, as the boundaries between species would be much more difficult to define. On the other hand, reducing the number of species could remove information about variants that can have clinical and epidemiological implications. A good example of the dilemma faced by taxonomists is *L. donovani* or the *L. donovani* complex. By eliminating species names such as *L. infantum*, the information that a group of isolates or lineages has dogs as reservoir is lost. But recognizing that subgroups with geographical associations, such as in Sudan/ Ethiopia and in India have quite different vectors and clinical traits, would increase dramatically the number of species, possibly to the point of being unusable.

One option is to apply the concept of sub-species, even if not in the strict sense of a geographically isolated group within a species. In that sense, *L. infantum* would become a sub-species of *L. donovani*, so *L. donovani infantum*. But, what to make of the South American variants of *L. donovani infantum*, which are effectively isolated geographically? Or of the two groups of *L. infantum*, as defined by microsatellite analyses, that have become known as (zymodeme Montpellier) "(MON)-1" and "non-MON-1" and that preferentially cause visceral or cutaneous disease, respectively? Should these also be given a taxonomic status or not?

It might be useful to apply the concept of superspecies and infra-species in this context (Mallet 2007). This concept was initially introduced by Mayr and Rensch and a notation proposed by Amadon (1966), in which the designation for the superspecies would appear in brackets. An example would be *Leishmania* (*Leishmania*) [*donovani*] *donovani* and *Leishmania* (*Leishmania*) [*donovani*] *infantum*.

Ultimately, taxonomic revisions should take into account the experience and needs of clinicians and laboratory scientists, and there should be greater communication between the two approaches. Taxonomy influences the choice of samples when studying parasite–host relationships, drug testing and, crucially, diagnostics. Are the samples representative of the species or of its diversity? Could some samples be too closely related that results obtained are too similar and not possible to extrapolate to the entire species?

As pointed out by Kuhn and Jahrling (2010), for the recent classification of viruses, most people, including researchers, are not fully equipped to deal with the intricacies of what constitutes different species, and many do not regularly read taxonomic revisions. As such, incorrect or outdated designations often persist in the literature. This issue is further confounded and maintained by taxonomic uncertainties. The next sections will review the consensus taxonomy for specific groups and the points of current debate.

2.4.1 A Visceral Question: *Leishmania donovani*

Leishmania donovani includes the aetiological agents of VL, a form that can be fatal without treatment. It has one of the widest geographical ranges of *Leishmania*, which only excludes Oceania and Antarctica, and it is the main species of *Leishmania* present in Europe. As such, it is perhaps not surprising that, despite its low intraspecific diversity, it is one of the most intensively sampled and genotyped taxa of the genus *Leishmania*. In spite, or because, of this, it has suffered extensive splitting into different species by some authors, although recognized to be close and considered by some authors as the *L. donovani* complex of species. Four species have been described: *L. donovani* (also often divided into sensu stricto, in India, and sensu lato, in East Africa), *L. infantum*, *L. archibaldi* and *L. chagasi* (Table 2.1). The initial support for such divisions was based on limited markers, particularly the description of *L. archibaldi* (Rioux et al. 1990) or a small number of strains, as reviewed previously regarding the *L. infantum*/*L. chagasi* debate (Mauricio et al. 2000). Indeed, in the past two decades, it has been consistently shown that it is not possible to distinguish populations from Europe and South America of, respectively, *L. infantum* and *L. chagasi*, using several techniques and targets, such as multilocus microsatellite typing (MLMT) (Kuhls et al. 2007) and multilocus sequence typing (MLST) (Zemanova et al. 2007; Mauricio et al. 2006), ribosomal internal transcribed spacer (ITS) and mini-exon PCR-RFLP (Mauricio et al. 2004) or random amplification of polymorphic DNA (RAPD) (Mauricio et al. 1999). These studies, which have included several strains from a wide geographical range, have supported synonymy. Considering that the epidemiology in the two regions is very similar, there are no molecular, clinical or epidemiological reasons that warrant separation into two species. However some researchers still claim they can be distinguished (e.g. Marcili et al. 2014), despite using restricted and unrepresentative sampling. In fact, an MLMT extensive study of 450 strains of the *L. donovani* complex had already shown that South American strains were most similar to those from Portugal and Spain and supported a recent introduction (in the past 500 years) consistent with the European colonization of South America (Leblois et al. 2011).

Table 2.1 Simplified nomenclature of the genus *Leishmania* (adapted from Fraga et al. 2010; Schönian et al. 2010)

Subgenus	Species	Other associated species names	Notes
L. (*Leishmania*)	*L. donovani*	*L. archibaldi*	
		L. chagasi	
		L. infantum	
	L. major	*L. arabica*	
		L. gerbilli	
		L. turanica	
	L. tropica	*L. aethiopica*	
		L. killicki	
	L. mexicana	*L. amazonensis*	
		L. aristidesi	
		L. garnhami	
		L. forattinii	
		L. pifanoi	
		L. venezuelensis	
		L. waltoni	
L. (*Sauroleishmania*)[a]	*L. tarentolae*		Telford (2009)
	L. adleri		
	L. gymnodactyli		
	L. hoogstraali		
	L. guliki[b]		
	L. zuckermani[c]		
	L. platycephala[c]		Telford (2009)
L. (*Viannia*)	*L. braziliensis*	*L. peruviana*	
	L. guyanensis	*L. panamensis*	
	L. lindenbergi	*L. shawi*	
	L. utingensis		
	L. lainsoni		
	L. naiffi		
Unnamed	*L. enrietti*		Harkins et al. (2016) and Kwakye-Nuako et al. (2015)
	L. sp. (Ghana)[d]		
	"*L. siamensis*"		No formal description
	L. martiniquensis		
	L. sp. (AM-2004)		
L. (*Endotrypanum*)[e]	*L. hertigi*	*L. deanei*	
	E. monterogeii[a]	*E. schaudinni*	
		E. sp	
		L. colombiensis	
		L. equatoriensis	

[a]Subject to confirmation

[b]Species description and one mitochondrial sequence available, with greatest similarity to *L. tarentolae*

[c]Species description, no molecular data

[d]Possible revision as *L. enrietti*

[e]Suggestion of subgenus, in alternative to section Paraleishmania or genus *Endotrypanum* for all species

Another taxonomic problem was the description of a separate species from *L. donovani*, *L. archibaldi* (Rioux et al. 1990), which has been shown not to be valid in its original definition based on multilocus enzyme electrophoresis (MLEE) analysis, but instead aspartate aminotransferase (ASAT) heterozygotes of *L. donovani*, and which thus led to misclassification of some *L. donovani* samples from East Africa as *L. infantum*, both demonstrated by further studies using markers such as ITS and the mini-exon (Mauricio et al. 2004), MLST (Mauricio et al. 2006; Zemanova et al. 2007) and MLMT (Kuhls et al. 2007) and, finally, in a joint analysis with a large number and variety of markers (Lukes et al. 2007).

It is of note that, although members of the *L. donovani* complex are agents of VL, a group within *L. infantum* (and referred to as non-MON-1) has been identified, through MLMT, that is mostly associated with CL in individuals without immune depressions (Kuhls et al. 2007). It would be more clinically useful to attribute to this group a sub-species status than to all South American *L. infantum*.

The genetic diversity within *L. donovani* can also have implications for control and diagnostics, such as it happened with rk39, which, despite successful implementation elsewhere, was found to have very low sensitivity when applied in Sudanese populations (Ritmeijer et al. 2006).

Despite some evidence for genetic groups and specific characteristics in some groups, genetic recombination has been shown to have occurred between *L. donovani* populations, including with *L. infantum* (Mauricio et al. 2006). As a result, it is very difficult to find consistent phylogenetic or diagnostic markers for any species, or subgroups, within the complex, and it is likely that it will become more difficult as more variants are found, such as in Sri Lanka and Cyprus (Alam et al. 2009). So, for diagnostic purposes and epidemiological studies, it seems more useful to recognize the existence of a single species, *L. donovani*, and analyse it as a single entity (Table 2.1), although research should take into account its full genetic diversity range.

2.4.2 A Major Issue: Parasite Species of Old World Rodents

L. major has been a non-controversial species, with limited genetic and clinical diversity, although subpopulations have been identified by MLMT with strong geographical associations (Al-Jawabreh et al. 2008). However, in the Old World, some *Leishmania* isolates from gerbils, the reservoir host of *L. major*, and so far not found to infect humans, had been found to be sufficiently distinct to be classified into different species: *L. turanica*, *L. arabica* and *L. gerbilli* (Table 2.1). Phylogenetic analyses that have included these species have shown that they form a monophyletic group with *L. major*, but that they are each sufficiently distant from *L. major*, by comparison with other species and complex of species, to warrant separate species status: a gp63 gene that included *L. turanica* and *L. arabica* (Mauricio et al. 2007), cytochrome b that included *L. turanica* and *L. arabica* (Asato et al. 2009), MLST of seven loci that included *L. turanica* and *L. gerbilli* (Auwera et al. 2014) or all three species (Baidouri et al. 2013). Such results, and considering the common association with gerbils, would support definition of a *L. major* species complex to include these four species to reflect a common origin and ecological similarities (Table 2.1).

2.4.3 Not a Minor Issue: The Case of *Leishmania tropica* and Hyraxes

Phylogenetics can be used to generate hypothesis regarding newly isolated species, for example, in terms of the reservoir hosts, vectors, possible clinical presentations, drug response, etc.

One such case is of *L. tropica* and *L. aethiopica*. *L. tropica*, once named as *L. tropica minor*, is considered to have an anthroponotic life cycle for most of its geographical range, whereas *L. aethiopica* has hyraxes as reservoirs, although also capable of infecting humans. Phylogenetic studies had showed that *L. aethiopica* and *L. tropica* were closely related species, suggesting that *L. tropica* could have evolved from a parasite of hyraxes and that it could still infect this species, a close relative of elephants. Indeed, *L. tropica* isolates were eventually found in hyraxes (Jacobson et al. 2003; Jaffe et al. 2004) leading to the hypothesis that differentiation between the *L. major* and the *L. tropica*/*L. aethiopica* lineages was driven by host associations, in particular by successful colonization of hyraxes by parasites originally associated with rodents (Mauricio et al. 2007).

A third related species, *L. killicki*, has been shown to be a small subgroup of *L. tropica* and should, thus, be considered synonymous (Baidouri et al. 2013; Chaara et al. 2015). It is not so consensual how to classify *L. aethiopica*, as it can appear to be very close to *L. tropica* (Fraga et al. 2010; Krayter et al. 2015) or to form a separated group from *L. tropica* in analyses of several strains (Asato et al. 2009; Baidouri et al. 2013; Auwera et al. 2014), although in an analysis of *hsp20*, they did not form a monophyletic cluster (Fraga et al. 2013). However, by comparison of distances between species, and for consistency and simplicity, it has been proposed that the entire complex is considered as a single species, *L. tropica* (Schönian et al. 2010) (Table 2.1).

2.4.4 From Mexico to the Amazon: Parasite Species of New World Rodents

Leishmania parasites of small rodents in the New World have been classified into the species *L. mexicana* (synonymous with *L. pifanoi*), *L. amazonensis* (synonymous with *L. garnhami*) (Asato et al. 2009) and *L. forattinii*, *L. venezuelensis* and *L. aristidesi* (Lainson 1997; Schönian et al. 2010) (Table 2.1). *Leishmania forattinii* was found to be closely related to *L. aristidesi* by Cupolillo et al. (1994). The ITS study by Berzunza-Cruz et al. (2002) found *L. venezuelensis* to be more closely related to *L. major*. Recently, the description of a new species (*L. waltoni*) has been published for a subset of strains within *L. mexicana* (Shaw et al. 2015) that is reportedly associated with diffuse CL in the Dominican Republic. The authors reported a single-nucleotide polymorphism among the five studied strains in 2.5 kbp of concatenated single gene sequences, which is very low, and only 37 polymorphic sites for the entire complex, which represents less than 1.5% genetic diversity across. It is more likely that this group of strains represents a clonal expansion or a geographically restricted group, and this new species name should, in fact, be considered synonymous with *L. mexicana*.

The overall phylogeny of this group has been less extensively studied than for other *Leishmania*, and *L. forattinii*, *L. venezuelensis* and *L. aristidesi* are seldom represented. The *hsp70* analysis by Fraga et al. (2010) failed to resolve between the species and respective strains of New World *Leishmania* parasites of rodents, thus concluding that it should be considered a single species. Other authors were able to separate the two main species within the group (*L. mexicana* and *L. amazonensis*) based on ITS (Berzunza-Cruz et al. 2002; Davila and Momen 2000) as well as by MLST and mini-exon (Auwera et al. 2014). In phylogenetic trees based on other markers, the genetic diversity within the group was low and comparable to that found within other species (Asato et al. 2009; Fraga et al. 2013; Kwakye-Nuako et al. 2015), including a recent genome-based tree (Harkins et al. 2016) (Fig. 2.1). Increased sampling of the least represented members of this group and more detailed analyses of all members with multilocus markers would further elucidate relationships and species status within it, although a conservative approach would group all species under *L. mexicana* (Table 2.1).

2.4.5 The *Viannia* Group

A group of *Leishmania* parasites that was found to colonize the hindgut of the sand fly vector was placed in the subgenus *L.* (*Viannia*), in contrast with the other known *Leishmania* that were only detected in the foregut of the vector (Lainson and Shaw 1987). Species in this subgenus have only been found in South America (Lainson and Shaw 1987). The most common species and the main agent of mucocutaneous leishmaniasis (MCL) is *L. braziliensis*, which is distributed throughout the endemic range of MCL in South America, with species with a more restricted geographical range, such as *L. peruviana* (in the Andes) and *L. guyanensis* (mostly in the tropical forest). Most strains cluster into two main groups that include, first, *L. braziliensis* and *L. peruviana* and, second, *L. guyanensis*, *L. panamensis* and *L. shawi*, according to multilocus analyses (Boité et al. 2012; Auwera et al. 2014). Other species, with fewer isolates obtained so far, are also placed in the subgenus *L. (Viannia)* (Table 2.1): *L. lindenbergi* and *L. utingensis*, which are phylogenetically closer to both *L. braziliensis* (Boité et al. 2012) and *L. guyanensis*, and *L. lainsoni* and *L. naiffi* (Fraga et al. 2010; Boité et al. 2012; Fraga et al. 2013; Auwera et al. 2014). Phylogenetic relationships within the subgenus *Leishmania* (*Viannia*), however, are not as well resolved as within the subgenus *Leishmania* (*Leishmania*), with lower bootstrap values in general and lack of resolution for cytochrome b-based phylogeny (Asato et al. 2009). An MLST network analysis suggests some level of recombination within and between groups (Boité et al. 2012), thus complicating phylogenetic inference.

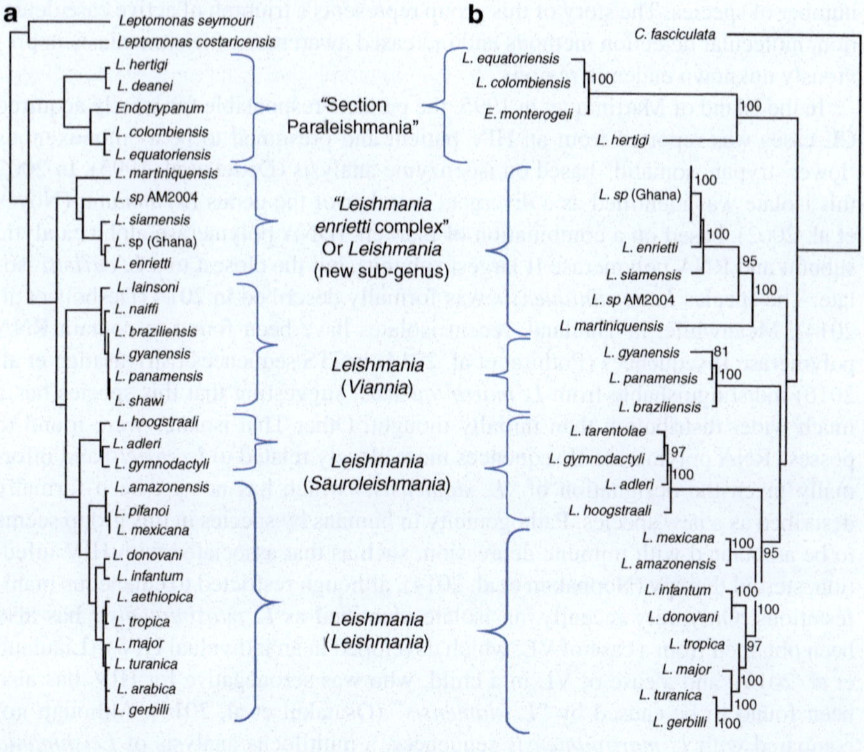

Fig. 2.1 Comparison of *Leishmania* phylogenies and corresponding taxa. (**a**) Phylogeny constructed in BEAST v1.8.2 using the loci available for isolate *L.* sp. AM-2004: 18S, ITS1/5.8S, RNA polymerase II large subunit partial sequences. Two species of *Leptomonas* were used as outgroup. Adapted from: Harkins et al. (2016) https://doi.org/10.1016/j.meegid.2015.11.030. (**b**) Maximum likelihood tree based on RNA polymerase II large subunit (RNAPolII) gene sequences, using *Chritidia fasciculata* as an outgroup. Bootstrap values above 80% are shown. Adapted from Kwakye-Nuako et al. (2015) https://doi.org/10.1016/j.ijpara.2015.05.001

2.4.6 A New Subgenus? The End of Solitude for *Leishmania enrietti*

While research on well-sampled taxa has been mostly concerned with clumping and reducing taxa diversity and complexity, better sampling and increased awareness of leishmaniasis has not only identified previously unknown *Leishmania* endemic regions, as it has uncovered new variants that are sufficiently genetically distant from known species to be awarded a separate species status. The most striking example has been the *L. enrietti* branch, for which four new groups of isolates, at genetic distances comparable to those between species in other groups, have been described recently. Once considered an "enigmatic" species (Lainson 1997), and neglected among the neglected, *L. enrietti* once stood isolated in its long branch in all *Leishmania* trees. At present, the group comprises at least five species-level groups, with a level of diversity and genetic distances suggestive of a much larger number of species. The story of this group represents a triumph of active case detection, molecular detection methods and increased awareness of leishmaniasis in previously unknown endemic regions.

In the island of Martinique, in 1995, the parasite responsible for locally acquired CL cases was reported from an HIV patient and presumed to be a "monoxenous 'lower' trypanosomatid" based on isoenzyme analysis (Dedet et al. 1995). In 2002 this isolate was identified as a divergent member of the genus *Leishmania* (Noyes et al. 2002), based on a combination of markers (DNA polymerase alpha catalytic subunit and RNA polymerase II largest subunit), but the closest to a *L. enrietti* isolate. The species *L. martiniquensis* was formally described in 2014 (Desbois et al. 2014). Meanwhile, in Thailand, recent isolates have been found to contain RNA polymerase II sequences (Pothirat et al. 2014) or ITS sequences (Siriyasatien et al. 2016) indistinguishable from *L. martiniquensis*, suggesting that this species has a much wider distribution than initially thought. Other Thai isolates were found to possess RNA polymerase II sequences more closely related to *L. enrietti* and informally given the designation of "*L. siamensis*," which has not yet been formally described as a new species. Pathogenicity in humans by species in this group seems to be associated with immune depression, such as that associated with HIV infection, steroid therapy (Noppakun et al. 2014), although restricted to cutaneous manifestations. Only very recently, an isolate identified as *L. martiniquensis* has also been obtained from a case of VL, which developed in an individual HIV+ (Liautaud et al. 2015), and a case of VL in a child, who was seronegative for HIV, has also been found to be caused by "*L. siamensis*" (Osatakul et al. 2014). Although not compared with *L. martiniquensis* sequences, a multilocus analysis of *Leishmania* samples from Thailand suggests that two distinct species of parasites circulate in that country (Leelayoova et al. 2013): one is "*L. siamensis*" and the other is likely to be *L. martiniquensis*. Similarly to *L. martiniquensis*, "*L. siamensis*" seems to have a wide global distribution, with closely related sequences isolated from Florida, USA (Reuss et al. 2012), from a horse, and in Central Europe, from a cow (Lobsiger et al. 2010) and from horses (Müller et al. 2009). The presence of these parasites in non-human hosts is highly suggestive of a zoonotic parasite, with occasional development in humans, particularly those with compromised or immature immune systems.

Furthermore, a third species (still unnamed) has been proposed for parasites isolated from Ghana (Kwakye-Nuako et al. 2015), also from human cutaneous cases.

All three of these species have the capacity to cause disease in humans. However, a related, although so far unnamed, group of isolates was obtained from kangaroos in Australia (Rose et al. 2004). Considering that other species in this group have been found capable of infecting humans, it can be considered that these parasites pose a risk for the human population in Australia, particularly if immunocompromised.

At the moment, based on ribosomal protein L23a intergenic region and RNA polymerase II large subunit gene sequences, it has been proposed (Kwakye-Nuako et al. 2015) that this group warrants a subgenus status, alongside the subgenera *Leishmania* (*Leishmania*), *Leishmania* (*Sauroleishmania*) and *Leishmania* (*Viannia*)

(Fig. 2.1). Such position of the *L. enrietti* branch at subgenus level is supported by a genome-based phylogeny (Harkins et al. 2016) (Fig. 2.1). However, the entire group is often referred to as "*L. enrietti* complex," which should be more accurately applied only to the group that includes *L. enrietti*, "I" and the Ghanaian samples, upon comparison of genetic distances with other *Leishmania* species (Fig. 2.1 and Table 2.1).

2.4.7 One Genus or Two Genera?

At the beginning of the twentieth century, an intraerythrocytic parasite resembling *Leishmania* was observed in sloths, and the genus *Endotrypanum* was created to accommodate these isolates (Mesnil and Brimont 1908). To date, two species have been described in this genus, *E. monterogeii* and *E. schaudinni*, with several isolates not assigned to a species (*Endotrypanum* sp). However, intracellular forms have not been observed in experimental infections, and as the isolated parasites grow well in standard *Leishmania* medium, it is possible that the isolates do not correspond to the originally observed forms, but to other parasites present in the host (Cupolillo et al. 2000). Indeed, phylogenetic analyses place these isolates in the same cluster as the *Leishmania* species *L. colombiensis*, *L. equatoriensis*, *L. hertigi*, *L. herreri* and *L. deanei* (Croan et al. 1997; Croan and Ellis 1996; Noyes et al. 1996, 1997; Cupolillo et al. 2000; Harkins et al. 2016; Kwakye-Nuako et al. 2015; Asato et al. 2009) (Fig. 2.1). This group is, genetically, quite distinct from other *Leishmania*, leading Cupolillo et al. (2000) to propose a division of the genus in two sections: section Paraleishmania to include these species and section Euleishmania the remainder.

Taxonomy of this group is still not agreed upon, although it can be argued that it should be revised to avoid polyphyly[1] of the genus *Endotrypanum* and paraphyly[2] of the genus *Leishmania*. In 2000, Cupolillo et al. proposed maintenance of the genus *Endotrypanum* for the existing isolates and pending revision of the genus with fresh isolates from sloths. However, more recently, Marcili et al. (2014) proposed that *L. hertigi* and *L. equatoriensis* (and presumably the other species in the same clade) should be renamed to become genus *Endotrypanum*, which would effectively form a sister genus to *Leishmania*. However, *L. colombiensis* and *L. equatoriensis* (Delgado et al. 1993; Rodriguez-Bonfante et al. 2003; Kreutzer et al. 1991; Ramírez et al. 2016) have reportedly been isolated from human CL and VL cases. Furthermore, no new descriptions of the elusive intraerythrocytic parasites of sloths have emerged. Finally, it has been reported that species of the subgenus *L. (Sauroleishmania)* can develop inside erythrocytes (Telford 2009). It is, thus, possible that either the first observations of intraerythrocytic parasites in sloths corresponded to non-isolated *L. (Sauroleishmania)* or that the two groups share this capacity. In any case, it is not a unique character within the genus *Leishmania*, and, therefore, it does not justify a

[1] Polyphyly: a group of organisms whose last common ancestor is not a member of the group.

[2] Paraphyly: a group of organisms that includes the last common ancestor, but not all of its descendants.

separate genus per se. Instead, this group could become a subgenus within the genus *Leishmania* to keep consistency across the genus *Leishmania*, to reflect the identity and the history of this group and to recognize the capacity of at least some species for causing pathology in humans. One possibility would be to name the subgenus as *L. (Paraleishmania)*. Alternatively, a subgenus *Leishmania* (*Endotrypanum*) could be proposed, considering priority, to keep the connection with a formal taxon name (*Endotrypanum*) (Table 2.1) and because it would replicate the process undergone for *Sauroleishmania*.

2.5 Perspectives

Several typing methodologies are available to researchers trying to unravel the taxonomy of *Leishmania*, many of which have been mentioned in this chapter, but mostly based on sequencing of conserved DNA regions, such as heat-shock protein 70 (Fraga et al. 2010, 2013), DNA and RNA polymerases (Croan et al. 1997), the ribosomal internal transcribed spacer (Berzunza-Cruz et al. 2002; Davila and Momen 2000; Mauricio et al. 2004), the mini-exon (Mauricio et al. 2004), the small subunit rDNA (Berzunza-Cruz et al. 2002; Marcili et al. 2014), GAPDH (Marcili et al. 2014), glycoprotein 63 (Mauricio et al. 2007), cytochrome b (Asato et al. 2009), multilocus sequences (Baidouri et al. 2013; Leelayoova et al. 2013; Mauricio et al. 2006; Zemanova et al. 2007) and entire genomes (Harkins et al. 2016), although other methods have been useful to resolve relationships between closely related species, such as multilocus microsatellite analysis (Al-Jawabreh et al. 2008; Alam et al. 2009; Kuhls et al. 2007).

Ideally, *Leishmania* taxonomy should be based on genomic data for all species. However, a case can be put forward for simplification and quick identification, such as the use of barcoding methods (Hebert et al. 2016). Barcoding is based on a small number of markers to identify species, typically mitochondrial targets, such as COI, or nuclear markers such as ribosomal RNA or spacer regions. However, *Leishmania* species can cross in nature to produce hybrids (Rougeron et al. 2015), and barcoding using only mitochondrial targets would not detect species hybrids due to uniparental transmission (Romano et al. 2014). As such, any barcoding system for *Leishmania* should be based on or include at least one nuclear region that could detect both parental sequences.

Phylogenetic inference methods have become quite sophisticated, and increased computational power allows application of complex and computer-intensive methods, such as maximum likelihood and Bayesian analyses, to larger number of samples and large volumes of genotyping data (Yang and Rannala 2012). However, good phylogenetic trees can only be obtained from adequate data, which should include neutral markers or markers shown to be good representatives of *Leishmania* genome evolution, from which robust alignments can be produced. Importantly, such trees should be based on sufficiently wide sampling of the biological and genetic diversity of the genus, which should include all known species of the genus *Leishmania*, as well as intraspecific diversity. Good sample representativity is crucial, as it has been quite rightly pointed out that "any sound taxonomy should take into account the full biological diversity of the group under study" (Auwera et al. 2011). From studies of well-represented species, such as *L. donovani*, *L tropica* or *L. braziliensis*, it has become clear that small numbers of isolates can lead to misleading divisions within those species. Most initial phylogenetic analyses of the genus *Leishmania* were based on a small number of samples, from a reduced number of locations and from markers with limitations, such as MLEE (as reviewed in the previous section). Such phylogenies have introduced taxonomic problems, such as the description of *L. archibaldi* as a separate species from *L. donovani* (Rioux et al. 1990). Efforts should, thus, be made to look for and study isolates related to new or poorly represented species or groups, even if from uncultured samples, and with a wide selection of markers, ideally, whole genome sequences.

Conclusions

Leishmania taxonomy remains complex and challenging. Some factors that have made a consensus difficult to reach by *Leishmania* taxonomists include lack of classic sexual recombination that precludes application of the biological concept of species, occasional recombination, including between different species, that blurs the boundaries between phylogenetic groups, the large number of described species that are now considered to be synonymous, lack of homologous genotyping data for all species as well as recent discoveries of new species or variants. However, as reviewed in this chapter, there is a clear case for taxonomic simplification at species level, as well as for a revision at genus and subgenus levels to reflect the now overwhelming molecular and phylogenetic data.

Acknowledgments I would like to thank all the researchers that have worked and published in the subject of *Leishmania* taxonomy, including any that have not been cited in this chapter.

This work has been written in the context of FCT (Portugal) financing of the GHTM centre, GHTM –UID/Multi/04413/2013.

References

Al-Jawabreh A, Diezmann S, Muller M, et al. Identification of geographically distributed subpopulations of *Leishmania (Leishmania) major* by microsatellite analysis. BMC Evol Biol. 2008;8:183.

Alam MZ, Haralambous C, Kuhls K, et al. The paraphyletic composition of *Leishmania donovani* zymodeme MON-37 revealed by multilocus microsatellite typing. Microbes Infect. 2009;11:707–15.

Amadon D. The superspecies concept. Syst Biol. 1966;15:245–9.

Asato Y, Oshiro M, Myint CK, et al. Phylogenic analysis of the genus *Leishmania* by cytochrome b gene sequencing. Exp Parasitol. 2009;121:352–61.

Auwera G, Fraga J, Montalvo AM, et al. *Leishmania* taxonomy up for promotion. Trends Parasitol. 2011;27:49–50.

Auwera G, Ravel C, Verweij JJ, et al. Evaluation of four single-locus markers for *Leishmania* species discrimination by sequencing. J Clin Microbiol. 2014;52:1098–104.

Baidouri F, Diancourt L, Berry V, et al. Genetic structure and evolution of the *Leishmania* genus in Africa and Eurasia: what does MLST tell us. PLoS Negl Trop Dis. 2013;7:e2255.

Berzunza-Cruz M, Cabrera N, Crippa-Rossi M, et al. Polymorphism analysis of the internal transcribed spacer and small subunit of ribosomal RNA genes of *Leishmania mexicana*. Parasitol Res. 2002;88:918–25.

Boité MC, Mauricio IL, Miles MA, Cupolillo E. New insights on taxonomy, phylogeny and population genetics of *Leishmania (Viannia)* parasites based on multilocus sequence analysis. PLoS Negl Trop Dis. 2012;6(11):e1888.

Chaara D, Ravel C, Bañuls A, Haouas N, Lami P, Talignani L, El Baidouri F, Jaouadi K, Harrat Z, Dedet JP, Babba H, Pratlong F. Evolutionary history of *Leishmania killicki* (synonymous *Leishmania tropica*) and taxonomic implications. Parasit Vectors. 2015;8:198.

Croan D, Ellis J. Phylogenetic relationships between *Leishmania*, *Viannia* and *Sauroleishmania* inferred from comparison of a variable domain within the RNA polymerase II largest subunit gene. Mol Biochem Parasitol. 1996;79:97–102.

Croan DG, Morrison DA, Ellis JT. Evolution of the genus *Leishmania* revealed by comparison of DNA and RNA polymerase gene sequences. Mol Biochem Parasitol. 1997;89:149–59.

Cupolillo E, Grimaldi G Jr, Momen H. A general classification of new world *Leishmania* using numerical zymotaxonomy. Am J Trop Med Hyg. 1994;50:296–311.

Cupolillo E, Medina-Acosta E, Noyes H, et al. A revised classification for *Leishmania* and *Endotrypanum*. Parasitol Today. 2000;16:142–4.

Davila AM, Momen H. Internal-transcribed-spacer (ITS) sequences used to explore phylogenetic relationships within *Leishmania*. Ann Trop Med Parasitol. 2000;94:651–4.

Dedet JP, Roche B, Pratlong F, et al. Diffuse cutaneous infection caused by a presumed monoxenous trypanosomatid in a patient infected with HIV. Trans R Soc Trop Med Hyg. 1995;89:644–6.

Delgado O, Castes M, White AC, et al. *Leishmania colombiensis* in Venezuela. Am J Trop Med Hyg. 1993;48:145–7.

Desbois N, Pratlong F, Quist D, et al. *Leishmania (Leishmania) martiniquensis* n. sp. (Kinetoplastida: Trypanosomatidae), description of the parasite responsible for cutaneous leishmaniasis in Martinique Island (French West Indies). Parasite. 2014;21:12.

Fraga J, Montalvo AM, De Doncker S, et al. Phylogeny of *Leishmania* species based on the heat-shock protein 70 gene. Infect Genet Evol. 2010;10:238–45.

Fraga J, Montalvo AM, Van der Auwera G, et al. Evolution and species discrimination according to the *Leishmania* heat-shock protein 20 gene. Infect Genet Evol. 2013;18:229–37.

Harkins KM, Schwartz RS, Cartwright RA, et al. Phylogenomic reconstruction supports supercontinent origins for *Leishmania*. Infect Genet Evol. 2016;38:101–9.

Hebert PDN, Hollingsworth PM, Hajibabaei M. From writing to reading the encyclopedia of life. Phil Trans R Soc B. 2016;371:20150321.

Jacobson R, Eisenberger CL, Svobodova M, et al. Outbreak of cutaneous leishmaniasis in northern Israel. J Infect Dis. 2003;188:1065–73.

Jaffe CL, Baneth G, Abdeen ZA, et al. Leishmaniasis in Israel and the Palestinian authority. Trends Parasitol. 2004;20:328–32.

Krayter L, Schnur LF, Schönian G. The genetic relationship between *Leishmania aethiopica* and *Leishmania tropica* revealed by comparing microsatellite profiles. PLoS One. 2015;10:e0131227.

Kreutzer RD, Corredor A, Grimaldi G, et al. Characterization of *Leishmania colombiensis* sp. n (Kinetoplastida: Trypanosomatidae), a new parasite infecting humans, animals, and phlebotomine sand flies in Colombia and Panama. Am J Trop Med Hyg. 1991;44:662–75.

Kuhls K, Keilonat L, Ochsenreither S, et al. Multilocus microsatellite typing (MLMT) reveals genetically isolated populations between and within the main endemic regions of visceral leishmaniasis. Microbes Infect. 2007;9:334–43.

Kuhn JH, Jahrling PB. Clarification and guidance on the proper usage of virus and virus species names. Arch Virol. 2010;155:445–53.

Kwakye-Nuako G, Mosore M-T, Duplessis C, et al. First isolation of a new species of *Leishmania* responsible for human cutaneous leishmaniasis in Ghana and classification in the *Leishmania enriettii* complex. Int J Parasitol. 2015;45:679–84.

Lainson R. On *Leishmania enriettii* and other enigmatic *Leishmania* species of the Neotropics. Mem Inst Oswaldo Cruz. 1997;92:377–87.

Lainson R, Shaw JJ. Evolution, classification and geographical distribution. In: Peters W, Killick-Kendrick R, editors. The leishmaniasis in biology and medicine. London: Academic Press Inc.; 1987.

Leblois R, Kuhls K, François O, Schönian G, Wirth T. Guns, germs and dogs: on the origin of *Leishmania chagasi*. Infect Genet Evol. 2011;11:1091–5.

Leelayoova S, Siripattanapipong S, Hitakarun A, et al. Multilocus characterization and phylogenetic analysis of *Leishmania siamensis* isolated from autochthonous visceral leishmaniasis cases, southern Thailand. BMC Microbiol. 2013;13:60.

Liautaud B, Vignier N, Miossec C, et al. First case of visceral leishmaniasis caused by *Leishmania martiniquensis*. Am J Trop Med Hyg. 2015;92:317–9.

Lobsiger L, Müller N, Schweizer T, et al. An autochthonous case of cutaneous bovine leishmaniasis in Switzerland. Vet Parasitol. 2010;169:408–14.

Lukes J, Mauricio IL, Schonian G, et al. Evolutionary and geographical history of the *Leishmania donovani* complex with a revision of current taxonomy. Proc Natl Acad Sci U S A. 2007;104:9375–80.

Mallet J. Subspecies, semispecies, superspecies. In: Levin SA, editor. Encyclopedia of biodiversity. New York: Elsevier; 2007.

Marcili A, Sperança MA, da Costa AP, et al. Phylogenetic relationships of *Leishmania* species based on trypanosomatid barcode (SSU rDNA) and gGAPDH genes: taxonomic revision of *Leishmania (L.) infantum chagasi* in South America. Infect Genet Evol. 2014;25:44–51.

Mauricio IL, Gaunt MW, Stothard JR, et al. Glycoprotein 63 (*gp63*) genes show gene conversion and reveal the evolution of old world *Leishmania*. Int J Parasitol. 2007;37:565–76.

Mauricio IL, Howard MK, Stothard JR, et al. Genetic diversity in the *Leishmania donovani* complex. Parasitology. 1999;119:237–46.

Mauricio IL, Stothard JR, Miles MA. The strange case of *Leishmania chagasi*. Parasitol Today. 2000;16:188–99.

Mauricio IL, Stothard JR, Miles MA. *Leishmania donovani* complex: genotyping with the ribosomal internal transcribed spacer and the mini-exon. Parasitology. 2004;128:1–5.

Mauricio IL, Yeo M, Baghaei M, et al. Towards multilocus sequence typing of the *Leishmania donovani* complex: resolving genotypes and haplotypes for five polymorphic metabolic enzymes (ASAT, GPI, NH1, NH2, PGD). Int J Parasitol. 2006;36:757–69.

Mesnil F, Brimont E. Sur un hematozoaire nouveau (*Endotrypanum* n.gen.) d'un edente de la Guyane. C R Soc Biol. 1908;65:581.

Müller N, Welle M, Lobsiger L, et al. Occurrence of *Leishmania* sp. in cutaneous lesions of horses in Central Europe. Vet Parasitol. 2009;166:346–51.

Noppakun N, Kraivichian K, Siriyasatien P. Disseminated dermal leishmaniasis caused by *Leishmania siamensis* in a systemic steroid therapy patient. Am J Trop Med Hyg. 2014;91:869–70.

Noyes H, Pratlong F, Chance M, et al. A previously unclassified trypanosomatid responsible for human cutaneous lesions in Martinique (French West Indies) is the most divergent member of the genus *Leishmania* ss. Parasitology. 2002;124:17–24.

Noyes HA, Arana BA, Chance ML, et al. The *Leishmania hertigi* (Kinetoplastida; Trypanosomatidae) complex and the lizard *Leishmania*: their classification and evidence for a neotropical origin of the *Leishmania-Endotrypanum* clade. J Eukaryot Microbiol. 1997;44:511–7.

Noyes HA, Belli AA, Maingon R. Appraisal of various random amplified polymorphic DNA-polymerase chain reaction primers for *Leishmania* identification. Am J Trop Med Hyg. 1996;55:98–105.

Osatakul S, Mungthin M, Siripattanapipong S, et al. Recurrences of visceral Leishmaniasis caused by *Leishmania* siamensis after treatment with amphotericin B in a Seronegative child. Am J Trop Med Hyg. 2014;90:40–2.

Pothirat T, Tantiworawit A, Chaiwarith R, et al. First isolation of *Leishmania* from Northern Thailand: case report, identification as *Leishmania martiniquensis* and phylogenetic position within the *Leishmania enriettii* complex. PLoS Negl Trop Dis. 2014;8:e3339.

Ramírez JD, Hernández C, León CM, et al. Taxonomy, diversity, temporal and geographical distribution of Cutaneous Leishmaniasis in Colombia: a retrospective study. Sci Rep. 2016;6:28266.

Ramírez JD, Llewellyn MS. Reproductive clonality in protozoan pathogens-truth or artefact? Mol Ecol. 2014;23:4195–202.

Reuss SM, Dunbar MD, Mays MBC, et al. Autochthonous *Leishmania* siamensis in horse, Florida, USA. Emerg Infect Dis. 2012;18:1545–7.

Rioux JA, Lanotte G, Serres E, et al. Taxonomy of *Leishmania*. Use of isoenzymes. Suggestions for a new classification. Ann Parasitol Hum Comp. 1990;65:111–25.

Ritmeijer K, Melaku Y, Mueller M, et al. Evaluation of a new recombinant K39 rapid diagnostic test for Sudanese visceral leishmaniasis. Am J Trop Med Hyg. 2006;74:76–80.

Rodriguez-Bonfante C, Bonfante-Garrido R, Grimaldi G Jr, et al. Genotypically distinct *Leishmania* colombiensis isolates from Venezuela cause both cutaneous and visceral leishmaniasis in humans. Infect Genet Evol. 2003;3:119–24.

Romano A, Inbar E, Debrabant A, et al. Cross-species genetic exchange between visceral and cutaneous strains of *Leishmania* in the sand fly vector. Proc Natl Acad Sci U S A. 2014;111:16808–13.

Rose K, Curtis J, Baldwin T, et al. Cutaneous leishmaniasis in red kangaroos: isolation and characterisation of the causative organisms. Int J Parasitol. 2004;34:655–64.

Rougeron V, De Meeûs T, Bañuls AL. A primer for *Leishmania* population genetic studies. Trends Parasitol. 2015;31:52–9.

Rougeron V, De Meeûs T, Bañuls AL. Reproduction in *Leishmania*: a focus on genetic exchange. Infect Genet Evol. 2017;50:128–32.

Schönian G, Mauricio I, Cupolillo E. Is it time to revise the nomenclature of *Leishmania*? Trends Parasitol. 2010;26:466–9.

Saf'janova VM, Aliev EI. Comparative study of biological characteristics of the causal agents of zoonotic and anthroponotic cutaneous leishmaniasis in the USSR. Bull World Health Organ. 1973;49:499–506.

Shaw J, Pratlong F, Floeter-Winter L, et al. Characterization of *Leishmania* (*Leishmania*) *waltoni* n. sp. (Kinetoplastida: Trypanosomatidae), the parasite responsible for diffuse cutaneous Leishmaniasis in the Dominican Republic. Am J Trop Med Hyg. 2015;93:552–8.

Siriyasatien P, Chusri S, Kraivichian K, et al. Early detection of novel *Leishmania* species DNA in the saliva of two HIV-infected patients. BMC Inf Dis. 2016;16:1–7.

Staley JT. Universal species concept: pipe dream or a step toward unifying biology? J Ind Microbiol Biotechnol. 2009;36:1331–6.

Telford SR. Hemoparasites of the reptilia: colour atlas and text. Boca Raton: CRC Press; 2009.

Yang Z, Rannala B. Molecular phylogenetics: principles and practice. Nat Rev Genet. 2012;13(5):303–14.

Zemanova E, Jirku M, Mauricio IL, et al. The *Leishmania donovani* complex: genotypes of five metabolic enzymes (ICD, ME, MPI, G6PDH, and FH), new targets for multilocus sequence typing. Int J Parasitol. 2007;37:149–60.

Parasite Biology: The Vectors

3

Vit Dvorak, Jeffrey Shaw, and Petr Volf

This chapter describes different aspects of phlebotomine sand flies (Diptera: Psychodidae, Phlebotominae) that are the only conclusively proven vectors of most disease-causing *Leishmania* species. However, biting midges are likely vectors (Dougall et al. 2011; Seblova et al. 2015) of some parasites belonging to the subgenus *Mundinia*. Sand flies also transmit other human and animal pathogens, namely, bacteria (*Bartonella*) and viruses (mainly *Bunyaviridae*) (Maroli et al. 2013). Life cycle and biology of sand flies are presented together with important stages of parasite development within the vector, immunomodulatory properties of sand fly saliva, a brief outline of taxonomy and their vectorial role of Old and New World leishmaniasis in different biomes.

3.1 Life Cycle

The life cycle of sand flies is fully terrestric. As holometabolic insects, after hatching from an egg, they undergo larval development that comprises four larval instars and a pupal period after which an adult emerges. Breeding sites of sand flies are generally poorly defined. Larvae were recovered from various sylvatic, peridomestic and domestic habitats (Fig. 3.1). While only a few species develop in an earth floor of human dwellings (*Phlebotomus* (*Euphlebotomus*) *argentipes* in India and *Phlebotomus* (*Adlerius*) *chinensis* in China), larvae of many species inhabit soil contaminated with

V. Dvorak • P. Volf (✉)
Department of Parasitology, Faculty of Science, Charles University,
Prague 2, CZ, Czech Republic
e-mail: icejumper@seznam.cz; volf@cesnet.cz

J. Shaw
Departamento de Parasitologia, Instituto de Ciências Biomédicas, Universidade de São Paulo,
São Paulo, SP, Brazil
e-mail: jayusp@hotmail.com

F. Bruschi, L. Gradoni (eds.), *The Leishmaniases: Old Neglected Tropical Diseases*, https://doi.org/10.1007/978-3-319-72386-0_3

Fig. 3.1 Typical breeding sites of important sand fly vectors in the Old World. (**a**) Termite mounds, breeding sites of *Phlebotomus martini* and *Ph. celiae*, south Ethiopia; (**b**) rural biotope in Acacia-Balanite woodlands growing on vertisols, breeding sites of *Phlebotomus orientalis*, north Ethiopia; (**c**) periurban rock boulders inhabited by hyraxes, breeding sites of *Phlebotomus arabicus*, northern Israel; (**d**) rural farm with stone walls and organic debris, breeding sites of *Phlebotomus neglectus* and *Ph. perfiliewi*, southern Macedonia (V. Dvorak)

dung, organic debris of animal shelters, and poultry coops. Sylvatic sites include forest litter, tree trunks, caverniculous biotopes, deep cracks of drying soils or rodent burrows (Feliciangeli 2004). Typical ecological affinities of several species to certain type of vegetation or habitat (i.e. termite hills) are described below.

Many breeding sites are also used as resting places by the adults, which due to their nocturnal activity hide in dark places during daylight hours. Several peridomestic endophilic species, many of them important *Leishmania* vectors, can be encountered in dark corners of houses, wall cracks and behind hanging clothes and furniture (*Ph.* (*Eu.*) *argentipes*, *Phlebotomus* (*Phlebotomus*) *papatasi*). Considerable densities of *Ph.* (*Ph.*) *papatasi* in agricultural fields in Israel (Orshan et al. 2016) or *Phlebotomus* (*Larroussius*) *orientalis* in intensively cultivated sesame fields in Ethiopia (Gebre-Michael et al. 2010) indicate that some can inhabit highly modified microhabitats.

Adult sand flies of both sexes feed on plant sap, nectar and honeydew; in addition, females feed on the blood of various vertebrate hosts. While a few are autogenous, the vast majority of species are anautogenous, requiring one or more blood meals to successfully produce progeny (Killick-Kendrick 1999).

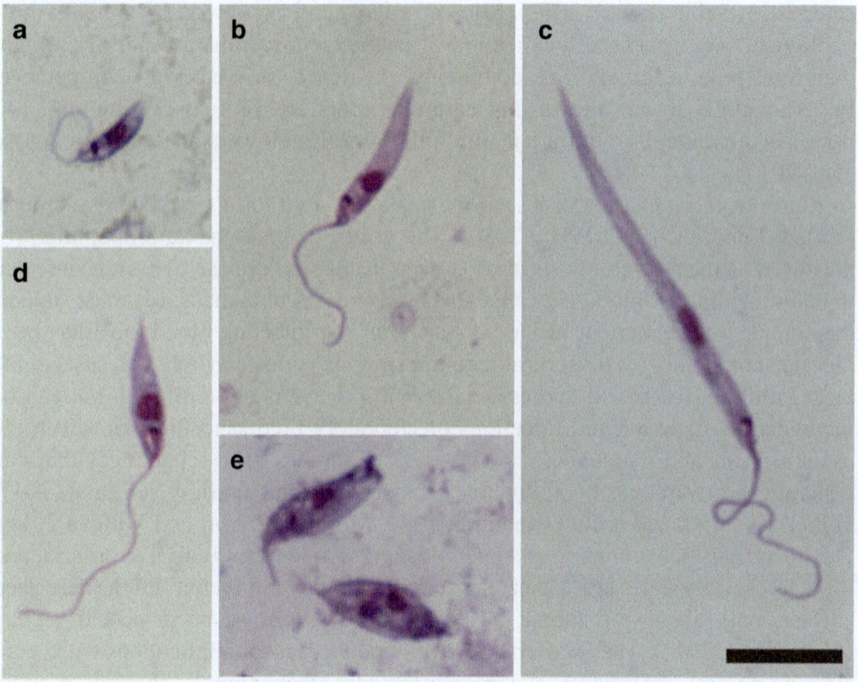

Fig. 3.2 Morphological forms of *Leishmania* occurring in sand fly midgut. *Leishmania major* in natural vector *Phlebotomus duboscqi*, scale bar 10 μm. (**a**) Procyclic promastigote; (**b**) leptomonad (=short nectomonads); (**c**) long nectomonad; (**d**) metacyclic promastigote; (**e**) two haptomonads (Courtesy of J. Sadlova)

3.2 Vector Competence and Its Epidemiological Consequences

During their life cycle, *Leishmania* parasites alternate between two major forms: in the sand fly's digestive tract, they develop as slender promastigotes that have a prominent flagellum and as rounded amastigotes that have no visible flagellum inside the vertebrate's phagocytic cells. Population of *Leishmania* promastigotes in sand fly gut is composed by several morphological forms (Fig. 3.2) that differ in cell shape, flagellum length, motility, surface molecules and other biochemical properties. These populations have specific roles during the vectorial part of the life cycle as they encounter various mechanical or biochemical barriers. These variable parasite forms and molecular factors affecting *Leishmania*-sand fly interaction are described in Sect. 3.3.

The *Leishmania* life cycle in the vector gut includes a complex series of interactions, which are in many cases species-specific. The early phase of infection in the vector is, however, nonspecific. Almost any *Leishmania* survive and divide within the sand fly blood meal, even in sand flies of genus *Sergentomyia* or in biting midges, but they thrive only until defecation (Lawyer et al. 1990; Sadlova et al. 2013; Seblova et al. 2012). Microscopical confirmation of *Leishmania* morphological stages and the localization of late-stage infections in sand fly midgut therefore remain crucial as convincing evidence of vectorial competence. Experimentally it was shown that natural vectors are very susceptible and 1–2 parasites are enough to initiate mature infections (Seblova et al. 2013; Pruzinova et al. 2015).

Rigorous incrimination of a sand fly species as a vector is based on five criteria (Killick-Kendrick 1990; WHO 2010). The vector (a) must be anthropophilic, (b) must feed on the reservoir hosts in zoonotic transmission cycles, (c) must be infected in nature with the same *Leishmania* species as occurs in humans, (d) must support the complete development of the parasite after the infecting blood meal has been digested and finally (e) must be able to transmit the parasite by bite to a susceptible host. Fulfilling the fourth and especially fifth criteria is often difficult but can be facilitated by experimental infections of laboratory bred sand flies. However, the establishment and maintenance of laboratory colonies poses challenges that can generally be overcome by following the correct protocols (Lawyer et al. 2017; Killick-Kendrick and Killick-Kendrick 1991; Modi 1997; Volf and Volfova 2011). Due to various constraints, it is sometimes impossible to fulfil all five criteria, and in such cases, a species is a "suspected vector" or "potential vector." Of the described species, only about 100 (about 10%) are listed as proven or suspected vectors (Maroli et al. 2013). These criteria have important practical implications when considering possible alternative vectors. For instance, *Sergentomyia* species are often abundant in leishmaniasis endemic foci, sometimes outnumbering *Phlebotomus* species. Originally, they were assumed to feed only on reptiles and were thus considered to transmit just parasites of subgenus *Sauroleishmania* (Lane 1993). However, many *Sergentomyia* species are known to feed on mammals, and indirect evidence suggests their possible involvement in the transmission of mammalian parasites (Maia and Depaquit 2016).

This chapter summarizes the most important Old and New World proven and suspected vectors focusing on the sand flies rather than on transmitted parasites. A too restrictive approach that insists on the fulfilment of all WHO criteria could eventually omit species which may have an important, albeit not yet recognized, epidemiological significance. Following this reasoning the inclusion of species is largely in accordance with the review of Maroli et al. (2013) that follows three minimal requirements for considering vectorial incrimination: (1) epidemiological evidence indicated by the overlapping of the geographical distributions of the vector and the human disease, (2) evidence that the vector feeds on humans and (3) evidence that the vector supports natural gut infections with promastigotes of the same *Leishmania* species as occurs in humans.

According to their ability to support late-stage development of different *Leishmania* species under experimental conditions, sand flies have been classified into two categories (Kamhawi 2006; Volf and Myskova 2007). Some species display strict specificity for *Leishmania* species, for example, *P.* (*Ph.*) *papatasi* supports late-stage development of *Leishmania* (*Leishmania*) *major* and *Leishmania* (*Leishmania*) *turanica* only (Pimenta et al. 1994; Chajbullinova et al. 2012). Such

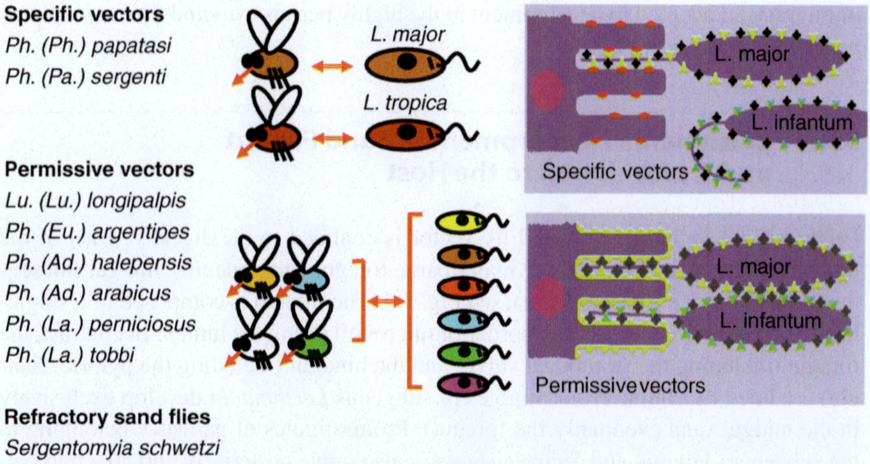

Fig. 3.3 Specificity of sand fly—*Leishmania* interaction: three categories of sand flies and proposed mechanisms of midgut attachment. Few sand fly species, namely, *P. papatasi* and *P. sergenti*, display remarkable specificity for *Leishmania* species they transmit. In these specific vectors midgut attachment requires specific binding of terminal sugars of lipophosphoglycan to lectin on midgut epithelium. These terminal sugars differ between *Leishmania* species and therefore *L. infantum* cannot attach to *P. papatasi* midgut. In contrast, most sand flies (e.g., *L. longipalpis*) are experimentally permissive for the development of a broad range of *Leishmania* species. Parasites attach due to unspecific binding to mucin-like molecules in sand fly midgut (P. Volf)

sand flies are called "specific" or "restrictive vectors." On the other hand, the majority of *Phlebotomus* and *Lutzomyia* species support late-stage development of multiple *Leishmania* species and are thus called "permissive vectors." In addition, there is also a third category, sand flies refractory to all *Leishmania* parasites are tested (Fig. 3.3).

Permissive sand fly species should be considered as potential vectors of various *Leishmania* species, and they could be responsible for establishing new leishmaniasis foci (Volf and Myskova 2007). In Israel, various *Leishmania (Leishmania) tropica* foci were described (Jacobson et al. 2003). Parasites possessing typical *L.* (*L.*) *tropica* lipophosphoglycan (LPG) terminated by glucose were transmitted by typical vector *Phlebotomus* (*Paraphlebotomus*) *sergenti*. In contrast, there is a focus in North Galilee (Svobodova et al. 2006) where *L.* (*L.*) *tropica* has modified LPG with atypical side chains (Soares et al. 2004). These parasites were not able to develop in the specific vector *Ph.* (*Pa.*) *sergenti* but were transmitted by the permissive species *Phlebotomus* (*Adlerius*) *arabicus* (Svobodova et al. 2006).

Probably the most important example of a permissive vector involved in the establishment of new disease foci is *Lutzomyia longipalpis*, the major New World vector of *L.* (*L.*) *infantum* (=*L.* (*L.*) *chagasi*). In the Old World around the Mediterranean Sea, *Leishmania infantum* is transmitted by *Ph. perniciosus* and other sand fly species belonging to subgenus *Larroussius* (Killick-Kendrick 1999). However, when imported into the New World by the dogs that accompanied Spanish and Portuguese immigrants, it adapted to development in the highly permissive sand fly *Lu. longipalpis* (Volf and Myskova 2007).

3.3 Leishmania Development in Sand Fly Gut and Transmission to the Host

Leishmania development in sand fly vector is confined to the digestive tract of the insect. This tract consists of three major parts, foregut (stomodeum), midgut (mesentheron) and hindgut (proctodeum), see Fig. 3.4. The midgut is composed of a single-layered epithelium with a brush border of microvilli lining the lumen. In contrast, the foregut (including the stomodeal valve) and the hindgut (including the pyloric triangle) are lined by chitin. Promastigotes of subgenus *Leishmania* develop exclusively in the midgut (and eventually the foregut). Promastigotes of parasites belonging to the subgenera *Viannia* and *Sauroleishmania* also move posteriorly and attach to chitin lining of the pylorus region (as haptomonads) (Lainson et al. 1977; Walters et al. 1993). The mechanisms of midgut attachment in members of subgenus *Viannia* remain to be studied in greater detail (Bates 2007; Wilson et al. 2010), but members of subgenus *Leishmania* develop exclusively in the midgut and foregut. Parasite development in the vector is summarized in Fig. 3.4, and in the text below.

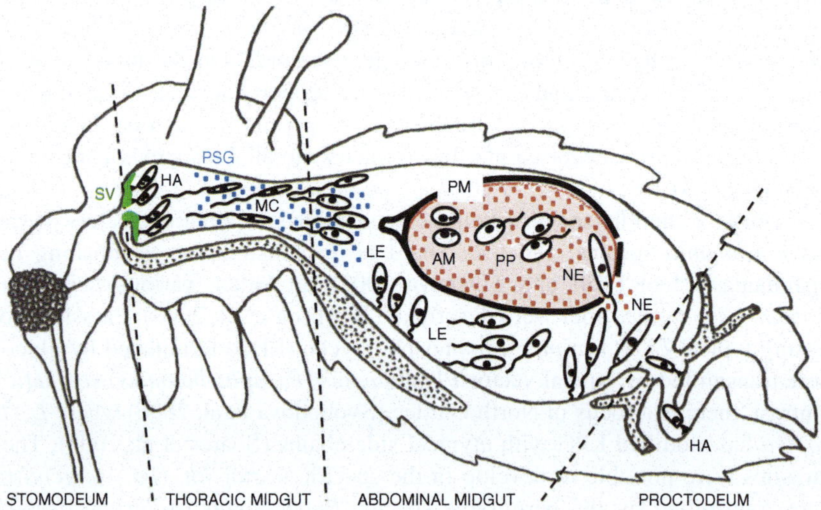

Fig. 3.4 Sand fly digestive tract and *Leishmania* development in the vector. Sand fly female ingest amastigotes (AM) together with the blood meal. In the abdominal midgut the blood meal is encased by peritrophic matrix (PM), and amastigotes transform to procyclic promastigotes (PP) which proliferate. After PM disintegrates, PP transform to long nectomonads (NE) which later transform to shorter form called leptomonads (*LE* = short nectomonads). These forms temporarily attach to midgut epithelium and proliferate. In subgenus *Viannia*, haptomonads (HA) attach to cuticular lining of proctodeum. In late-stage infections, leptomonads migrate to thoracic midgut, produce promastigote secretory gel (PSG) and transform either to metacyclics (MC) or haptomonads (HA) attaching to cuticular lining of stomodeal valve (SV) (J. Sadlova and P. Volf)

3.3.1 Early-Stage Development Within the Blood Meal

While feeding on the vertebrate, sand fly females take a blood meal that is directed to the abdominal midgut and surrounded by peritrophic matrix (PM) in so-called peritrophic sac or endoperitrophic space. The amount of the blood meal varies from 0.6 μL in *Ph. argentipes* or *Ph.* (*La.*) *orientalis* to 0.9 μL in *Ph.* (*Ph.*) *papatasi* and *Se. schwetzi* (Pruzinova et al. 2015). Amastigotes ingested along with a blood meal transform first into procyclic promastigotes. Within the peritrophic sac, these forms are exposed to nutrient-rich as well as relatively harsh environment full of proteolytic enzymes which are excreted in response to blood meal intake (Dillon and Lane 1993; Telleria et al. 2010, reviewed by Dostalova and Volf 2012). Some earlier studies have described that parasites transforming from amastigotes to promastigotes are susceptible to sand fly proteases, but recent experiments did not show any direct negative effect of these enzymes on transforming parasites (Pruzinova et al. 2017). *Leishmania* are protected against proteolytic damage by surface glycoconjugates called phosphoglycans; in the natural parasite-vector combination, these molecules enable promastigotes to thrive in an environment full of digestive enzymes (Secundino et al. 2010; Svarovska et al. 2010).

The PM is the acellular layer composed of proteins, glycoproteins and chitin. It is secreted by midgut epithelium in response to feeding, and it surrounds the blood in midgut lumen within few hours after the meal (Fig. 3.5) (Lehane 1997). It represents an important mechanical barrier to promastigotes and prevents their escape from endoperitrophic space. This may result in defecation of parasites together with remnants of the meal (reviewed by Bates 2008). At the end of digestion process, sand fly chitinases (Ramalho-Ortigao and Traub-Cseko 2003; Ramalho-Ortigao et al. 2005) disintegrate the

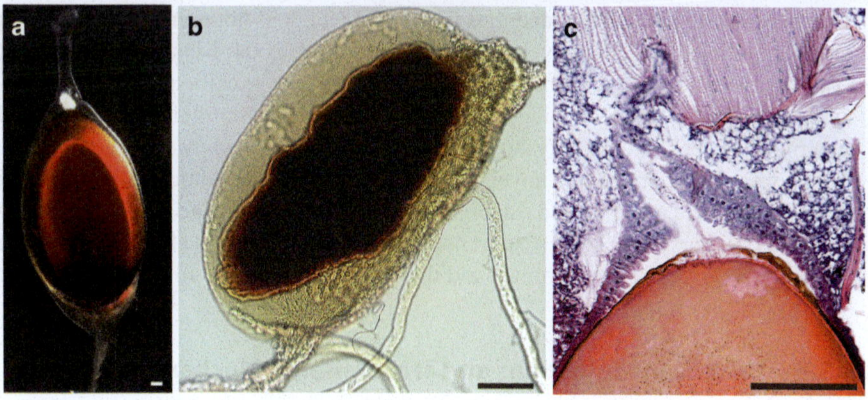

Fig. 3.5 The peritrophic matrix (PM). (**a**) The PM surrounding the blood meal, its anterior and posterior ends darkened by heme incrustations. Erythrocytes are agglutinated into a compact central bolus. Midgut of *P. papatasi* dissected 24 h post blood meal. (**b**) Thick and dark PM with folded surface at the end of digestive process, *S. schwetzi* midgut dissected by day 3 post blood meal. (**c**) Anterior end of PM with anterior plug, protrusion of the PM formed in the thoracic midgut which remains transparent as it is not in contact with heme. Histological section of *P. papatasi* by 24 h PBM stained with hematoxylin and eosine. Scale bars = 100 μm (Courtesy J. Sadlova)

PM on its posterior end (Sadlova and Volf 2009), but *Leishmania* chitinase does not seem to participate in this process (Schlein and Jacobson 1994; Sadlova and Volf 2009). The kinetics of PM synthesis and disintegration differs between sand fly species (Walters et al. 1993; Pruzinova et al. 2015) as does the period between PM breakdown and defecation. After the PM disintegrates, the promastigotes must attach themselves to the midgut wall to avoid being expelled with the blood meal remnants. The duration of this phase and the mechanisms controlling it determine whether a *Leishmania* will become established in the sand fly's midgut. In some species, like *Sergentomyia schwetzi*, this period is extremely short which results in this sand fly being refractory to various *Leishmania* parasites tested (Sadlova et al. 2013; Pruzinova et al. 2015).

3.3.2 Leishmania Attachment in the Midgut

In the midgut, promastigotes attach by flagella, inserting them between microvilli (Fig. 3.6), and this binding is the prerequisite for further development (reviewed by Bates 2008). The midgut binding is stage-dependent, being limited to the forms found in the middle phase of development (nectomonad and leptomonad forms), but absent in procyclics and metacyclics (Wilson et al. 2010).

However, the binding mechanism differs between sand fly species (Fig. 3.3). In *P.* (*Ph.*) *papatasi*, the well-known example of specific vector, the attachment to midgut microvilli is controlled by galectin (Kamhawi et al. 2004) which serves as a receptor for terminal galactose present on *L.* (*L.*) *major* and *L.* (*L.*) *turanica* lipophosphoglycan (LPG), the main surface glycoconjugate on parasite surface (Pimenta et al. 1997; Volf et al. 2014). More recently, partial involvement of the flagellar protein FLAG1/SMP1 was demonstrated in *P.* (*Ph.*) *papatasi*-*L.* (*L.*) *major* pair (Di-Blasi et al. 2015).

In contrast, in permissive sand flies the attachment does not require LPG, as even parasites deficient in LPG develop well (Rogers et al. 2004; Myskova et al. 2007; Svarovska et al. 2010). A correlation was found between specificity versus permissivity and glycosylation of midgut proteins; O-linked glycoproteins were detected in permissive species only (Myskova et al. 2007), and in *Lutzomyia longipalpis* this protein was characterized as a novel mucin localized on midgut microvilli (Myšková et al. 2016). Once attached and established in the midgut, parasite forms called leptomonads (Rogers et al. 2002) (=short nectomonads by Walters et al. 1993; Cihakova and Volf 1997) replicate vigorously and then migrate anteriorly to thoracic midgut.

3.3.3 Late-Stage Infections and Transmission Mechanisms

Leptomonad forms accumulate in large numbers in the thoracic part of the midgut (Fig. 3.6), and produce promastigote secretory gel (PSG) containing filamentous proteophosphoglycan (fPPG) (Stierhof et al. 1999). This PSG, together with parasite masses, physically obstructs the gut creating a gel-like plug (Rogers et al. 2002). Another parasite form found in matured infections is metacyclics: small, highly

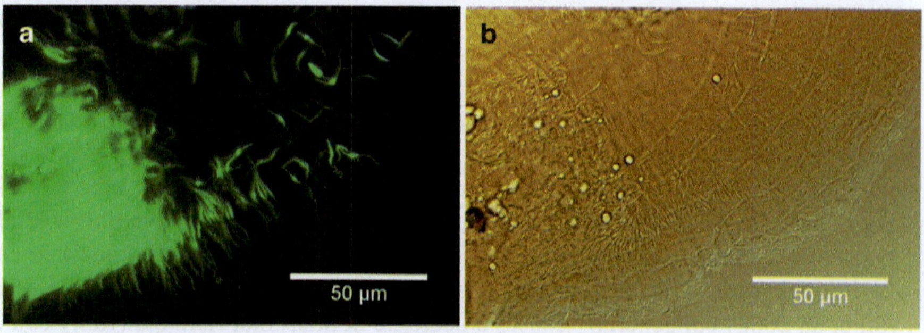

Fig. 3.6 Late-stage development of *Leishmania* in sand fly midgut. GFP transfected *L. donovani* in natural vector *P. orientalis*, 6 days post-infected blood meal. Same dissected midgut under fluorescent (**a**) and light microscopy (**b**). Promastigotes insert flagella between gut microvilli in the abdominal part of midgut, and mass of parasites form a plug in the thoracic part of midgut (Courtesy J. Sadlova)

motile forms possessing long flagella (Sacks and Perkins 1985; Rogers et al. 2002). They have a dense coat of long LPG molecules (Sacks et al. 1990), some stage-specific proteins, like HASPs and SHERP (Sadlova et al. 2010) and are highly infective for vertebrate hosts (Sacks and Perkins 1985).

In late-stage infections, *Leishmania* colonize the stomodeal valve. This valve is the junction between anterior midgut and foregut and in non-infected sand flies ensures one-way flow of the food. It is composed by cylindrical cells covered by chitin lining (Schlein et al. 1992; Volf et al. 2004). Promastigote forms called haptomonad attach to cuticular lining of the stomodeal valve through an expanded flagellum containing hemidesmosomal structures and cause damage to the structure of the stomodeal valve, interfering with its function and facilitating reflux of parasites from the midgut (Schlein et al. 1992). Stomodeal valve of heavily infected sand flies seems to be permanently open, the shape of cells is changed and the chitin lining is destroyed (Volf et al. 2004) due to the action of *Leishmania* chitinase (Rogers et al. 2008). The so-called blocked sand flies have problem in taking a blood meal, bite repeatedly, increasing the chance of *Leishmania* transmission (Rogers and Bates 2007). Clearly, pathological changes of the stomodeal valve, together with obstruction of the thoracic midgut by PSG, are important for parasite transmission.

In matured infections, metacyclic promastigotes are concentrated in the anterior part of midgut, on the edge of PSG plug (Rogers et al. 2002), sometimes invading also the pharynx, cibarium and proboscis. Occasionally *Leishmania* metacyclics were found in salivary glands of sand flies (Killick-Kendrick et al. 1996) or in urine discharged by infected females during blood feeding (Sadlova and Volf 1999). However, two principal mechanisms were described: either small numbers of metacyclics present in the proboscis are deposited into the skin during feeding or parasites masses residing in thoracic midgut behind the stomodeal valve are regurgitated with a backflow of ingested blood (reviewed by Bates 2008). The parasite numbers transmitted by individual sand flies are very variable. In *L.* (*L.*) *major-P.(Ph.) duboscqi* model, numbers of promastigotes inoculated by individual sand flies ranged between 10 and 10^5 (Kimblin et al. 2008) and in *L. infantum* combinations with *P.(La) perniciosus* or *Lu. (Lu.) longipalpis* from 4 up to 4×10^4 (Maia et al. 2011).

The development of *Leishmania* is also affected by bacteria, which usually inhabit the sand fly's midgut (Dillon and Dillon 2004; Sant'Anna et al. 2014; Kelly et al. 2017). Most of these bacteria are acquired by sand fly females from environment or with a contaminated sugar meal, but few specific ones survive transstadial passage from larval stage (Volf et al. 2002), like strain AK recently identified as *Ochrobactrum* sp. Blood feeding supports growth of aerobically cultivating bacteria (Volf et al. 2002) but generally causes a decrease in microbial richness (Kelly et al. 2017). Some bacteria species, like *Serratia marcescens*, were found to diminish subsequent colonization of *Lu. longipalpis* midgut with *Leishmania mexicana*, and conversely, established *L. mexicana* infection protected sand flies from *Serratia*-induced mortality (Sant'Anna et al. 2014). However, other bacteria are required for full *Leishmania* development, and antibiotic-mediated perturbation of the midgut microbiome rendered sand flies unable to support *Leishmania* metacyclogenesis (Kelly et al. 2017). Manipulating or inhibiting the sand fly's gut bacterial fauna represents a novel approach to controlling *Leishmania* transmission (Kelly et al. 2017).

3.3.4 Sex in the Vector

Hybrid parasites have been observed in wild caught sand flies for several decades in both the Old World and the Americas, and in some cases, they were found widespread (reviewed by Rougeron et al. 2010). However, only recently has genetic exchange, sometimes referred to as a sexual cycle, been attained in experimentally infected laboratory bred sand flies (Akopyants et al. 2009; Sadlova et al. 2011). Hybridization occurs within the natural as well as unnatural parasite-vector pairs (Inbar et al. 2013) and even between various *Leishmania* species (Romano et al. 2014; Volf et al. 2007). These studies provided the evidence that genetic exchange is responsible for the generation of hybrids observed in natural populations. No morphologically distinct sexual *Leishmania* stages were observed, but the early timing of hybrid recovery (Sadlova et al. 2011; Inbar et al. 2013) suggests that nectomonads are the most likely sexually competent stage of the parasite (Inbar et al. 2013). Importantly, interspecific hybridization may increase the fitness of parasite progeny, enabling them to develop in new sand fly vectors (Volf et al. 2007) and modify their ability to produce pathology in the skin or disseminate to and grow in visceral organs (Romano et al. 2014). In nature, this may lead to newly emerged foci, where all isolates are derived from a single cross of two diverse strains with rare subsequent recombination with the population reproducing primarily clonally (Rogers et al. 2014).

3.3.5 Immunomodulation of the Host During *Leishmania* Transmission

In the host skin, metacyclic promastigotes are swallowed by the phagocytic cells, neutrophils and macrophages (Peters et al. 2008), and inside them transform to non-flagellated amastigotes. It is, however, important to note that *Leishmania* are co-inoculated into the vertebrate host together with sand fly saliva and midgut content that also includes PSG, microbiota (see above) and vesicles called exosomes (Atayde et al. 2015). All these components immunomodulate the host and enhance the virulence of the parasite. The effect of saliva on *Leishmania* infection in the mammalian host is called "enhancing effect" or "exacerbating effect" as it results in more severe disease accompanied by larger lesions when compared to the controls receiving parasites alone (Titus and Ribeiro 1988; Theodos et al. 1991). There is a vast literature about the molecular mechanisms of this effect on mammalian immune cells (reviewed by Abdeladhim et al. 2014; Lestinova et al. 2017). Here we decided to give only a brief summary as the detailed overview is beyond the scope of this chapter.

Saliva is chemotactic for different immune cells. The influx of phagocytes is beneficial to parasites because of their earlier entry into these cells that protect them from hostile extracellular environment in the mammalian host. First host cells recruited to the feeding site and entered by *Leishmania* are neutrophils (Silva et al. 2005; Peters et al. 2008). *Leishmania* survive temporarily inside these phagocytic cells; moreover, in the presence of sand fly saliva, infected neutrophils produce chemokines which attract macrophages. These cells phagocyte the infected neutrophils, and *Leishmania* silently enter the macrophage either inside apoptotic neutrophils (van Zandbergen et al. 2004) or after escaping from dying neutrophils (Peters et al. 2008).

In parallel, saliva modifies the inflammatory process at the site of bite (Anjili et al. 1995; Zer et al. 2001) and T-cell response to *Leishmania* antigen (Theodos and Titus 1993). It enhances the production of anti-inflammatory cytokines (Rogers and Titus 2003; Costa et al. 2004) and inhibits various functions of macrophages, reducing expression of inducible nitric oxide synthase (iNOS) (Mbow et al. 1998) and production of nitric oxide (Hall and Titus 1995) which facilitates *Leishmania* survival. In *Lu. (Lu.) longipalpis*, the immunomodulating component was identified as a potent vasodilator called maxadilan (Morris et al. 2001).

However, mice repeatedly exposed to sand fly saliva or to uninfected sand flies were protected against *Leishmania* infection (Belkaid et al. 1998; Kamhawi et al. 2000). This is due to induction of delayed-type hypersensitivity and inflammatory cytokines elicited by repeated exposure to sand fly salivary antigens, which create an inhospitable environment for *Leishmania* injected by the vector (Belkaid et al. 1998). This saliva-elicited immunity in repeatedly bitten host is sand fly species-specific (Thiakaki et al. 2005) and limited to short-term exposures (Rohousova et al. 2011), but it led to development of transmission blocking vaccines using saliva-derived proteins or plasmids coding these proteins (Valenzuela et al. 2001).

Another potent factor enhancing *Leishmania* infection is PSG, particularly glycan moieties of its main bioactive component fPPG (Rogers et al. 2004). Similarly to saliva, the enhancing role of PSG was demonstrated for different cutaneous and visceral parasites (Rogers 2012). PSG is chemotactic to macrophages, and its exacerbative effect is probably related to its modulation of macrophage activity (Rogers et al. 2009). Finally, infected sand flies regurgitate exosomes that are liberated by the parasite inside the sand fly into the feeding pool lesion. These vesicles secreted by *Leishmania* into midgut lumen of infected sand fly female contain parasite DNA and protein. They were proven to regulate the immune response of the host and exacerbate disease outcome as well (Atayde et al. 2015).

3.4 Taxonomy and Species Identification of Sand Flies

Taxonomy of phlebotomine sand flies underwent a long and complex evolution and still fails to bring universal agreement on the generic and higher classification that would truly reflect the presumed phylogeny of the group. Conservative approach based on practical criteria and involvement of certain groups in the transmission cycles of leishmaniases recognizes six genera, three in the Old World (*Phlebotomus*, *Sergentomyia* and *Chinius*) and three in the New World (*Lutzomyia*, *Brumptomyia* and *Warileya*). Traditionally, vectors of human leishmaniases are grouped in two sand fly genera only, Old World sand flies in genus *Phlebotomus* and New World sand flies in genus *Lutzomyia*. However, species of each genus are now further divided into numerous subgenera and genera, some of which have a practical importance as they harbour vectors that transmit different *Leishmania* species.

In this chapter, we retained the single genus *Phlebotomus* denomination for the Old World vectors and refer to its subgenera. American species names follow an updated nomenclature that raises many *Lutzomyia* subgenera to the generic level (Galati 2003). It is beyond the scope of this chapter to further discuss this taxonomic arrangement; however, it appears inevitable that similar revision will be adopted also for the subgenera of the Old World genus *Phlebotomus*. Future phylogenies of the genera and subgenera of the Phlebotominae subfamily will require both nuclear and mitochondrial genes combined with other criteria related to ecology and biology of these groups (Akhoundi et al. 2016).

The Old World genus *Sergentomyia* has an extensive geographical distribution and contains ten subgenera and many ungrouped taxa (Seccombe et al. 1993); many of the latter are poorly described and known from only a few specimens. However, some like *Sergentomyia garnhami* are found in many geographical regions. Unlike the medically important taxonomic groups of *Phlebotomus* and the American genera, the *Sergentomyia*'s taxonomy and phylogeny are less studied. Their role in the transmission of *Leishmania* parasites remains unproven, as discussed later.

Species identification of sand flies traditionally utilizes morphological characters mainly on the head (cibarium and pharyngeal armature, antennal and palpal formula) and genitalia (arrangements of aedeagi and appendages in males, morphology of internal genitalia in females) (Killick-Kendrick 1990). Advances of molecular techniques enabled their deployment in species identification, relying first on isoenzymes and cuticular hydrocarbons but nowadays mainly on DNA markers (Depaquit 2014). DNA barcoding approach that uses a standard, single-locus marker for species identification was successfully applied also on sand flies to identify species at different geographical scales, including countrywide inventories (Contreras Gutiérrez et al. 2014; Nzelu et al. 2015). It helped in revealing cryptic species ("reverse taxonomy") as in description of two new species of the subgenus *Transphlebotomus*, *Ph. (Tr.) anatolicus* and *Ph. (Tr.) killicki*, where molecular evidence preceded morphological differences (Kasap et al. 2015).

Further savings of financial and time investment offer high-throughput metabarcoding of sand fly mixtures as demonstrated on correct species-level assignations of neotropical sand flies in French Guiana (Kocher et al. 2017). Wider availability of next-generation sequencing (NGS) techniques will facilitate species identification of large-scale entomological surveys. Beside DNA-based methods, MALDI-TOF protein profiling that identifies sand fly specimens (Dvorak et al. 2014) upon species-specific protein fingerprints is an attractive alternative for its simple, rapid and low-cost sample preparation. The aforementioned molecular methods enable vouchering (i.e. preservation of morphological vouchers; in case of sand flies, relevant body parts mounted on slides) and represent complementary approaches that supplement rather than substitute traditional morphological techniques.

3.5 *Phlebotomus* Species as *Leishmania* Vectors in the Old World

In the Old World, over 40 sand fly species within the genus *Phlebotomus* are regarded as proven or probable vectors, depending on applied criteria. Most of them are implicated in the transmission of *Leishmania* (*Leishmania*) *infantum*/*donovani* complex (26); others are vectors of predominantly CL agents *L.* (*L.*) *major* (7), *L.* (*L.*) *tropica* (7) and *L.* (*L.*) *aethiopica* (3). With few exceptions described below, each serves as a natural vector of a single *Leishmania* species, mainly due to geographical distribution, ecology and behaviour (e.g. host preferences), while many were found permissive to various *Leishmania* under laboratory conditions (see Sect. 3.2).

Together with the decision on the stringency of "proven" and "suspected" vector categories, it is necessary to understand the complex and more than often unresolved taxonomy of many sand fly species throughout most of Old World subgenera. This is not a theoretical issue confined to taxonomists; understanding evolution of a given taxon in time is necessary to correctly assess historical records and epidemiological significance. Traditional taxonomy is developed based largely on a taxonomic species concept. With the boom of molecular approaches that use the genetic properties of organisms, the phylogenetic species concept has stepped in, revealing putative cryptic species, some of which may have distinct epidemiological significance. Classification of Old World species that is applied below is rather conservative not to blur their epidemiological and medical significance; however, future reclassification based mainly on complementary use of morphological and phylogenetic approaches may enrich the list of species incriminated in transmission cycles.

3.5.1 The Subgenus *Phlebotomus* (*Ph.*)

The importance of *Phlebotomus* (*Ph.*) *papatasi* is manifold: it was the first sand fly described formally as a species; it has one of the widest distributions in the Old World, covering three zoogeographical regions of Africa, Europe and Asia; and it serves as a principal vector of *L.* (*L.*) *major*, causative agent of zoonotic cutaneous leishmaniasis, in many regions. In some areas, its ecology is closely connected with the reservoir hosts, as it uses burrows of colonial gregarious rodents of the genera *Rhombomys*, *Psammomys* or *Meriones* as breeding sites. The species is anthropophilic and readily takes blood meals on humans. Despite its vast geographical distribution, the study on mitochondrial and nuclear markers including microsatellites showed notably low degree of intraspecific variability and population structuring, suggesting that the absence of *L.* (*L.*) *major* transmission in some areas where *Ph.* (*Ph.*) *papatasi* is present is probably due to the lack of reservoir hosts rather than different vectorial capacities of local sand fly populations (Depaquit et al. 2008; Hamarsheh et al. 2009).

Ph. (*Ph.*) *duboscqi* is a proven vector of *L.* (*L.*) *major* in Senegal and is a suspected one in most Western African countries including Mali, where endemic foci are known for more than a century. Several autochthonous rodent species were found infected with *L.* (*L.*) *major* (*Arvicanthis niloticus*, *Mastomys erythroleucus*, *Tatera gambiana*), and sand flies breed in these same burrows. Seasonal study in Mali demonstrated its anthropophily (Anderson et al. 2011). *P.* (*Ph.*) *bergeroti* is a suspected vector of *L.* (*L.*) *major* in several regions of Africa and Middle East; *P.* (*Ph.*) *salehi* is the only species of the subgenus restricted to occur only in Asia and was proven to transmit *L.* (*L.*) *major* in Iran, India and probably also Pakistan (Maroli et al. 2013).

3.5.2 The Subgenus *Paraphlebotomus* (*Pa.*)

Within this subgenus, a principal vector species is *Ph.*(*Pa.*) *sergenti*, which has a broad range of distribution, stretching from Canary Islands and Madeira over Portugal, Spain, Maghreb, Sicily, Turkey, Egypt, Middle East up to Pakistan and India in west-east direction, from southern France to Kenya in north-south direction. It is a proven and specific vector of *L.* (*L.*) *tropica* causing mainly anthroponotic cutaneous leishmaniasis in many countries of North Africa, Middle East, Central and East Asia. Unlike *Ph.* (*Ph.*) *papatasi*, within this wide range it shows population structuring that reflects proposed scenario of species expansion from ancestral area (Depaquit et al. 2002). It is anthropophilic in most places studied. However, in some foci it was shown to feed also on a wide range of both domestic and wild animals (Ajaoud et al. 2015). Low rates of feeding on humans were detected in Israel where *Ph.* (*Pa.*) *sergenti* takes blood meals mainly on rock hyraxes which serve as reservoir hosts of *L. (L.) tropica* transmission cycle that is here zoonotic (Svobodova et al. 2006). Study on breeding sites in a focus of CL in Judean desert showed that both natural (caves) and man-made (support walls) habitats are utilized (Moncaz et al. 2012). While proven vector of *L.* (*L.*) *tropica* in many countries, *Ph.* (*Pa.*) *sergenti* was also demonstrated to host *L. aethiopica* in a focus of leishmaniasis in the Awash Valley in northern Ethiopia (Gebre-Michael et al. 2004).

Closely related *Ph.* (*Pa.*) *similis* may serve as a vector of *L. (L.) tropica* in eastern Mediterranean, including a reactivated historical focus in Crete; however, conclusive evidence is not available (Christodoulou et al. 2012). Three species of the *P.* (*Pa.*) *caucasicus* complex (*Ph.* (*Pa.*) *caucasicus*, *Ph.* (*Pa.*) *mongolensis* and *Ph.* (*Pa.*) *andrejevi*) that exhibit different degrees of xerophily were shown to transmit *L.* (*L.*) *major* to animal hosts in Central Asia but probably do not participate in the transmission to humans (Strelkova 1996). The latter two play the same role in some regions of China (Lun et al. 2015).

Phlebotomus (Pa.) alexandri has a very wide range of distribution that stretches from the western Mediterranean and east Africa across towards China. Nevertheless, throughout this vast distribution, it was only proven as a vector of VL due to *L.* (*L.*) *infantum* in north-western China where it readily takes blood meals on humans and inhabits stony deserts at the foot of mountains at altitudes of 500–1750 m with sparse vegetation (Guan et al. 2016). However, in other countries like Iran, *Ph. (Pa.) alexandri* is suspected vector of CL, occurs in different biotopes and is known to be more synanthropic (Jahanifard et al. 2014).

3.5.3 The Subgenus *Synphlebotomus* (*Sy.*)

Only two species of this subgenus are incriminated in the transmission; however, both *Ph.* (*Sy.*) *martini* and *Ph.* (*Sy.*) *celiae* are significant proven vectors of anthroponotic VL caused by *L.* (*L.*) *donovani* in several countries of East Africa (Sudan, Ethiopia, Kenya and Uganda), one of the regions worst affected by kala-azar. They both show a strong and distinctive ecological association with termite mounds, which provide suitable breeding and resting sites inside their ventilation shafts (Hoogstraal and Heyneman 1969). Peak of their abundance is typically associated with high humidity and moderate temperature (Gebre-Michael et al. 2004), unlike *Ph. (La.) orientalis*, another principal vector of *L.* (*L.*) *donovani* in East Africa, which exhibits different ecological associations and climatic preference.

3.5.4 The Subgenus *Larroussius* (*La.*)

This subgenus harbours many important, mainly permissive vectors that contribute to transmission cycles of several *Leishmania*. About 14 members of this subgenus are regarded as proven or suspected vectors of *L.* (*L.*) *infantum* in the Mediterranean basin (southern Europe, northern Africa and adjacent regions) where principal reservoir host is a dog, while the role of cats as well as various wild canids is also expected. In the western Mediterranean, *Ph. (La.) ariasi* and *Ph. (La.) perniciosus* are proven vectors, while *Ph. (La.) neglectus*, *Ph. (La.) tobbi* and *Ph. (La.) perfiliewi* transmit VL in the east Mediterranean. Some form different morphotypes and constitute species complexes: *P. (La.) perniciosus*, *Ph. (La.) longicuspis*, *Ph. (La.) major* and *Ph. (La.) perfiliewi* (Pesson et al. 2004; Kasap et al. 2013; Depaquit et al. 2013); there may be variations in ecological preferences within the complexes and species in reference to climate, altitude and other characteristics as shown for *Ph. (La.) perniciosus* complex in Morocco (Kahime et al. 2015).

Recent northward spread of *Ph. (La.) neglectus* was demonstrated in Italy, followed by expansion of canine leishmaniasis (Maroli et al. 2008). Several modelled scenarios based on anticipated changes of climatic variables in Europe suggest a possible spread of other *Larroussius* species into recently non-endemic regions (Fischer et al. 2011; Haeberlein et al. 2013). Outside the Mediterranean region, *Ph. (La.) kandelakii* was proven to transmit *L. infantum* in urban foci of VL in Georgia, and it is suspected to facilitate the transmission also in neighbouring countries (Giorgobiani et al. 2012).

In some foci, while being caused by *L. (L.) infantum* or *L. infantum/L. donovani* hybrids, the clinical manifestation of the disease is nevertheless cutaneous; *Ph. (La.) tobbi* was identified as vector of CL caused by *L. infantum/L. donovani* hybrids in Anatolia, Turkey (Svobodova et al. 2009; Rogers et al. 2014), while *Ph. (La.) wui* was demonstrated as a vector of CL caused by *L. (L.) infantum* in north-western China (Guan et al. 2013, 2016).

Vast data were accumulated over four decades about the ecology of *Ph. (La.) ariasi*, showing that the species prefers subhumid and humid bioclimatic zones in altitudes often exceeding 800 m, while sympatrically present *P. perniciosus* preferentially occurs in lower altitudes (Rioux et al. 1980; Ballart et al. 2014; Kahime et al. 2015; Prudhomme et al. 2016). In southern France *Ph. (La.) perniciosus* is an opportunistic feeder with wide host preferences (Cotteaux-Lautard et al. 2016), while in Greece *Ph. (La.) perfiliewi* and *Ph. (La.) tobbi* readily feed on humans despite availability of animal hosts (Chaskopoulou et al. 2016).

In eastern Africa, *Ph. (La.) orientalis* is a proven vector of *L. (L.) donovani* (Elnaiem 2011; Seblova et al. 2013) causing VL in highly endemic foci of Sudan, South Sudan, Ethiopia and northern Kenya. It shows periodic seasonality pattern following the dry and rainy seasons and distinct ecological association with Acacia-Balanite woodlands (Hoogstraal and Heyneman 1969; Elnaiem 2011). In Ethiopia, the species also adapted to open agricultural fields, and therefore, apart from permanent residents of the foci, also migrant workers bitten during seasonal agricultural campaigns are affected. In both habitats, *Ph. (La.) orientalis* breeds in deep cracks of vertisols (Moncaz et al. 2014). Low numbers of specimens collected indoors suggest exophily (Yared et al. 2017). Experiments testing the feeding habits revealed that the species is primarily zoonotic and attracted to different domestic and wild animals (Gebresilassie et al. 2015).

Beside their profound importance as vectors of *L. (L.) donovani/infantum*, two *Larroussius* species, namely, *Ph. (La.) pedifer* and *Ph. (La.) longipes*, are key vectors of *L. aethiopica*, a parasite that occurs mostly in the highlands of Ethiopia and causes zoonotic cutaneous leishmaniasis with hyraxes serving as reservoir hosts. These sand fly species live in high altitudes (1700–2700 m) being typically rural; they readily feed on humans and cattle although their primary blood meal source is the hyrax. However, *Ph.* (*La.*) *pedifer* occurs also in Kenya, the only other country where *L. aethiopica* is autochthonous. It may occur in suburban habitats and was described as the vector in a VL focus in residential districts of Addis Ababa (Lemma et al. 2009).

The role of *Larroussius* species in the transmission of *L.* (*L.*) *tropica* is not pivotal; however, in Kenya, *Ph.* (*La.*) *guggisbergi* is a proven vector, while *Ph.* (*Ph.*) *aculeatus* is suspected to participate in the transmission in other Kenyan foci (Lawyer et al. 1991).

3.5.5 The Subgenus *Adlerius* (*Ad.*)

This subgenus has very complex taxonomy, and conclusive identification of its species is a significant challenge, especially in females where we lack robust and stable morphological species-specific characters. The phylogenetic relations within the subgenus are not well resolved, and species status of several taxons is unclear, partly because the core of distribution is in Asia, including regions poorly accessible due to remoteness or political instability. Nevertheless, several members of this subgenus are important proven or suspected vectors of VL caused by *L.* (*L.*) *infantum*, and one species serves as an alternative vector of *L.* (*L.*) *tropica*.

In China, *Phlebotomus* (*Adlerius*) *chinensis* is the principal VL vector, a species with an extensive geographical distribution that stretches from the Yangtze river up to over 20 provinces in the northern China (Wang et al. 2012), indicating a high ecological plasticity and adaptation to markedly different habitats in coastal plains, mountains and plateaus (Guan et al. 2016). Consequently, it serves as a vector of two distinct transmission VL cycles: zoonotic VL in mountainous regions of Sichuan with dogs and wild canids as reservoirs and anthroponotic VL in Henan with human reservoirs (Lun et al. 2015). *Ph.* (*Ad.*) *sichuanensis* is a sister species which occurs at higher altitudes sympatrically with *Ph. chinensis* that is also incriminated in VL transmission (Zhang et al. 2013).

In Georgia, *Ph.* (*Ad.*) *balcanicus* was proven to be a vector of VL due to *L.* (*L.*) *infantum* in urban foci, together with *Ph.* (*Larroussius*) *kandelakii*. Sand flies of both species successfully utilize urban habitats including the capital Tbilisi where they are found on the slopes in courtyards, orchards and vegetable gardens (Giorgobiani et al. 2012). *Ph.* (*Ad.*) *longiductus* is a dominant species in VL foci in Kazakhstan and was shown to transmit *L.* (*L.*) *infantum* to golden hamsters (Dergacheva and Strelkova 1985). It is also suspected to transmit *L.* (*L.*) *infantum* in western provinces of China (Guan et al. 2016) and *L.* (*L.*) *donovani* in Himalayan regions of India (Sharma et al. 2009). Several other species, including *Ph.* (*Ad.*) *brevis*, *Ph.* (*Ad.*) *halepensis* and *Ph.* (*Ad.*) *turanicus*, are suspected vectors of *L.* (*L.*) *infantum* in Middle East and Central Asia mainly upon their geographical distribution and ecological observations (Maroli et al. 2013; Strelkova et al. 2015).

Ph. (*Ad.*) *arabicus* was shown to be an efficient vector in a focus of zoonotic cutaneous leishmaniasis in northern Israel. Two markedly different transmission cycles were characterized in closely adjacent foci near the Lake Kinneret (Jacobson et al. 2003). In a southern focus, *L.* (*L.*) *tropica* strains with typical LPG were transmitted by *Ph.*(*Pa.*) *sergenti*. In a northern focus, however, *L.* (*L.*) *tropica* strains with a different LPG were transmitted by *Ph.* (*Ad.*) *arabicus*. Experimentally, *Ph.* (*Pa.*) *sergenti* was shown to be refractory to these "nontypical" *L.* (*L.*) *tropica* strains because of rearrangement of LPG structure (Svobodova et al. 2006). Rock hyraxes (*Procavia capensis*) were incriminated as reservoir hosts in both foci.

3.5.6 The Subgenus *Euphlebotomus* (*Eu.*)

While most species of this subgenus are not incriminated in transmission of leishmaniases, *Ph.* (*Eu.*) *argentipes* is of a paramount importance being the proven vector of kala-azar due to *L.* (*L.*) *donovani* in Indian subcontinent. This region is one of the world's worst VL foci where two thirds of all human cases occur (Chappuis et al. 2007; Alvar et al. 2012). It is a synanthropic species that utilizes microhabitats in houses, animal sheds and mixed dwellings and shows endophilic and endophagic behaviour. In addition, a capability of notable vertical dispersal was demonstrated by collecting the specimens in canopies of palmyra palm trees (*Borassus flabellifer*) up to 18 m above the ground in Bihar (Poché et al. 2012), suggesting that a proportion of the population shows also consistent exophilic behaviour with implications on vector control measures. Preferentially a cattle feeder, the species also readily takes mixed blood meals and shows anthropophilic behaviour in foci where no cattle is available (Srinivasan et al. 2015).

Exhaustive morphological reassessment revealed the existence of three sibling species within *Ph.* (*Eu.*) *argentipes* complex in India: *Ph.* (*Eu.*) *argentipes* sensu stricto, *Ph.* (*Eu.*) *annandalei* and *Ph.* (*Eu.*) *glaucus* (Illango 2010). A study combining morphological and molecular data revealed the presence of two undetermined sibling species named A and B in Sri Lanka (Gajapathy et al. 2013).

Despite Indian subcontinent being a highly endemic region for visceral leishmaniasis, sporadic foci of cutaneous leishmaniasis are reported. In a CL focus in Kerala, the causative agent was *L.* (*L.*) *donovani* (Kumar et al. 2015). Subsequent studies using multiple genetic markers on pools of local sand fly species incriminated *Ph.* (*Eu.*) *argentipes* in the transmission of *L.* (*L.*) *donovani* CL. Interestingly, DNA barcoding of *Ph.* (*Eu.*) *argentipes* population revealed presence of all the sibling species of the complex, two lineages corresponding to species A and B from Sri Lanka, suggesting that they may be the vectors of *L.* (*L.*) *donovani* in this focus (Srinivasan et al. 2016).

3.5.7 Species of Other Old World Subgenera

In eastern Sudan, females of *Phlebotomus* (*Anaphlebotomus*) *rodhaini* were found infected by *L.* (*L.*) *donovani* in a zoonotic focus of VL where *Ph.* (*La.*) *orientalis* is a proven vector (Elnaiem et al. 2011). *Ph.* (*An.*) *rodhaini* is considered a rare species that is not anthropophilic, but it was suggested to contribute to the transmission among the animal reservoir hosts.

One or more species of the subgenus *Transphlebotomus* may be incriminated in transmission of *L.* (*L.*) *infantum* in some parts of Europe. While they are generally regarded as caverniculous, recently described *Ph.* (*Tr.*) *anatolicus* was recorded from microhabitats near domestic animals and human dwellings (Kasap et al. 2015). *Ph.* (*Tr.*) *mascittii* has an extensive distribution that extends from the Mediterranean basin northwards into Germany, Austria and southern Slovakia (Dvorak et al. 2016). It was suggested as a vector in the autochthonous canine leishmaniasis focus in southern Germany's Rhine valley (Naucke et al. 2008).

3.6 Vectors in the American Continent

Some 535 phlebotomine sand fly species have been described in the Americas (Galati 2015), and of these approximately 45 species (WHO 2010) have in one way or another been implicated as vectors. The greatest challenge is to grade the vectorial capacity of a species. In some cases, the supporting evidence is irrefutable as to its importance as a consecrated vector, but in others it may merely refer to an infection of a *Leishmania* parasite determined by a molecular test. The keys to this problem lie in understanding the zoonotic cycle, the fly's biology and its susceptibility to infection (Shaw 1999). Unlike the Old World vectors that are often found in arid and semiarid areas, the vast majority of American sand flies are closely associated with forests. A level of adaptation to other habitats and food sources varies from species to species. Some have left their native environment, particularly in the drier areas to take up residence in peridomestic and urban sites. This has led to exceptional increases in population sizes that have resulted in greater levels of transmission. An excellent example of this is the major vector of visceral leishmaniasis (VL), *Lutzomyia longipalpis* is now found in many major Brazilian cities where the disease has become endemic (Costa 2008). These same trends can be found throughout the Americas from Argentina to Mexico as well as for other species.

However, generalizations can be dangerous and as more populations of a species are examined by sophisticated molecular methods what was once considered to be a species is found to be a species complex. Again, *Lu. longipalpis* is an excellent example of this. Biological studies showed mating barriers (Ward et al. 1983) that were fully confirmed by later genetic studies (Bauzer et al. 2002). The idea of species complexes raises the fascinating question of possible variations in vector capacity.

The literature citations that are used are those that we consider best illustrate different epidemiological situations. The abbreviations of the sand fly genera follow those suggested by Marcondes (2007). All the information on distributions and infections is based on a careful review of the literature, but it is impossible to cite each one given the limited space. We have prioritized certain references that show particular trends that may not be obvious to the general reader.

3.7 Interspecific Genetic Diversity and Species Complexes of American Sand Flies

Genetic diversity increases when a founding population expands, and as biological and geographical barriers are crossed, genetically distinct populations arise resulting in species and species complexes. Hybridization and introgression between genetically distinct groups occur, and the permeability of the species barrier (gene exchange) depends on a genome region's function (Harrison and Larson 2014). American *Leishmania* vectors, such as *Nyssomyia umbratilis*, *Ny. whitmani* and *Lu. longipalpis*, have extensive geographical distributions in different biomes. Are there genetic differences between these populations, and if so, are they associated with variations of vectorial capacity?

In Panama and neighbouring Pacific Coast, *Ny. trapidoi* is the principal vector of *L.* (*Viannia*) *panamensis*. Populations of *Ny. trapidoi* from two Ecuadorian locations proved to be sympatric species (Dujardin et al. 1996), but so far potential differences in their vectorial capacity have not been investigated. *Psychodopygus carrerai* was similarly found to be two cryptic species (Caillard et al. 1986), and the one from the Bolivian highlands was named *Ps. yucumensis*. Both are considered to be vectors of *L.* (*V.*) *braziliensis*.

Ny. whitmani sensu stricto has been recorded in practically every Brazilian state, extending into the northeastern tip of Argentina and eastern Paraguay in the south, as well as in countries bordering Amazonia such as Bolivia, Peru, Suriname, Guyana and French Guiana (Young et al. 1994). It is associated with *L.* (*V.*) *guyanensis* transmission in Amazonia and elsewhere with both peridomestic and sylvatic *L.* (*V.*) *braziliensis* and *L.* (*V.*) *shawi* transmissions. Preliminary molecular studies (Rangel et al. 1996) showed that there were two populations. One occurred on both sides of the Amazon River and the other in Ceará and Bahia. A more detailed study (Ready et al. 1997) supported three or four mitochondrial lineages. One from the eastern Atlantic forest habitat, a second central north-south lineage stretching from Teresina to São Paulo and a less well-defined Amazon group consisting of two sublineages north and south of the Amazon river. Behavioural studies (Campbell-Lendrum et al. 1999, 2000) showed that the degree of anthropophily was varied and was greatest in populations from Paraná where *L.* (*V.*) *braziliensis* is endemic. It is well known that *Ny. whitmani* feeds more avidly on humans outdoors in the domestic environment than indoors. This is due to its reluctance to enter sheds or houses with greater levels of closure (Campbell-Lendrum et al. 2000). These observations draw attention to the difficulties of associating genetic and behavioural differences. For instance, in hotter regions dwellings are more open, but in colder ones they tend to be more closed. The fact that *Ny. whitmani* bites humans inside the house more in northern Brazil than in the south is due to the construction of the houses and not behavioural differences. Clearly this affects its intradomiciliary vectorial capacity.

Another vector with an extensive geographical distribution is *Ny. umbratilis*. It is a vector of parasites of the *guyanensis* complex and is found throughout Brazilian, Bolivian, Colombian, Peruvian and Venezuelan Amazonia, French Guiana and Surinam extending in the east to the Atlantic forest. This a tree trunk loving species that comes down to rest at the base of trees during the day (Ready et al. 1986). It is when humans disturb these populations that they become infected. Initial morphological studies (Azevedo et al. 2002) failed to detect significant variations between Venezuelan and Brazilian populations. However, a molecular analysis (Scarpassa and Alencar 2012) showed that in Amazonia there were two mitochondrial lineages separated by the Amazon and Rio Negro Rivers. A later publication (de Souza Freitas et al. 2015) found that a population from the Atlantic forest region near Recife was closer to the lineage north of the Rio Negro/Amazon rivers than to that from the south. So far no behavioural differences have been detected between these clades, but epidemiologically the incidence of *L. (V.) guyanensis* is greater in the region of the clade north of the Negro/Amazon rivers.

Without doubt the most extensively studied group is the *Lu. longipalpis* complex that was initially considered to be composed of at least three cryptic species (Lanzaro et al. 1993). Subsequent molecular and pheromone studies (Araki et al. 2009; Bauzer et al. 2002; Hamilton et al. 2005) indicated further cryptic species within the complex. In spite of these extensive studies, the status of the complex remains controversial. There is an overall agreement between pheromone and period gene typing although it is not 100%. Some authors (Hamilton et al. 2005) consider that pheromone typing should preempt molecular studies. What is surprising is that in Sobral, located in the Brazilian State of Ceará, there are three cryptic species that do not interbreed, each being characterized by distinct pheromones, cembrene-1, 9MGBþ ((S)-9-methylgermacrene-B) and 3MaH (3-methyl-ahimachalene). A combination of molecular and pheromone results (Arrivillaga et al. 2003; Lima Costa et al. 2015) shows that there are at least nine cryptic species in Central and South America of which two have been named (*Lu. pseudolongipalpis*, *Lu. cruzi*). However, no convincing difference in vectorial capacity has been shown for these genetically distinct species.

3.8 Transmission in Different Neotropical Biomes

3.8.1 Tropical Forest Environments

The American tropical rain forests are divided by the Andean mountains. On the western side, you have the Chocó-Darien forest that is continuous with those of Central America. In the east, there is the more extensive Amazonian rain forest that extends from the Andean foothills to the boarders of the Orinoco river in Venezuela and to the edge of the cerrado in eastern Brazil. All are endemic for cutaneous leishmaniasis, but with deforestation visceral leishmaniasis is becoming endemic.

3.8.1.1 Chocó-Darien and Central American Forests

The Chocó-Darien forest is contiguous with those of the Panamanian isthmus that continues northwards to Mexico. On the Atlantic side, the forests are humid but those on the Pacific side are dry. In this ecoregion, the dominant parasite found in humans in Ecuadorian and Colombian Pacific coastal forest is *L. (V.) panamensis* followed by *L.(V.) braziliensis*. The proportion of these infections is in the order of 6:1 (Weigle et al. 1986). In Panama and Costa Rica, *L.(V.) panamensis* predominates, and in Honduras and Nicaragua it is sympatric with *L.(V.) braziliensis*. The finding of *braziliensis/panamensis* hybrids in Nicaragua suggests that the same sand fly species was or is involved in the transmission of both parasites. The northerly limit of *L.(V.) panamensis* seems to be somewhere between Nicaragua and Guatemala. In the Petén region of Mexico and Guatemala, the number of the infections of *L.(V.) braziliensis* and *L.(Leishmania) mexicana* is more or less equal in patients, but in Mexico the latter parasite dominates. These different epidemiological profiles reflect a variety of vectorial patterns associated with different ecologies. In the forests of the Chocó-Darien and Panamanian isthmus, *Ny. trapidoi* is the principal vector of *L.(V.) panamensis*. Other sand flies, in which infections have been found, such as *Lu. gomezi*, *Lu. sanguinaria* and *Ps. panamensis*, may be involved in enzootic and zoonotic transmission. At the present time, it is unclear what species are involved in the transmission of *L.(V.) braziliensis* in the Chocó-Darien forests. Human infections of *mexicana* complex in this biozone are very rare, but its vector *Bichromomyia olmeca* is present. The distribution of the major vectors in Costa Rica, Nicaragua, Honduras and Guatemala is very similar, although vector infections of the four major parasites have not been found in all these countries. In the cutaneous leishmaniasis endemic Petén region of Guatemala, *L.(V.) braziliensis* has been found in *Ny. ylephilator*, *Ps. panamensis* and *Pintomyia ovallesi*. The principal vectors of *L.(V.) panamensis* (*Ny trapidoi*) and *L.(L.) mexicana* (*Bi. olmeca*) are both present in all four countries. Except for *Ny. trapidoi*, these same sand fly species also occur in the Yucatán Peninsula (Mexico and Belize) where only two *Leishmania* species, *L.(L.) mexicana* and *L.(V.) braziliensis*, have been found in humans. Infections by *L.(L.) mexicana* have consistently been found by different authors in *Bi. olmeca*, but more recently this same parasite has been detected in *Ps. panamensis* and *Psathyromyia shannoni*. Attempts to associate vectors with human infection using ecological modelling are beginning to be used as more data becomes available. However, interpreting the results depends on the accuracy of this data. Recently a study showed that in Mexico the distribution of *Bi. olmeca* did not accompany that of cutaneous leishmaniasis (Gonzalez et al. 2011) leading the authors to suggest that other species were also vectors. Unfortunately, no parasites were identified so the question is vectors of what parasite since both *L.(L.) mexicana* and *L.(V.) braziliensis* are present? On the drier Pacific side of the above-mentioned Central American countries, there are populations of *Lu. longipalpis* and *Pi. evansi* in visceral leishmaniasis (VL) areas.

3.8.1.2 Amazonian Forests

In a recently published study, 70 sand fly species were captured in the Tapajós forest reserve, south of the Amazon River (de Souza et al. 2016), of which 16 are implicated as *Leishmania* vectors in Amazonian forests (see Table 3.1). North of the river in Amapá state, where "pian bois" (CL by *L. (V.) guyanensis*) is highly endemic, 46 species were captured (Freitas et al. 2002). Of these 14 are vectors or suspected (potential) vectors, but *Ny. umbratilis*, *Ps. squamiventris* and *Trichophoromyia*

Table 3.1 Sand fly species captured in the Tapajós forest reserve, Pará State, Brazil, that are implicated as vectors of cutaneous leishmaniasis in Amazonia

Leishmania species	Sand flies implicated as vectors
Leishmania (Leishmania) amazonensis	*Bichromomyia flaviscutellata*
Leishmania (*Viannia*) *braziliensis*	*Psychodopygus carrerai*
	Psychodopygus complexus
	Psychodopygus davisi
	Psychodopygus hirsutus
	Psychodopygus squamiventris
	Psychodopygus wellcomei
	Nyssomyia shawi
Leishmania (*Viannia*) *naiffi*	*Psychodopygus ayrozai*
	Psychodopygus davisi
	Psychodopygus hirsutus
	Psychodopygus paraensis
	Psychodopygus squamiventris
Leishmania (*Viannia*) *guyanensis*	*Nyssomyia umbratilis*
	Nyssomyia anduzei
Leishmania (*Viannia*) *lainsoni*	*Trichophoromyia ubiquitalis*
Leishmania (*Viannia*) *shawi*	*Lutzomyia gomezi*
	Nyssomyia whitmani
Leishmania (*Viannia*) *utingensis*	*Viannamyia tuberculata*

ubiquitalis were dominant. An overall comparison shows that *Ny. umbratilis* was present on both sides of the river, but in the south *Ps. complexus/wellcomei* replaces *Ps. squamiventris*. Also, *Psychodopygus* species were significantly ($p = <0.0001$) more prevalent as a group south of the river. In the higher ground of the lower Amazon, such as in the Serra das Carajás, *Ps. wellcomei* is the vector of *L.* (*V.*) *braziliensis*, while in the lowlands of this region, it is replaced by *Ps. complexus*. Figures on the prevalence of the different *Leishmania* species in humans are controversial due to the method of collection that is primarily passive rather than active, but they reflect different vectorial situations. However, the literature is consistent in recording that in humans, north of the Amazon river, *L.* (*V.*) *guyanensis* is very common and *L.* (*V.*) *braziliensis* is rare, but south of the river, the situation is reversed. This strongly supports the hypothesis that *Psychodopygus* species are the primary vectors of sylvatic *L.* (*V.*) *braziliensis* and that in Amazonia *Nyssomyia* species are the primary vectors of parasites of the *guyanensis* clade that includes *L.* (*V.*) *guyanensis* and *L.* (*V.*) *shawi*.

Cases of CL of the Amazon forest from the southeastern regions of Colombia, eastern Ecuador and northern Peru are predominantly *L.*(*V.*) *guyanensis*. Infections due to *L.*(*V.*) *braziliensis* are rarer as are those due to *L.* (*L.*) *amazonensis*. In these biotas, the sand fly vectors predominantly belong to the genera *Nyssomyia* and *Lutzomyia*. However, in the Ecuadorian Amazon forest zone, the dominance of *L.* (*V.*) *guyanensis* infections in the absence of *Ny. umbratilis* suggests that another species is transmitting this parasite. Moving south, *L.* (*V.*) *braziliensis* becomes the dominant pathogen. In the Bolivian sub-Andean lowlands (Le Pont and Desjeux 1986), *Psychodopygus* species dominated, whereas in the *L.* (*V.*) *guyanensis* endemic zone of Ecuador to the north, they only represented 5% of the sand fauna (Gomez et al. 2014). The incriminated *L.(V.) braziliensis* vectors in the above-mentioned Bolivian foci were *Ps. yucumensis* and *Ps. llanosmartinsi*, but other important potential vectors are present, such as *Bi. flaviscutellata*, *Ny. umbratilis*, *Ps. carrerai* and *Ps. hirsutus*.

Across the border from Bolivia in the Amazon forest of Rondônia State, Brazil, 78 species have been captured. Of the total number of flies caught, 70% belonged to the genus *Psychodopygus* (Gil et al. 2003). Here and in the neighbouring State of Acre the dominant parasite in humans is again *L.* (*V.*) *braziliensis*. Further studies are needed to find its vector, but the most likely candidate is *Ps. carrerai* or *Ps. yucumensis*. Other species, such as *L.* (*V.*) *shawi*, *L.*(*V.*) *lainsoni*, *L.* (*V.*) *naiffi* and *L.* (*L.*) *amazonensis*, also occur in this western Amazonian region. Although many potential vectors of these parasites exist in this area, there is strong evidence indicating that *Ps. davisi* and *Ps. paraensis* are vectors of *L.* (*V.*) *naiffi* and *Bi. reducta* is the vector *L.*(*L.*) *amazonensis*.

Throughout the Amazon forest (sensu stricto) there have been increasing records of human infections by *L.* (*V.*) *lainsoni*, whose principal vector is *Th. ubiquitalis* (Lainson et al. 1992). As its specific name suggests, this is a species that has an extensive geographical distribution that covers the whole of Amazonia. It rarely represents more than 8% of the fauna in a given locality, but it may rank as high as third based on its SISA index. The SISA index is the Standardized Index of Species Abundance used to analyse entomological abundance data (Roberts and Hsi 1979). So far *Th. ubiquitalis* is the only known vector of *L.* (*V.*) *lainsoni*, but more recently this parasite has been detected by the PCR method in *Pi. nuneztovari* and *Th. auraensis* from Amazonia. Are these infections dead-ends or do they represent enzootic cycles involving other sand flies? This is an example of the complexity of determining the importance of infections without broader ecological and epidemiological studies. This applies to many records of *Leishmania* in sand flies that are based on molecular methods.

3.8.1.3 Atlantic Forest

Before deforestation the Atlantic Forest occupied a coastal strip from northeastern Brazil to northern Argentina and across into southeastern Paraguay, but today less than 15% has survived, having been replaced by enormous anthromes that are biomes created by man such as cities and farm lands. This ecoregion consists of a humid tropical coastal forest that inland gives way to semi-deciduous forests. It is endemic for CL where it reached epidemic levels in the late 1800s and early 1900s during the intense deforestation periods that accompanied the colonization of southern Brazil. The principal parasite is *L. (V.) braziliensis.* The number of sand fly species in the humid Atlantic forest is less than in Amazonia but may be as many as 30. Domination of different sand fly genera appears to vary according to the location. The proportion of *Psychodopygus* species (*Ps. amazonensis*, *Ps. ayrozai, Ps. bispinosa*, *Ps. carrerai*, *Ps. davisi*, *Ps. complexus*, *Ps. hirsutus*, *Ps. wellcomei*) may vary from 19 to 89%, the latter being very similar to that of area of Amazonia south of the river. Infections of *L. (V.) naiffi* have been identified in *Ps. ayrozai* and *P.s hirsutus* captured in Atlantic forest remnants of Bahia and Rio de Janeiro. Other important *L. (V.) braziliensis* vectors, such as *Ny. intermedia*, *Ny umbratilis*, *Ny. whitmani*, *Mi. migonei*, *Ps. complexus*, *Ps wellcomei* and *Pi. fischeri*, have been found in these forests but generally in small numbers. At the time of writing, infections of *L. (V.) braziliensis* have not been found in examples of species captured in climax Atlantic forest. However, finding *L. (V.) braziliensis* in small sylvatic mammals captured in this forest (Brandão-Filho et al. 2003) indicates the presence of active enzootic cycles, but however so far we do not know what the vectors are. Perhaps it is maintained, as in Amazonia, by *Psychodopygus* species. However, their absence in the semi-deciduous forest suggests that species such as *Ny. intermedia/neivai*, *Ny. whitmani* and *Mi. migonei* that are dominant (Teodoro et al. 1991), or *Pi. fischeri* and *Pi. pessoai* that are present in smaller numbers, maintain the *L. (V.) braziliensis* enzootic. In the southern most regions of the Atlantic humid and semi-deciduous forests, *Ny. intermedia* is replaced by its sister species *Ny. neivai.*

Infections from these patches of Atlantic forest reach the peridomestic and domestic habitats where parasites are transmitted by *Ny. whitmani* or in the drier areas in Minas Gerais to the south and east by *Ny. intermedia.* When the coverage is less dense, such as the gallery forests that run along river courses in the savannah (cerrado/cerradão), the number of species falls to below 20 (Galati et al. 1996), and in residual forest it may be fewer than ten species (Gomes et al. 1989). In spite of the reduction of the fauna, important vectors such as *Ny. intermedia*, *Ny. whitmani* and *Mi. migonei* are the principal species and appear to be a source of flies for the peridomestic ecotope in these rural areas.

3.8.2 Montane Environments

L. (V.) peruviana is responsible for a form of CL known as "uta" that is endemic in the western slopes of the Andes between 1000 and 3000 m. At the lower altitudes, *Pi. verrucarum* dominates giving way in the higher altitudes to *Lu. peruensis* that is considered to be *L. (V.) peruviana*'s principal vector. Transmission is mostly during the warmer wetter seasons of the year, and there is a correlation between disease incidence and the intradomiciliary abundance of *Lu. peruensis*. The population appears to be a single unit whose intradomiciliary presence is greater in the rainy season, presumably avoiding worsening extradomiciliary conditions (Villaseca et al. 1993). It is possible that other sand flies are also vectors since the parasite has been detected in *Lu. ayacuchensis* and *Pi. verrucarum*. The latter species has been shown to transmit this parasite experimentally (Davies et al. 1993) and is more endophilic than *Lu. peruensis*. However, there was a positive correlation between the number of new cases and the number of intradomiciliary *Lu. peruensis* indicating that it was the principal vector (Villaseca et al. 1993).

A clinical form considered to be uta also occurs in Ecuador, at altitudes between 2300 and 2500 m, but the dominant parasite in humans was identified as

L. (*L.*) *mexicana* and the suspected vector was *Lu. ayacuchensis* (Hashiguchi et al. 1991), transmission probably being intradomiciliary. In both Peru and Ecuador intradomiciliary transmission is attributed to the flies preferring the warmer interior temperatures of the houses during the cold nights, when outside temperatures can fall to between 6 and 8 °C.

3.8.3 Drylands and Savannah

The vegetation in the drier regions of the Andes and Savannah consists mostly of shrubs and small trees. There are wetter seasons during the warmer periods of the year both in the northern and southern hemispheres. The colder periods are less humid and may be associated with periods of drought.

CL is recorded in the USA, but most cases are imported; however in the dryland biome of eastern Texas that is congruent with the same biome in Mexico, there are autochthonous cases. The parasite is *L.* (*L.*) *mexicana*, and there is an enzootic cycle involving rodents and *Dampfomyia anthophora* which has been found naturally infected and has transmitted this parasite experimentally (Endris et al. 1987). Cases have been recorded in Texas since 1942 and more recently in the Dallas-Fort Worth metropolitan area, but in 2004 and 2005, two cases were reported from southeastern Oklahoma (Clarke et al. 2013). It is suggested the appearance of cases outside the old endemic areas of southern Texas is due to the northerly spread of sand flies, including *Da. anthophora.*

So far there is no evidence of vectorial VL transmission in the USA. Autochthonous cases of canine VL have been recorded in many states including Oklahoma, Ohio and New York and are the result of vertical transmission (Petersen and Barr 2009).

Moving from the Amazonian forest of the Venezuelan States of Bolivar across the Orinoco river in a northwesterly direction, there is a drier grassland biome that stretches for the northeastern State of Monagas down into Colombia. Here VL is endemic (De Lima et al. 2009), and suspected vectors are *Lu. longipalpis*, *Lu. pseudolongipalpis* and *Pi. evansi.* The northeastern, northern and north-western rim of this grassland transforms into drier woodlands and highland mountainous vegetation. In the lower foothills, VL dominates giving way to the cutaneous disease as the altitude increases in a band that continues into northern-western Colombia. In the northern region Anzoátegui State, Venezuela, *Pi. evansi* is present in the foothills and is replaced by cutaneous vectors such as *Ps. panamensis* at lower altitudes and *Pi. ovallesi* at altitudes above 300 m (Gonzalez et al. 2002). Throughout this whole northern region of Venezuela, there is very strong evidence that the major vector of *L.* (*V.*) *braziliensis* is *Pi. ovallesi* rather than *Lu. gomezi* (Bonfante-Garrido et al. 1991; Feliciangeli and Rabinovich 1998). Differences in altitude moderate the relative abundance of different species resulting in a greater variety of vectors in these hilly and mountainous regions of northern Colombia and Venezuela.

This same northern region is the home of parasites belonging to the *mexicana* clade, such as *L.* (*L.*) *venezuelensis* in the Venezuelan States of Lara, Merida and Yaracuy and *L.* (*L.*) *pifanoi* in Miranda. The first species has been principally linked to keloid-like lesions and together with the second with diffuse CL. The suspected vector of both is possibly *Bi. olmeca bicolor* that is recorded as occurring throughout the region.

There is a very extensive drier region situated between the Amazon and Atlantic forests that stretches from northeastern Brazil to the Argentinian Chaco, ending in the Patagonian Desert in the south. In Brazil cerrado refers to an ecoregion of the central Brazilian highlands that includes forested and grassland savanna as well as gallery forests. The term cerradão is used for a similar biome in which the woodlands of the forested savanna are taller and denser. The original biomes are the very dry Caatinga in Brazil's northeast followed by the cerrado and then the Chaco subtropical grasslands in the south. In the cerrado of Mato Grosso State, Brazil, 15 different species were found, and in the same region in cerradão, which has denser vegetation, there were 20 species (Galati et al. 1996). These biomes are endemic for *L.* (*V.*) *braziliensis* tegumentary leishmaniasis. Although the number of sand fly species is relatively high, only 1–5 individuals were taken for 50% of the species during a 2-year study period. In another area of cerrado further to the north in Minas Gerais, 20 species were captured including two important vectors, *Ny. whitmani* and *Lu. longipalpis*, but again 50% of the species were represented by only 1–5 specimens (Nascimento et al. 2013). *L.* (*V.*) *braziliensis* has been detected by molecular methods in nine species (*Ev. termitophila*, *Mt. minasensis*, *Micropygomyia capixaba*, *Mi. quinquefer*, *Ny. intermedia*, *Ny. whitmani*, *Psathyromyia aragaoi*, *Ps. claustrei*) that were captured in the woodlands of this environment. Given such a variety of infected species, it is possible that the sylvatic enzootic is maintained by a mosaic of sand fly species.

Human disease caused by *L.* (*V.*) *braziliensis* is endemic in the Chaco bioregion that is divided into a wetter eastern zone and a drier one in the west. The number of species present is less than ten, but the enzootic cycle has survived. The eastern zone is at the edge of the semi-deciduous Atlantic forest biome, and here the dominant species in the natural habitats is *Ny. neivai*. In the drier eastern area, the landscape is xeric and the dominant species is *Mi. migonei* (Salomon et al. 2008b).

3.9 Domestication and Urbanization of American Sand Flies

The above-mentioned natural biomes no longer have their original geographical extensions, having been replaced in many regions by human biomes or anthromes. These range from farmland to cities and are associated with an increase of humans and animals that represent increased food sources and humid ecotopes that favour some but not all sand fly species. Indigenous civilizations have occupied Latin America for thousands of years so it is feasible that in some areas sand fly adaptation predates European colonization. But what is the epidemiological importance of capturing sand flies in the vicinity or within human dwellings? Many studies have addressed this complex question and below we give a few examples of their conclusions.

Table 3.2 Published records of proven* and putative vectors of the New World collected close to or inside houses

Visceral Leishmaniasis	Cutaneous leishmaniasis
Evandromyia cortelezzii	*Bichromomyia olmeca**
Evandromyia sallesi	*Evandromyia cortelezzii*
Evandromyia lenti	*Evandromyia lenti*
Lutzomyia almerio	*Lutzomyia gomezi**
Lutzomyia ayacuchensis	*Lutzomyia longipalpis*
*Lutzomyia cruzi**	*Lutzomyia renei*
Lutzomyia ischnacantha	*Lutzomyia peruensis**
*Lutzomyia longipalpis**	*Nyssomyia intermedia**
*Migonemyia migonei**	*Nyssomyia neivai*
Nyssomyia whitmani	*Nyssomyia shawi**
*Pintomyia evansi**	*Nyssomyia whitmani**
	Nyssomyia yuilli
	*Migonemyia migonei**
	Micropygomyia ferreirana
	*Pintomyia fischeri**
	Pintomyia longiflocosa
	Pintomyia nuneztovari
	*Pintomyia ovallesi**
	*Pintomyia spinicrassa**
	*Pintomyia verrucarum**
	*Pintomyia youngi**
	Trichophoromyia auraensis

The term putative vector is used for species in which parasites have been identified by molecular methods

There are published records of vectors and putative VL and of CL vectors being collected close to or inside houses (Table 3.2). *Putative vector refers to a sand fly species in which a *Leishmania* species has only been detected by a molecular method. It's a surprisingly long list and field studies are needed to determine the epidemiological importance of many of these species.

All domestic sand fly populations are originally derived from feral ones, but some are more successful at adapting to these anthromes than others. In some situations, peridomestic population of a species may be greater than those of the sylvatic one. Understanding and predicting where and how this happens underpins any control strategy. To produce prediction maps for CL control in Colombia using land-cover classes, 167 classes were reduced to 25 (King et al. 2004). This emphasizes the extreme difficulty of assessing domestication in so many different situations. There are various degrees of domestication ranging from a sylvatic species merely entering a house that is close to its natural habitat too actually breeding within the peridomestic and domestic environment.

An outbreak of *L. (V.) guyanensis* CL in 16% of the inhabitants of a newly settled suburb of Manaus, Brazil, in 1980 initially suggested that the vector, *Ny. umbratilis*, may have adapted to the domestic environment. A study (Ready et al. 1985) showed however that this was not the case. Small pieces of forest were cleared to accommodate houses that were literally inside the forest. People were being infected inside their houses at night or when they went into the surrounding forest during the day. A similar situation occurred in French Guiana (Le Pont and Pajot 1981) when a small outbreak of CL occurred in the newly established village of Cacao. As the village developed and proximity to the forest diminished, so did the numbers of *Ny. umbratilis* (Esterre et al. 1986) and human cases. At distances of over 200 m from the forest edge no flies were captured. These observations and others have led to the recommendation that in areas where there are sylvatic vectors dwellings should be built around 400 m from the forest edge. However, in Chaparral, located in the sub-Andean Amazon forest region of Colombia, there was strong evidence that there was domiciliary transmission by *Pi. longiflocosa* (Ferro et al. 2011). In this situation, it will be interesting to know what the reservoir is.

In western Panama 24 different sand fly species were collected in and around houses (Calzada et al. 2013), and of these eight (*Bi. olmeca, Lu. gomezi, Lu. sanguinaria, Ny. trapidoi, Ny. ylephilator, Pi. ovallesi, Pa. shannoni, Ps. panamensis*) were potential vectors due to their anthropophilic habit. In another study (Saldana et al. 2013) in the same region, the abundance of *Lu. gomezi* inside the houses was positively associated with household CL rates. These rates increased by 6% when catches of other species, such as *Ny. trapidoi* or *Ps. panamensis* that were also captured inside the same houses, were incorporated in the analysis. Similar situations associating peridomiciliary and domiciliary sand fly abundance to human cases have been found in north-central Venezuela (Feliciangeli and Rabinovich 1998). This data suggests that transmission occurs inside the dwellings, but the question remains as to where the flies are becoming infected.

In the coffee and cacao plantations of Colombia, Bolivia and Venezuela (Alexander et al. 1995; Le Pont et al. 1989; Maingon et al. 1994; Mouchet et al. 1994), there is evidence of endophyly. Here houses are located within the plantations where tall native trees protect the coffee bushes from the sun. The following species have been found in the plantations and also in their houses: *Pintomyia spinicrassa, Pi. ovallesi, Lu. gomezi, Pi. nuneztovari, Ny. trapidoi, Lu. hartmanni* and *Pi. youngi*. The *Pintomyia* species and *Lu. gomezi* are associated with the transmission of *L. (V.) braziliensis, Ny. trapidoi* with *L. (V.) panamensis* and *Lu. hartmanni* with *L. colombiensis*.

Convincing evidence of *Pi. ovallesi* endophyly has been shown in the El Ingenio village located in the foothills of Venezuela's Cordillera de la Costa (Feliciangeli and Rabinovich 1998). There was a clear correlation between the sizes of the outside and indoor populations that led to the assumption that transmission was inside the houses. This suggested that combating the indoor population would reduce the number of cases but it did not resolve the problem as to where they were breading or becoming infected. This is a recurrent problem with all endophyly studies.

There is no doubt that many CL vectors have successfully adapted to the human-made environments. Some of the best documented vectors of *L.* (*V.*) *braziliensis* are *Ny. whitmani* and *Ny. intermedia* in Brazil and *Pi. ovallesi* and *Lu. gomezi* in Venezuela. However, besides these there is also good evidence indicating that other species belonging to the genera *Nyssomyia*, *Lutzomyia* and *Pintomyia* are responsible for the transmission of "viannian" CL within the peridomestic environment. The situation of the transmission of "leishmanian" CL caused by such parasites as *L* (*L.*) *amazonensis* and *L* (*L.*) *mexicana* in anthromes is less clear. A possible candidate for this is *Lu. cruciata* that was captured in large numbers both outside and inside houses and recently was found infected with *L* (*L.*) *mexicana* in Campeche, Mexico, in dry forest.

A difficulty in assessing the degree of adaptation to new environments is the denomination used for the capture site. Some are described as peridomestic or domestic but occur in rural or urban settings, and urban varies from a village to a town. However, ecological studies (Salomon et al. 2008a) show that periurban vegetation is present to varying degrees even in the towns. In Minas Gerais State, Brazil there were clear differences (Nascimento et al. 2013) in the sand fly fauna of the urban and green park areas. Ninety-nine percent of the *Lu. longipalpis* were captured in the peridomestic urban zone, whereas 62% of the *Ny. whitmani* were collected in the green areas. These two flies are respectively confirmed vectors of VL and CL. Similar situations are recorded throughout Latin America. Patches of urban vegetation may serve as sources or refuges for sand flies. The Mexican State of Chiapas is endemic for both CL and to a lesser extent VL and is the principal coffee producer of the country. Ninety-nine percent of the sand flies captured in an urban area, in which there was a patch of woodland, and in coffee plantations (Perez et al. 2014) were *Lu. cruciata.* The vectorial importance of this species is uncertain but the mere fact that it has been found infected with *L.* (*L.*) *mexicana* and is present in such large numbers clearly indicates its potential importance.

The level of Latin American peridomestic transmission of VL far exceeds that of CL and has been increasing steadily in recent years. As mentioned above 12 species have been found infected in peridomestic situations with *L.* (*L.*) *infantum chagasi*, but there is little doubt that the principal vectors belong to the *Lu. longipalpis* complex. The first records of rural peridomestic populations of *Lu. longipalpis* date back to the 1930s in the Brazilian States of Sergipe and Pará States (Deane 1956), but obviously adaptation to this environment must have occurred long before. None of the records of this period refer to urban situations. However, in 1954 Pessoa (Pessoa 1954) drew attention to the occurrence of cases of VL in the urban environment of Sobral in Ceará State, which implied the presence of the vector. In 1962 a focus was found in the Amazonian town of Santarem that reached epidemic status in 1984 (Lainson et al. 1985), and in this same period, another epidemic was recorded in the capital of Piaui State, Teresina (Costa 2008). Outbreaks had been recorded since the 1950s in smaller towns, but the 1980s heralded a new situation—transmission in cities with populations over 100,000 such as Belo Horizonte, Montes Claros (Minas Gerais State), Araguaína, Palmas (Tocantins State), Campo Grande, Três Lagoas (Mato Grosso do Sul State), Araçatuba, Bauru (São Paulo State), Cametá (Pará

State) and Rondonópolis (Mato Grosso State) (Werneck 2014). This epidemiological picture clearly shows that over a period of 50 years, *Lu. longipalpis* moved from a rural to a truly urban environment.

Before 1998 VL was not endemic in São Paulo State, Brazil (da Costa et al. 1997), but by 2011 the human cases had been recorded in 64 municipalities (Cardim et al. 2013). The apparent cause of this was the rapid expansion of *Lu. longipalpis*. A comparison of the chemotypes (Casanova et al. 2015) showed that the expanding population were 9MGBþ flies. However, the native cembrine-1 population, present in the eastern region of the State had not expanded. Some environmental factor presumably favoured the 9MGBþ population. This emphasizes that vectorial potential is also associated with a vector's response to environmental conditions as well as parasitological parameters related to infection and transmission.

In a rural region of Mato Grosso do Sul, *L.* (*L.*) *i. chagasi* infection levels of *Lu. longipalpis* and *Lu. almerio* were respectively 0.3% and 0.5% (Savani et al. 2009). What was surprising was that 1.25% of the *Lu. longipalpis* infections were *L.* (*L.*) *amazonensis*. However, the proportion of peridomiciliary versus domiciliary for *Lu. almerio* was 7 to 1 and for Lu. *longipalpis* 18 to 1 showing that *Lu. almerio* was more endophilic and thus, potentially, a more important VL vector. There is strong epidemiological evidence that other species are also VL vectors, such as *Mi. migonei* in the coastal region of the Brazilian State of Pernambuco. Its vectorial capacity was recently confirmed experimentally (Guimarães et al. 2016), so it must now be considered as a proven VL vector.

In areas of Costa Rica, Colombia and Venezuela (Travi et al. 1990) *Pi. evansi* is considered to be an important peridomestic VL vector. Both *Lu. longipalpis* and *Pi. evansi* are found in anthromes located in the tropical dry forest biome. They are sympatric in the Magdalena valley of Colombia and in Central American Pacific coast (Zeledón et al. 1984). However, *Pi. evansi* predominates in the Colombian and Venezuelan Atlantic Coast (Gonzalez et al. 2006).

3.10 Evidence for Permissive and Specific Vectors in the Americas

The experimental studies on sand fly susceptibility to various *Leishmania* led to the hypothesis that there are specific and permissive vectors (Kamhawi 2006; Volf and Myskova 2007). Field and experimental data confirms this idea and indicates that it is a crucial factor in delineating epidemiological patterns. The finding of different *Leishmania* species in the same sand fly species is a circumstantial support to classify the vector as a permissive vector. On the other hand, constantly finding only one parasite in a sand fly may suggest a specific vector. Clearly this situation is tempered by feeding preferences.

The most well-known American permissive vector is *Lu. longipalpis* which has been found naturally infected with *L.* (*L.*) *infantum*, *L.* (*L.*) *amazonensis* and an unidentified *L.* (*Viannia*) parasite. It was one of the first sand flies to be successfully colonized that led to be being used as a laboratory model worldwide. It has been experimentally infected with different *L. (Viannia)* and *L. (Leishmania)* parasites as well as being used for xenodiagnoses to evaluate reservoir infectiousness and transmission efficiency (Andrade et al. 2015; Courtenay et al. 2002; Maia et al. 2011; Stamper et al. 2011). Comparisons of the development of different *Leishmania* in *Lu. longipalpis* led to the establishment of a robust taxonomy for the leishmanial parasites (Lainson and Shaw 1987). It is arguable that choosing a specific vector for these studies might have impaired the development of this classification and that true permissiveness is the capacity to transmit rather than merely support development. Nevertheless, *Lu. longipalpis*'s permissiveness is responsible for the establishment and spread of VL in the Americas (Volf and Myskova 2007), and it was extremely unfortunate that *L.(L.) infantum* found such an excellent vector.

Natural infections of *L.* (*V.*) *braziliensis* and *L.* (*L.*) *infantum* have been found in *Mi. migonei*, and it has also been experimentally infected with *L.* (*V.*) *braziliensis* and *L.* (*L.*) *amazonensis* (Nieves and Pimenta 2000), suggesting that it is a permissive vector. Further experimental evidence (Guimarães et al. 2016) showed that it was also susceptible to different strains of *L.*(*L.*) *infantum* and led to its classification as a permissive vector (Guimarães et al. 2016). In different regions *Ny. whitmani* has been found infected with *L.* (*V.*) *braziliensis* and *L.* (*L.*) *infantum.* Recently it was infected experimentally (Fonteles et al. 2016) with *L.* (*L.) amazonensis*, adding to the idea it is a permissive vector. Similar criteria indicate that *Lu. almerio*, *Lu. gomezi*, *Lu. renei*, *Ev. cortelezzii*, *Ny. intermedia*, *Ny. yuilli*, *Pi. nuneztovari* and *Pi. ovallesi* may also be permissive vectors. Of these species, *Lu. renei* has been recorded (Coelho and Falcão 1962) as transmitting *L.* (*L.*) *mexicana* experimentally.

The genus *Viannamyia* presently has only four species and is seldom considered of any vectorial importance. *Vi. tuberculata* is the natural host of *L.* (*V.*) *utingensis*, and it was noted (Braga et al. 2003) that this parasite also developed in an identical manner in its sister species, *Vi. furcata*. This species has been experimentally infected with *L.* (*V.*) *braziliensis*, *L.* (*L.*) *infantum* and *Endotrypanum* (Braga et al. 2003; Ryan et al. 1987; Shaw 1981), and it transmitted the visceral parasite. The infection patterns of the other two parasites were similar, both having fore and hind gut infections, but there were no attached forms of *L.* (*V.*) *braziliensis* in the pylorus. From the above one must consider that *Vi. furcata* is a permissive vector for the subgenera *L.* (*Leishmania*) and *L.* (*Viannia*). Both the above-mentioned vectors have a very extensive geographical distribution in forested areas, and recently of 68 species captured in the Tapajós National Forest Reserve, *Vi. furcata* ranked 8th in the SISA index (de Souza et al. 2016). These species could be important vectors in enzootic cycles.

The *Evandromyia* species are typical of the drier savannah regions where they can be found in the peridomestic environments and may outnumber (Pinheiro et al. 2016) classical vectors such as *Lu. longipalpis.* From the literature, one might consider species of *Evandromyia* as being permissive. Infections of *L.* (*V.*) *braziliensis* have been recorded in *Ev. apurinan*, *Ev. cortelezzii*, *Ev. edwardsi* and *L.* (*L.*) *infantum* in *Ev. lenti*, *Ev. sallesi* and *Ev. termitophila* by molecular methods. All the infections were found in groups of pooled flies, and no flagellates were seen in dissections. Were these simply lingering parasites from abortive infections? *Ev. carmolinoi* taken on a dog with visceral leishmaniasis were not infected and experimental infections with different *Leishmania* did not develop in laboratory bred *Ev. carmolinoi* (Ryan et al. 1986). These results add weight to the hypothesis that the infections in the above mentioned evandromyias were in fact abortive ones. However, it is possible that permissiveness varies within a species group. The 44 *Evandromyia* species are divided into three subgenera (Galati 2015), and only *E. lenti* is in the same subgenus as *Ev. carmolinoi*. Clearly at the time of writing, there is not enough information to determine the vectorial status of *Evandromyia* species except to say that the only species investigated in sufficient detail is refractive to infection.

Finding evidence to support the existence of specific American sand fly vectors is more difficult. *Bichromomyia* species have been dissected in large numbers throughout the Americas and, so far, have only been found infected with parasites belonging to the *mexicana*/*amazonensis* complexes. *Bi. flaviscutellata* has been colonized successfully. Experimentally it was found to be refractory to *L.* (*V.*) *braziliensis* (Ready et al. 1984), but when fed on an *Endotrypanum*-infected two-toed sloth, surprisingly 80% were positive (Shaw 1981). The only parasites seen were attached forms in the hind gut with no evidence of any in the foregut or midgut. This may suggest that *Bichromomyia* species are probably specific for *L.*(*L.*) *mexicana* and *L.*(*L.*) *amazonensis*. Clearly further experimental studies are necessary to prove such a hypothesis and the fact that *Endotrypanum* developed in *Bi. flaviscutellata* does not favour it as being a specific vector.

3.11 Non-phlebotomine Vectors

There is a group of parasites belonging to the subgenus *L.* (*Mundinia*) Shaw, Camargo & Texeira 2016 (Espinosa et al. 2016) that occur on both sides of the Atlantic, including even Australia. They are an ancient phylogenetic group that predates the other three leishmanial subgenera (Harkins et al. 2005). The most well-known member of the subgenus is the enigmatic guinea pig parasite *Leishmania* (*Mundinia*) *enriettii*. More recently another parasite belonging to the subgenus *L.* (*M.*) *martiniquensis* was described from a patient who contracted the infection on the island of Martinique (Desbois et al. 2014). Their vectors are unknown. However, ceratopogonid midges (Dougall et al. 2011) are the suspected vector of another closely related parasite, *L.* (*M.*) *macropodum*, found in Australian red kangaroos. Related parasites have been found associated with human VL in Thailand (Pothirat et al. 2014) and with CL in Ghana (Kwakye-Nuako et al. 2015). It is possible that these phylogenetically ancient parasites are transmitted by midges rather than sand flies. Seblova et al. (2015) demonstrated that *L.*(*M.*) *enriettii* develops late-stage infections in *Culicoides sonorensis*. In contrast, the same biting midge and *C. nubeculosus* were refractory to the development of the principal Old World parasites, namely, *L. major*, *L. infantum* and *L. donovani* (Seblova et al. 2012, 2015). These authors suggest that midges should be assessed in the field when searching for the vectors of *L. enriettii* and related parasites, including *L. martiniquensis* and the mundinian parasite causing VL in Thailand.

3.12 Concluding Remarks

In the Old World, the transmission of main *Leishmania* agents of CL is biased towards certain sand fly subgenera, while parasites of *L. donovani* complex, usually responsible for VL, are transmitted by four different subgenera. Members of subgenus *Phlebotomus* are associated with the transmission of *L.* (*L.*) *major* throughout this parasite's geographical distribution, while the other cutaneous parasite, *L.* (*L.*) *tropica*, is transmitted principally by species of subgenus *Paraphlebotomus*. Sand flies of the subgenus *Larroussius* are key vectors of VL in the Mediterranean and East African regions and of CL in Ethiopia and Kenya. Just two *Synphlebotomus* species are VL vectors in East Africa. Few species of the subgenus *Euphlebotomus* are involved in the transmission of leishmaniasis with the notable exception of the Indian subcontinent's principal VL vector *Ph.* (*Eu.*) *argentipes*. Finally, several members of subgenus *Adlerius* are suspected or proven vectors of *L.* (*L.*) *infantum*, and one appears to be an alternative *L.* (*L.*) *tropica* vector.

Advances in the taxonomy (Galati 2015) of the American sand flies have revealed some important epidemiological tendencies and patterns. In Amazonia ten *Psychodopygyus* species as compared to four *Nyssomyia* are the principal CL vectors (Table 3.1). As we move eastwards and southwards the importance of *Psychodopygyus* diminishes to just one or two as does the forest coverage. Deforestation has resulted in VL becoming endemic in many areas of Amazonia. In these more open drier regions flies belonging to the genera *Lutzomyia* and *Nyssomyia* are the main leishmaniasis vectors. Moving back westwards to the sub-Andean lowlands that run down into Amazonia and the Andean mountain chain, the principal vectors are species of *Pintomyia* and *Lutzomyia* where they transmit both VL and CL parasites. In Colombia *Pi. evansi* surpasses *Lu. longipalpis*'s role as the vector of VL in the more northerly region of the dry tropical forest (Gonzalez et al. 2006) close to the Venezuelan border. In Central America, the principal vectors of both forms of leishmaniasis belong to the genera *Lutzomyia*, *Nyssomyia* and *Psychodopygus*. The subgenus *Bichromomyia* has an extensive geographical distribution in both central and south America, and its species (*Bi. nociva, Bi. olmeca, Bi. flaviscutellata*) are intimately linked to the transmission of two closely related parasites, *L.* (*L.*) *mexicana* and *L.* (*L.*) *amazonensis*.

One of the biggest difficulties in assessing the vectorial importance of a sand fly species is defining and predicting the biomes in which it occurs besides determining its feeding habits, population dynamics and susceptibility to *Leishmania* infection. In American sand flies, many papers have shown that the quantity and type of vegetation and its distance from houses, irrespective of the area being considered as rural or urban, profoundly affects a sand fly's presence in the domiciliary environment. A house vegetation index could help resolve this problem and give a more accurate picture of comparative distributions.

Greater understanding of the vector/parasite interphase will undoubtedly contribute to the development of innovative control measures as well as generating epidemiological insights. For instance, recent understating of parasite sand fly specificity led to the concept of specific and permissive vectors. This has important consequences. The expansion of a permissive vector can completely change the gravity of an epidemiological situation.

With the ever-increasing arsenal of molecular techniques, more infections are being found, and sand flies are being identified with greater precision. However, carefully executed field studies are needed to understand the vectorial importance of these infections in the disease's epidemiology.

References

Abdeladhim M, Kamhawi S, Valenzuela JG. What's behind a sand fly bite? The profound effect of sand fly saliva on host hemostasis, inflammation and immunity. Infect Genet Evol. 2014;28:691–703.

Ajaoud M, Es-Sette N, Charrel R, Laamrani-Idrissi A, Nhammi H, Riyad M, et al. Phlebotomus sergenti in a cutaneous leishmaniasis focus in Azilal province (High Atlas, Morocco): molecular detection and genotyping of Leishmania tropica, and feeding behavior. PLoS Negl Trop Dis. 2015;93:e0003687.

Akhoundi M, Kuhls K, Cannet A, Votýpka J, Marty P, Delaunay P, et al. A historical overview of the classification, evolution, and dispersion of *Leishmania* parasites and sandflies. PLoS Negl Trop Dis. 2016;103:e0004349.

Akopyants N, Kimblin N, Secundino N, Patrick R, Peters N, Lawyer P, et al. Demonstration of genetic exchange during cyclical development of *Leishmania* in the sand fly vector. Science. 2009;324:265–8.

Alexander B, Usma MC, Cadena H, Quesada BL, Solarte Y, Roa W, et al. Phlebotomine sandflies associated with a focus of cutaneous leishmaniasis in Valle del Cauca, Colombia. Med Vet Entomol. 1995;9:273–8.

Alvar J, Velez I, Bern C, Herrero M, Dejeux P, Cano J, et al. Leishmaniasis worldwide and global estimates of its incidence. PLoS One. 2012;7:e35671.

Anderson J, Samake S, Jaramillo-Gutierrez G, Sissoko I, Coulibaly CA, Traoré B, et al. Seasonality and prevalence of *Leishmania major* infection in *Phlebotomus duboscqi* Neveu-Lemaire from two neighboring villages in central Mali. PLoS Negl Trop Dis. 2011;5:e1139.

Andrade MS, Courtenay O, Brito ME, Carvalho FG, Carvalho AW, Soares F, et al. Infectiousness of sylvatic and synanthropic small rodents implicates a multi-host reservoir of *Leishmania* (*Viannia*) *braziliensis*. PLoS Negl Trop Dis. 2015;9:e0004137.

Anjili CO, Mbati PA, Mwangi RW, Githure JI, Olobo JO, Robert LL, et al. The chemotactic effect of Phlebotomus duboscqi (Diptera, Psychodidae) salivary gland lysates to murine monocytes. Acta Trop. 1995;60:97–100.

Araki AS, Vigoder FM, Bauzer LG, Ferreira GE, Souza NA, Araujo IB, et al. Molecular and behavioral differentiation among Brazilian populations of *Lutzomyia longipalpis* (Diptera: Psychodidae: Phlebotominae). PLoS Negl Trop Dis. 2009;3:e365.

Arrivillaga J, Mutebi JP, Pinango H, Norris D, Alexander B, Feliciangeli MD, et al. The taxonomic status of genetically divergent populations of *Lutzomyia longipalpis* (Diptera: Psychodidae) based on the distribution of mitochondrial and isozyme variation. J Med Entomol. 2003;40:615–27.

Atayde V, Aslan H, Townsend S, Hassani K, Kamhawi S, Olivier M. Exosome secretion by the parasitic protozoan *Leishmania* within the sand fly midgut. Cell Rep. 2015;13:957–67.

Azevedo AC, Lainson R, Souza AA, Fe NF, Feliciangeli DM, Meneses CR, Rangel EF. Comparative studies of populations of *Lutzomyia umbratilis* (Diptera: Psychodidae) in Brazil and Venezuela. J Med Entomol. 2002;39:587–600.

Ballart C, Guerrero I, Castells X, Barón S, Castillejo S, Alcover M, et al. Importance of individual analysis of environmental and climatic factors affecting the density of *Leishmania* vectors living in the same geographical area: the example of *Phlebotomus ariasi* and *P. perniciosus* in northeast Spain. Geospat Health. 2014;8:389–403.

Bates P. Transmission of Leishmania metacyclic promastigotes by phlebotomine sand flies. Int J Parasitol. 2007;37:1097–106.

Bates P. Leishmania sand fly interaction: progress and challenges. Curr Opin Microbiol. 2008;11:340–4.

Bauzer LG, Gesto JS, Souza NA, Ward RD, Hamilton JG, Kyriacou CP, Peixoto AA. Molecular divergence in the period gene between two putative sympatric species of the *Lutzomyia longipalpis* complex. Mol Biol Evol. 2002;19:1624–7.

Belkaid Y, Kamhawi S, Modi G, Valenzuela J, Noben-Trauth N, Rowton E, et al. Development of a natural model of cutaneous leishmaniasis: powerful effects of vector saliva and saliva preexposure on the long-term outcome of *Leishmania major* infection in the mouse ear dermis. J Exp Med. 1998;188:1941–53.

Bonfante-Garrido R, Urdaneta R, Urdaneta I, Alvarado J. Natural infection of *Lutzomyia ovallesi* (Diptera: Psychodidae) with *Leishmania* in Duaca, Lara State, Venezuela. Trans R Soc Trop Med Hyg. 1991;85:61.

Braga RR, Lainson R, Ishikawa EA, Shaw JJ. *Leishmania* (*Viannia*) *utingensis* n. sp., a parasite from the sandfly *Lutzomyia* (*Viannamyia*) *tuberculata* in Amazonian Brazil. Parasite. 2003;10:111–8.

Brandão-Filho SP, Brito ME, Carvalho FG, Ishikawa EA, Cupolillo E, Floeter-Winter L, Shaw JJ. Wild and synanthropic hosts of *Leishmania* (*Viannia*) *braziliensis* in the endemic cutaneous leishmaniasis locality of Amaraji, Pernambuco State, Brazil. Trans R Soc Trop Med Hyg. 2003;97:291–6.

Caillard T, Tibayrenc M, Le Pont F, Dujardin JP, Desjeux P, Ayala FJ. Diagnosis by isozyme methods of two cryptic species, *Psychodopygus carrerai* and *P. yucumensis* (Diptera: Psychodidae). J Med Entomol. 1986;23:489–92.

Calzada JE, Saldana A, Rigg C, Valderrama A, Romero L, Chaves LF. Changes in phlebotomine sand fly species composition following insecticide thermal fogging in a rural setting of Western Panama. PLoS One. 2013;8:e53289.

Campbell-Lendrum DH, Pinto MC, Brandão-Filho SP, de Souza AA, Ready PD, Davies CR. Experimental comparison of anthropophily between geographically dispersed populations of *Lutzomyia whitmani* (Diptera: Psychodidae). Med Vet Entomol. 1999;13:299–309.

Campbell-Lendrum DH, Brandão-Filho SP, Pinto MC, Vexenat AJ, Ready PD, Davies CR. The domesticity of *Lutzomyia whitmani* populations: field experiments indicate behavioural differences. Bull Entomol Res. 2000;90:41–8.

Cardim MF, Rodas LA, Dibo MR, Guirado MM, Oliveira AM, Chiaravalloti-Neto F. Introduction and expansion of human American visceral leishmaniasis in the state of São Paulo, Brazil, 1999–2011. Rev Saúde Pública. 2013;47:691–700.

Casanova C, Colla-Jacques FE, Hamilton JG, Brazil RP, Shaw JJ. Distribution of *Lutzomyia longipalpis* chemotype populations in Sao Paulo state, Brazil. PLoS Negl Trop Dis. 2015; 9:e0003620.

Chajbullinova A, Votypka J, Sadlova J, Kvapilova K, Seblova V, Kreisinger J, et al. The development of *Leishmania turanica* in sand flies and competition with L. major. Parasit Vectors. 2012;5:219.

Chappuis F, Sundar S, Hailu A, Ghalib H, Rijal S, Peeling R, et al. Visceral leishmaniasis: what are the needs for diagnosis, treatment and control? Nat Rev Microbiol. 2007;5:873–82.

Chaskopoulou A, Giantsis I, Demir S, Bon M. Species composition, activity patterns and blood meal analysis of sand fly populations (Diptera: Psychodidae) in the metropolitan region of Thessaloniki, an endemic focus of canine leishmaniasis. Acta Trop. 2016;158:170–6.

Christodoulou V, Antoniou M, Ntais P, Messaritakis I, Ivovic V, Dedet J, et al. Re-emergence of visceral and cutaneous leishmaniasis in the Greek Island of Crete. Vector Borne Zoonotic Dis. 2012;12:214–22.

Cihakova J, Volf P. Development of different Leishmania major strains in vector sandflies *Phlebotomus papatasi* and *P. duboscqi*. Ann Trop Med Parasitol. 1997;91:267–79.

Clarke CF, Bradley KK, Wright JH, Glowicz J. Case report: emergence of autochthonous cutaneous leishmaniasis in northeastern Texas and southeastern Oklahoma. Am J Trop Med Hyg. 2013;88:157–61.

Coelho MV, Falcão AR. Transmissão experimemtal de *Leishmania braziliensis*. II. Tranmissão de amostra mexicana pela picada de *Phlebotomus longipalpis* e de *Phlebotomus renel*. Rev Inst Med Trop Sao Paulo. 1962;4:220–4.

Contreras Gutiérrez M, Vivero R, Vélez I, Porter C, Uribe S. DNA barcoding for the identification of sand fly species (Diptera, Psychodidae, Phlebotominae) in Colombia. PLoS One. 2014;9:e85496.

Costa CH. Characterization and speculations on the urbanization of visceral leishmaniasis in Brazil. Cad Saude Publica. 2008;24:2959–63.

Costa D, Favali C, Clarencio J, Afonso L, Conceicao V, Miranda J, et al. *Lutzomyia longipalpis* salivary gland homogenate impairs cytokine production and costimulatory molecule expression on human monocytes and dendritic cells. Infect Immun. 2004;72:1298–305.

Cotteaux-Lautard C, Leparc-Goffart I, Berenger J, Plumet S, Pages F. Phenology and host preferences *Phlebotomus perniciosus* (Diptera: Phlebotominae) in a focus of Toscana virus (TOSV) in South of France. Acta Trop. 2016;153:64–9.

Courtenay O, Quinnell RJ, Garcez LM, Shaw JJ, Dye C. Infectiousness in a cohort of brazilian dogs: why culling fails to control visceral leishmaniasis in areas of high transmission. J Infect Dis. 2002;186:1314–20.

da Costa AI, Casanova C, Rodas LA, Galati EA. Atualização da distribuição geografica e primeiro encontro de *Lutzomyia longipalpis* em area urbana no Estado de São Paulo, Brasil. Rev. Saúde Pública. 1997;31:632–3.

Davies CR, Fernandez M, Paz L, Ronca N, Llanos-Cuentas A. *Lutzomyia verrucarum* can transmit *Leishmania peruviana*, the aetiological agent of Andean cutaneous leishmaniasis. Trans R Soc Trop Med Hyg. 1993;87:603–6.

De Lima H, Rodriguez N, Feliciangeli MD, Barrios MA, Sosa A, Agrela I, et al. Cutaneous leishmaniasis due to *Leishmania chagasi/Le. infantum* in an endemic area of Guarico State, Venezuela. Trans R Soc Trop Med Hyg. 2009;103:721–6.

de Souza Freitas MT, Rios-Velasquez CM, Costa CR Jr, Figueiredo CA Jr, Aragão NC, da Silva LG, et al. Phenotypic and genotypic variations among three allopatric populations of *Lutzomyia umbratilis*, main vector of *Leishmania guyanensis*. Parasit Vectors. 2015;8:448.

de Souza AA, Dos Santos TV, Jennings YL, Ishikawa EA, Barata ID, Silva MD, et al. Natural *Leishmania (Viannia)* spp. infections in phlebotomine sand flies (Diptera: Psychodidae) from the Brazilian Amazon region reveal new putative transmission cycles of American cutaneous leishmaniasis. Parasite. 2016;23:22.

Deane LM. Leishmaniose Visceral no Brasil. Rio de Janeiro: Serviço Nacional de Educação Sanitária; 1956.

Depaquit J. Molecular systematics applied to Phlebotomine sandflies: review and perspectives. Infect Genet Evol. 2014;28:744–56.

Depaquit J, Ferté H, Léger N, Lefranc F, Alves-Pires C, Hanafi H, et al. ITS2 sequences heterogeneity in *Phlebotomus sergenti* and *Phlebotomus similis* (Diptera, Psychodidae): possible consequences in their ability to transmit Leishmania tropica. Int J Parasitol. 2002;32:1123–31.

Depaquit J, Lienard E, Verzeaux-Griffon A, Ferté H, Bounamous A, Gantier J, et al. Molecular homogeneity in diverse geographical populations of *Phlebotomus papatasi* (Diptera, Psychodidae) inferred from ND4 mtDNA and ITS2 rDNA Epidemiological consequences. Infect Genet Evol. 2008;8:159–70.

Depaquit J, Bounamous A, Akhoundi M, Augot D, Sauvage F, Dvorak V, et al. A taxonomic study of Phlebotomus (Larroussius) perfiliewi s. l. Infect Genet Evol. 2013;20:500–8.

Dergacheva T, Strelkova M. Epidemiological role of sandflies *Phlebotomus smirnovi* Perfiliev, 1941 and *P. longiductus* Parrot, 1928 in visceral leishmaniasis foci in the Kazakh SSR. Trans R Soc Trop Med Hyg. 1985;79:34–6.

Desbois N, Pratlong F, Quist D, Dedet JP. *Leishmania (Leishmania) martiniquensis* n. sp. (Kinetoplastida: Trypanosomatidae), description of the parasite responsible for cutaneous leishmaniasis in Martinique Island (French West Indies). Parasite. 2014;21:12.

Di-Blasi T, Lobo A, Nascimento L, Cordova-Rojas J, Pestana K, Marin-Villa M, et al. The flagellar protein FLAG1/SMP1 is a candidate for Leishmania-sand fly interaction. Vector Borne Zoonotic Dis. 2015;15:202–9.

Dillon R, Dillon V. The gut bacteria of insects: nonpathogenic interactions. Annu Rev Entomol. 2004;49:71–92.

Dillon R, Lane R. Bloodmeal Digestion in the midgut of *Phlebotomus papatasi* and *Phlebotomus langeroni*. Med Vet Entomol. 1993;7:225–32.

Dostalova A, Volf P. *Leishmania* development in sand flies: parasite-vector interactions overview. Parasit Vectors. 2012;5:276.

Dougall A, Alexander B, Holt D, Harris T, Sultan A, Bates P, Rose K, Walton SF. Evidence incriminating midges (Diptera: Ceratopogonidae) as potential vectors of *Leishmania* in Australia. Int J Parasitol. 2011;41:571–9.

Dujardin P, Le Pont F, Cruz M, Leon R, Tarrieu LF, Guderian R, Echeverria R, Tibayrenc M. Cryptic speciation in *Lutzomyia (Nyssomyia) trapidoi* (Fairchild & Hertig) (Diptera: Psychodidae) detected by multilocus enzyme electrophoresis. Am J Trop Med Hyg. 1996;54:42–5.

Dvorak V, Halada P, Hlavackova K, Dokianakis E, Antoniou M, Volf P. Identification of phlebotomine sand flies (Diptera: Psychodidae) by matrix-assisted laser desorption/ionization time of flight mass spectrometry. Parasit Vectors. 2014;7:21.

Dvorak V, Hlavackova K, Kocisova A, Volf P. First record of Phlebotomus (Transphlebotomus) mascittii in Slovakia. Parasite 2016;23:48.

Elnaiem D. Ecology and control of the sand fly vectors of *Leishmania donovani* in East Africa, with special emphasis on *Phlebotomus orientalis*. J Vector Ecol. 2011;36(Suppl 1):S23–31.

Elnaiem D, Hassan H, Osman O, Maingon R, Killick-Kendrick R, Ward R. A possible role for *Phlebotomus* (*Anaphlebotomus*) *rodhaini* (Parrot, 1930) in transmission of *Leishmania donovani*. Parasit Vectors. 2011;4:238.

Endris RG, Young DG, Perkins PV. Experimental transmission of *Leishmania mexicana* by a North American sand fly, *Lutzomyia anthophora* (Diptera: Psychodidae). J Med Entomol. 1987;24:243–7.

Espinosa OA, Serrano MG, Camargo EP, Teixeira MM, Shaw JJ. An appraisal of the taxonomy and nomenclature of trypanosomatids presently classified as *Leishmania* and *Endotrypanum*. Parasitology. 2016;15:1–13. https://doi.org/10.1017/S0031182016002092.

Esterre P, Chippaux JP, Lefait JF, Dedet JP. Evaluation d'un programme de lutte contre la leishmaniose cutanée dans un village forestier de Guyane francaise. Bull Org Mond Sant. 1986;64:559–65.

Feliciangeli M. Natural breeding places of phlebotomine sandflies. Med Vet Entomol. 2004;18:71–80.

Feliciangeli MD, Rabinovich J. Abundance of *Lutzomyia ovallesi* but not *Lu. gomezi* (Diptera: Psychodidae) correlated with cutaneous leishmaniasis incidence in north-central Venezuela. Med Vet Entomol. 1998;12:121–31.

Ferro C, Marin D, Gongora R, Carrasquilla MC, Trujillo JE, Rueda NK, et al. Phlebotomine vector ecology in the domestic transmission of American cutaneous leishmaniasis in Chaparral, Colombia. Am J Trop Med Hyg. 2011;85:847–56.

Fischer D, Moeller P, Thomas S, Naucke T, Beierkuhnlein C. Combining climatic projections and dispersal ability: a method for estimating the responses of sandfly vector species to climate change. PLoS Negl Trop Dis. 2011;5:e1407.

Fonteles RS, Pereira Filho AA, Moraes JL, Kuppinger O, Rebelo JM. Experimental Infection of *Lutzomyia (Nyssomyia) whitmani* (Diptera: Psychodidae: Phlebotominae) With *Leishmania (Viannia) braziliensis* and Leishmania (L.) amazonensis, Etiological Agents of American Tugumentary Leishmaniasis. J Med Entomol. 2016;53:206–9.

Freitas RA, Naiff RD, Barrett TV. Species diversity and flagellate infections in the sand fly fauna near Porto Grande, State of Amapa, Brazil (Diptera: Psychodidae. Kinetoplastida: Trypanosomatidae). Mem Inst Oswaldo Cruz. 2002;97:53–9.

Gajapathy K, Peiris L, Goodacre S, Silva A, Jude P, Surendran S. Molecular identification of potential leishmaniasis vector species within the *Phlebotomus (Euphlebotomus) argentipes* species complex in Sri Lanka. Parasit Vectors. 2013;6:302.

Galati EAB. Morfologia e Taxonomiia: 2.1 Classificção de Phlebotominae. In: Rangel E, Lainson R, editors. Flebotomíneos do Brasil. Rio de Janeiro: Editora FIOCRUZ; 2003. p. 23–51.

Galati EAB. Phlebotominae (Diptera, Psychodidae) Classificação, morfologia, terminologia e identificação de Adultos, vol. 1; 2015. p. 136. http://www.fsp.usp.br/egalati/ApostilaPhlebotominae_2015_vol_I.pdf.

Galati EAB, Nunes VLB, Dorval MEC, Oshiro ET, Cristaldo G, Espíndola MA, et al. Estudo dos flebotomíneos (Diptera, Pychodidae), em área de leishmaniose tegumentar, no Estado de Mato Grosso do Sul, Brasil. Rev Saúde Pública. 1996;30:115–28.

Gebre-Michael T, Balkew M, Ali A, Ludovisi A, Gramiccia M. The isolation of *Leishmania tropica* and *L. aethiopica* from Phlebotomus (Paraphlebotomus) species (Diptera: Psychodidae) in the Awash Valley, northeastern Ethiopia. Trans R Soc Trop Med Hyg. 2004;98:64–70.

Gebre-Michael T, Balkew M, Berthe N, Hailu A, Mekonnen Y. Further studies on the phlebotomine sandflies of the kala-azar endemic lowlands of Humera-Metema (north-west Ethiopia) with observations on their natural blood meal sources. Parasit Vectors. 2010;3:6.

Gebresilassie A, Yared S, Aklilu E, Kirstein O, Moncaz A, Tekie H, et al. Host choice of *Phlebotomus orientalis* (Diptera: Psychodidae) in animal baited experiments: a field study in Tahtay Adiyabo district, northern Ethiopia. Parasit Vectors. 2015;8:190.

Gil LH, Basano SA, Souza AA, Silva MG, Barata I, Ishikawa EA, Camargo LM, Shaw JJ. Recent observations on the sand fly (Diptera: Psychodidae) fauna of the State of Rondonia, Western Amazonia, Brazil: the importance of *Psychdopygus davisi* as a vector of zoonotic cutaneous leishmaniasis. Mem Inst Oswaldo Cruz. 2003;98:751–5.

Giorgobiani E, Lawyer P, Babuadze G, Dolidze N, Jochim RC, Tskhvaradze L, et al. Incrimination of *Phlebotomus kandelakii* and *Phlebotomus balcanicus* as vectors of Leishmania infantum in Tbilisi, Georgia. PLoS Negl Trop Dis. 2012;6:e1609.

Gomes AC, Barata JM, Rocha e Silva EO, Galati EA. Aspectos ecologicos da leishmaniose tegumentar americana. 7. Capacidade vetorial flebotominea em ambiente florestal pimario do sistema da Serra do Mar, região do Vale do Ribeira, Estado de São Paulo, Brasil. Rev Inst Med Trop Sao Paulo. 1989;31:32–9.

Gomez EA, Kato H, Hashiguchi Y. Man-biting sand fly species and natural infection with the Leishmania promastigote in leishmaniasis-endemic areas of Ecuador. Acta Trop. 2014;140:41–9.

Gonzalez R, Jorquera A, De Sousa L, Ledezma E, Devera R. Sandfly fauna of endemic leishmaniasis foci in Anzoategui State, Venezuela. Trans R Soc Trop Med Hyg. 2002;96:57–9.

Gonzalez C, Cabrera OL, Munstermann LE, Ferro C. Distribución de los vectores de *Leishmania infantum* (Kinetoplastida: Trypanosomatidae) en Colombia. Biomedica. 2006;26(Suppl 1):64–72.

Gonzalez C, Rebollar-Tellez EA, Ibanez-Bernal S, Becker-Fauser I, Martinez-Meyer E, Peterson AT, Sanchez-Cordero V. Current knowledge of Leishmania vectors in Mexico: how geographic distributions of species relate to transmission areas. Am J Trop Med Hyg. 2011;85:39–46.

Guan L, Yang Y, Qu J, Ren H, Chai J. Discovery and study of cutaneous leishmaniasis in Karamay of Xinjiang, West China. Infect Dis Poverty. 2013;2:20.

Guan L, Zhou Z, Jin C, Fu Q, Chai J. Phlebotomine sand flies (Diptera: Psychodidae) transmitting visceral leishmaniasis and their geographical distribution in China: a review. Infect Dis Poverty. 2016;5:15.

Guimarães VC, Pruzinova K, Sadlova J, Volfova V, Myskova J, Brandão Filho SP, Volf P. *Lutzomyia migonei* is a permissive vector competent for *Leishmania infantum*. Parasit Vectors. 2016;9:159. https://doi.org/10.1186/s13071-016-1444-2.

Haeberlein S, Fischer D, Thomas SM, Schleicher U, Beierkuhnlein C, Bogdan C. First assessment for the presence of phlebotomine vectors in Bavaria, Southern Germany, by combined distribution modeling and field surveys. PLoS One. 2013;8:e81088.

Hall L, Titus R. Sand fly vector saliva selectively modulates macrophage functions that inhibit killing of *Leishmania major* and nitric oxide production. J Immunol. 1995;155:3501–6.

Hamarsheh O, Presber W, Yaghoobi-Ershadi M, Amro A, Al-Jawabreh A, Sawalha S, et al. Population structure and geographical subdivision of the *Leishmania major* vector *Phlebotomus papatasi* as revealed by microsatellite variation. Med Vet Entomol. 2009;23:69–77.

Hamilton JG, Maingon RD, Alexander B, Ward RD, Brazil RP. Analysis of the sex pheromone extract of individual male *Lutzomyia longipalpis* sandflies from six regions in Brazil. Med Vet Entomol. 2005;19:480–8.

Harkins KM, Schwartz RS, Cartwright RA, Stone AC. Phylogenomic reconstruction supports supercontinent origins for *Leishmania*. Infect Genet Evol. 2005;38:101–9.

Harrison RG, Larson EL. Hybridization, introgression, and the nature of species boundaries. J Hered. 2014;105(Suppl 1):795–809.

Hashiguchi Y, Gomez EA, de Coronel VV, Mimori T, Kawabata M, Furuya M, Nonaka S, Takaoka H, Alexander JB, Quizhpe AM, et al. Andean leishmaniasis in Ecuador caused by infection with Leishmania mexicana and L. major-like parasites. Am J Trop Med Hyg. 1991;44:205–17.

Hoogstraal J, Heyneman D. Leishmaniasis in the Sudan Republic. Am J Trop Med Hyg. 1969;18:1091–210.

Illango KA. Taxonomic reassessment of the *Phlebotomus argentipes* species complex (Diptera: Psychodidae: Phlebotominae). J Med Entomol. 2010;47:1–15.

Inbar E, Akopyants NS, Charmoy M, Romano A, Lawyer P, Elnaiem D, et al. The mating competence of geographically diverse *Leishmania major* strains in their natural and unnatural sand fly vectors. PLoS Genet. 2013;9:e1003672.

Jacobson RL, Eisenberger CL, Svobodova M, Baneth G, Sztern J, Carvalho J, et al. Outbreak of cutaneous leishmaniasis in northern Israel. J Infect Dis. 2003;188:1065–73.

Jahanifard E, Yaghoobi-Ershadi MR, Akhavan AA, Akbarzadeh K, Hanafi-Bojd AA, Rassi Y, et al. Diversity of sand flies (Diptera, Psychodidae) in southwest Iran with emphasis on synanthropy of *Phlebotomus papatasi* and *Phlebotomus alexandri*. Acta Trop. 2014;140:173–80.

Kahime K, Boussaa S, El Mzabi A, Boumezzough A. Spatial relations among environmental factors and phlebotomine sand fly populations (Diptera: Psychodidae) in central and southern Morocco. J Vector Ecol. 2015;40:342–54.

Kamhawi S. Phlebotomine sand flies and *Leishmania* parasites: friends or foes? Trends Parasitol. 2006;22:439–45.

Kamhawi S, Belkaid Y, Modi G, Rowton E, Sacks D. Protection against cutaneous Leishmaniasis resulting from bites of uninfected sand flies. Science. 2000;290:1351–4.

Kamhawi S, Ramalho-Ortigao M, Pham V, Kumar S, Lawyer P, Turco SJ, et al. A role for insect galectins in parasite survival. Cell. 2004;119:329–41.

Kasap O, Votýpka J, Alten B. The distribution of the *Phlebotomus major* complex (Diptera: Psychodidae) in Turkey. Acta Trop. 2013;1273:204–11.

Kasap O, Dvorak V, Depaquit J, Alten B, Votypka J, Volf P. Phylogeography of the subgenus *Transphlebotomus* Artemiev with description of two new species, Phlebotomus anatolicus n. sp. and Phlebotomus killicki n. sp. Infect Genet Evol. 2015;34:467–79.

Kelly P, Bahr S, Serafim T, Ajami N, Petrosino J, Meneses C. The gut microbiome of the vector *Lutzomyia longipalpis* is essential for survival of *Leishmania infantum*. MBio. 2017;8:e01121–16.

Killick-Kendrick R. Phlebotomine vectors of the leishmaniases: a review. Med Vet Entomol. 1990;4:1–24.

Killick-Kendrick R. The biology and control of phlebotomine sand flies. Clin Dermatol. 1999;17:279–89.

Killick-Kendrick M, Killick-Kendrick R. The initial establishment of sand fly colonies. Parassitologia. 1991;33(Suppl 1):321–33.

Killick-Kendrick R, Killick-Kendrick M, Tang Y, Bastien P. Metacyclic promastigotes of *Leishmania* in the salivary glands of experimentally infected phlebotomine sandflies. Parasite. 1996;3:55–60.

Kimblin N, Peters N, Debrabant A, Secundino N, Egen J, Lawyer P, et al. Quantification of the infectious dose of *Leishmania major* transmitted to the skin by single sand flies. Proc Natl Acad Sci U S A. 2008;105:10125–30.

King RJ, Campbell-Lendrum DH, Davies CR. Predicting geographic variation in cutaneous leishmaniasis, Colombia. Emerg Infect Dis. 2004;10:598–607. https://doi.org/10.3201/eid1004.030241.

Kocher A, Gantier J, Gaborit P, Zinger L, Holota H, Valiere S, et al. Vector soup: high-throughput identification of Neotropical phlebotomine sand flies using metabarcoding. Mol Ecol Resour. 2017;17:172–82.

Kumar N, Srinivasan R, Anish T, Nandakumar G, Jambulingam P. Cutaneous leishmaniasis caused by *Leishmania donovani* in tribal population ofthe Agasthyamala biosphere reserve forest, Western Ghats, Kerala, India. J Med Microbiol. 2015;64:157–63.

Kwakye-Nuako G, Mosore MT, Duplessis C, Bates MD, Puplampu N, Mensah-Attipoe I, et al. First isolation of a new species of *Leishmania* responsible for human cutaneous leishmaniasis in Ghana and classification in the *Leishmania enriettii* complex. Int J Parasitol. 2015;45:679–84.

Lainson R, Shaw JJ. Evolution, classification and geographical distribution. In: Peters W, Killick-Kendrick R, editors. The Leishmaniases in biology and medicine: volume I biology and epidemiology, vol. I. London: Academic Press Inc; 1987. p. 1–120.

Lainson R, Ward R, Shaw J. *Leishmania* in phlebotomid sandflies: VI. Importance of hindgut development in distinguishing between parasites of the *Leishmania mexicana* and *L. braziliensis* complexes. Proc R Soc Lond B Biol Sci. 1977;199:309–20.

Lainson R, Shaw JJ, Ryan L, Ribeiro RS, Silveira FT. Leishmaniasis in Brazil. XXI. Visceral leishmaniasis in the Amazon Region and further observations on the role of *Lutzomyia longipalpis* (Lutz & Neiva, 1912) as the vector. Trans R Soc Trop Med Hyg. 1985;79:223–6.

Lainson R, Shaw JJ, Souza AA, Silveira FT, Falqueto A. Further observations on *Lutzomyia ubiquitalis* (Psychodidae: Phlebotominae), the sandfly vector of *Leishmania (Viannia) lainsoni*. Mem Inst Oswaldo Cruz. 1992;87:437–9.

Lane RP. Sandflies (Phlebotominae). In Lane RP, Crosskey LW, editors. Medical Insects and Arachnids, Dordrecht: Springer; 1993. p. 78–119.

Lanzaro GC, Ostrovska K, Herrero MV, Lawyer PG, Warburg A. *Lutzomyia longipalpis* is a species complex: genetic divergence and interspecific hybrid sterility among three populations. Am J Trop Med Hyg. 1993;48:839–47.

Lawyer P, Ngumbi P, Anjili C, Odongo S, Mebrahtu Y, Githure J, et al. Development of *Leishmania major* in *Phlebotomus duboscqi* and *Sergentomyia schwetzi* (Diptera: Psychodidae). Am J Trop Med Hyg. 1990;43:31–43.

Lawyer P, Mebrahtu Y, Ngumbi P, Mwanyumba P, Mbugua J, Kiilu G, et al. *Phlebotomus guggisbergi* (Diptera: Psychodidae), a vector of *Leishmania tropica* in Kenya. Am J Trop Med Hyg. 1991;44:290–8.

Lawyer P, Killick-Kendrick M, Rowland T, Rowton E, Volf P. Laboratory colonization and mass rearing of phlebotomine sand flies (Diptera, Psychodidae). Parasite 2017;24:42.

Le Pont F, Desjeux P. Leishmaniasis in Bolivia. II. The involvement of *Psychodopygus yucumensis* and *Psychodopygus llanosmartinsi* in the selvatic transmission cycle of *Leishmania braziliensis braziliensis* in a lowland sub-Andean region. Mem Inst Oswaldo Cruz. 1986;81:311–8.

Le Pont F, Pajot FX. La leishmaniose en Guyane francaise. 2. Modalites de la transmission dans un village forestie: Cacao. Cah ORSTOM ser Ent Med Parasitol. 1981;19:223–31.

Le Pont F, Mouchet J, Desjeux P, Torres Espejo JM, Richard A. Epidemiology of cutaneous leishmaniasis in Bolivia. 2. Transmission patterns. Ann Soc Belg Med Trop. 1989;1989(69):307–12.

Lehane M. Peritrophic matrix structure and function. Annu Rev Entomol. 1997;42:525–50.

Lemma W, Erenso G, Gadisa E, Balkew M, Gebre-Michael T, Hailu A. A zoonotic focus of cutaneous leishmaniasis in Addis Ababa, Ethiopia. Parasit Vectors. 2009;2:60.

Lestinova T, Rohousova I, Sima M, de Oliveira CI, Volf P. Insights into the sand fly saliva: blood-feeding and immune interactions between sand flies, hosts and *Leishmania*. PLoS Negl Trop Dis. 2017;11:e0005600.

Lima Costa CR Jr, Freitas MT, Santiago Figueiredo CA Jr, Aragão NC, da Silva LG, Marcondes CB, et al. Genetic structuring and fixed polymorphisms in the gene period among natural populations of *Lutzomyia longipalpis* in Brazil. Parasit Vectors. 2015;8:193. https://doi.org/10.1186/s13071-015-0785-6.

Lun Z, Wu M, Chen Y, Wang J, Zhou X, Liao L, et al. Visceral Leishmaniasis in China: an endemic disease under control. Clin Microbiol Rev. 2015;28:987–1004.

Maia C, Depaquit J. Can *Sergentomyia* (Diptera, Psychodidae) play a role in the transmission of mammal-infecting *Leishmania*? Parasite. 2016;23:55.

Maia C, Seblova V, Sadlova J, Votypka J, Volf P. Experimental transmission of *Leishmania infantum* by two major vectors: a comparison between a viscerotropic and a dermotropic strain. PLoS Negl Trop Dis. 2011;5:e1181.

Maingon R, Feliciangeli D, Guzman B, Rodriguez N, Convit J, Adamson R, et al. Cutaneous leishmaniasis in Tachira State, Venezuela. Ann Trop Med Parasitol. 1994;88:29–36.

Marcondes CB. A proposal of generic and subgeneric abbreviations for phlebotomine sandflies (Diptera: Psychodidae: Phlebotominae) of the world. Entomol News. 2007;118:361–56.

Maroli M, Rossi L, Baldelli R, Capelli G, Ferroglio E, Genchi C, et al. The northward spread of leishmaniasis in Italy: evidence from retrospective and ongoing studies on the canine reservoir and phlebotomine vectors. Trop Med Int Health. 2008;13:256–64.

Maroli M, Feliciangeli M, Bichaud L, Charrel R, Gradoni L. Phlebotomine sand flies and the spreading of leishmaniases and other diseases of public health concern. Med Vet Entomol. 2013;27:123–47.

Mbow M, Bleyenberg J, Hall L, Titus R. *Phlebotomus papatasi* sand fly salivary gland lysate down-regulates a Th1, but up-regulates a Th2, response in mice infected with *Leishmania major*. J Immunol. 1998;161:5571–7.

Modi GB. Care and maintenance of phlebotominae sandfly colonies. In: Crampton JM, Beard CB, Louis C, editors. Molecular biology of insect disease vectors. London: Chapman & Hall; 1997. p. 21–30.

Moncaz A, Faiman R, Kirstein O, Warburg A. Breeding sites of *Phlebotomus sergenti*, the sand fly vector of cutaneous leishmaniasis in the Judean desert. PLoS Negl Trop Dis. 2012;6:e1725.

Moncaz A, Kirstein O, Gebresellassie A, Lemma W, Yared S, Gebre-Michael T, Hailu A, et al. Characterization of breeding sites of *Phlebotomus orientalis* - the vector of visceral leishmaniasis in northwestern Ethiopia. Acta Trop. 2014;139:5–14.

Morris R, Shoemaker C, David J, Lanzaro G, Titus R. Sandfly maxadilan exacerbates infection with *Leishmania major* and vaccinating against it protects against *L. major* infection. J Immunol. 2001;167:5226–30.

Mouchet J, Le Pont F, Leon R, Echeverria R, Guderian RH. Leishmaniasis in Ecuador. 5. Leishmaniasis and anthropization on the Pacific coast. Ann Soc Belg Med Trop. 1994;74:35–41.

Myskova J, Svobodova M, Beverley S, Volf P. A lipophosphoglycan-independent development of *Leishmania* in permissive sand flies. Microbes Infect. 2007;9:317–24.

Myšková J, Dostalova A, Penickova L, Halada P, Bates PA, Volf P. Characterization of a midgut mucin-like glycoconjugate of Lutzomyia longipalpis with a potential role in Leishmania attachment. Parasit Vectors 2016;9:413.

Nascimento BW, Saraiva L, Neto RG, Meira PC, Sanguinette CC, Tonelli GB, et al. Study of sand flies (Diptera: Psychodidae) in visceral and cutaneous leishmaniasis areas in the central-western state of Minas Gerais, Brazil. Acta Trop. 2013;125:262–8. https://doi.org/10.1016/j.actatropica.2012.11.005.

Naucke T, Menn B, Massberg D, Lorentz S. Sandflies and leishmaniasis in Germany. Parasitol Res. 2008;103:65–8.

Nieves E, Pimenta PF. Development of *Leishmania (Viannia) braziliensis* and *Leishmania (Leishmania) amazonensis* in the sand fly *Lutzomyia migonei* (Diptera: Psychodidae). J Med Entomol. 2000;37:134–40.

Nzelu C, Cáceres A, Arrunátegui-Jiménez M, Lañas-Rosas M, Yañez-Trujillano H, Luna-Caipo D, et al. DNA barcoding for identification of sand fly species (Diptera: Psychodidae) from leishmaniasis-endemic areas of Peru. Acta Trop. 2015;145:45–51.

Orshan L, Elbaz S, Ben-Ari Y, Akad F, Afik O, Ben-Avi I, et al. Distribution and Dispersal of *Phlebotomus papatasi* (Diptera: Psychodidae) in a Zoonotic Cutaneous Leishmaniasis Focus, the Northern Negev, Israel. PLoS Negl Trop Dis. 2016;10:e0004819.

Perez J, Virgen A, Rojas JC, Rebollar-Tellez EA, Alfredo C, Infante F, et al. Species composition and seasonal abundance of sandflies (Diptera: Psychodidae: Phlebotominae) in coffee agroecosystems. Mem Inst Oswaldo Cruz. 2014;109:80–6.

Pessoa SB. Surto epidêmico de calazar no município de Sobral (Estado do Ceará). Rev Paulista Med. 1954;44:141–2.

Pesson B, Ready J, Benabdennbi I, Martín-Sánchez J, Esseghir S, Cadi-Soussi M, et al. Sandflies of the *Phlebotomus perniciosus* complex: mitochondrial introgression and a new sibling species *of P. longicuspis* in the Moroccan Rif. Med Vet Entomol. 2004;18:23–37.

Peters NC, Egen J, Secundino N, Debrabant A, Kimblin N, Kamhawi S, et al. In vivo imaging reveals an essential role for neutrophils in leishmaniasis transmitted by sand flies. Science. 2008;321:970–4.

Petersen CA, Barr SC. Canine leishmaniasis in North America: emerging or newly recognized? Vet Clin North Am Small Anim Pract. 2009;39:1065–74.

Pimenta P, Saraiva E, Rowton E, Modi G, Garraway L, Beverley S, et al. Evidence that the vectorial competence of phlebotomine sand flies for different species of *Leishmania* is controlled by structural polymorphisms in the surface lipophosphoglycan. Proc Natl Acad Sci U S A. 1994;91:9155–9.

Pimenta P, Modi G, Pereira S, Shahabuddin M, Sacks D. A novel role for the peritrophic matrix in protecting *Leishmania* from the hydrolytic activities of the sand fly midgut. Parasitology. 1997;115:359–69.

Pinheiro MP, Silva MM, Junior JB, da Silva JH, Alves ML, Ximenes MF. Sand flies (Diptera, Psychodidae, Phlebotominae), vectors of *Leishmania* protozoa, at an Atlantic Forest Conservation Unit in the municipality of Nisia Floresta, Rio Grande do Norte state, Brazil. Parasit Vectors. 2016;9:83. https://doi.org/10.1186/s13071-016-1352-5.

Poché R, Garlapati R, Elnaiem D, Perry D, Poché D. The role of Palmyra palm trees (Borassus flabellifer) and sand fly distribution in northeastern India. J Vector Ecol. 2012;37:148–53.

Pothirat T, Tantiworawit A, Chaiwarith R, Jariyapan N, Wannasan A, Siriyasatien P, et al. First isolation of *Leishmania* from Northern Thailand: case report, identification as *Leishmania martiniquensis* and phylogenetic position within the *Leishmania enriettii* complex. PLoS Negl Trop Dis. 2014;8:e3339.

Prudhomme J, Cassan C, Hide M, Toty C, Rahola N, Vergnes B, et al. Ecology and morphological variations in wings of *Phlebotomus ariasi* (Diptera: Psychodidae) in the region of Roquedur (Gard, France): a geometric morphometrics approach. Parasit Vectors. 2016;9:578.

Pruzinova K, Sadlova J, Seblova V, Homola M, Votypka J, Volf P. Comparison of bloodmeal digestion and the peritrophic matrix in four sand fly species differing in susceptibility to *Leishmania donovani*. PLoS One. 2015;10:e0128203.

Pruzinova K, Sadlova J, Myskova J, Lestinova T, Janda J, Volf P. *Leishmania* mortality in sand fly bloodmeal is not species-specific and does not result from direct effect of proteinases. Parasit Vectors. 2017;10:276.

Ramalho-Ortigao J, Traub-Cseko Y. Molecular characterization of Llchit1, a midgut chitinase cDNA from the leishmaniasis vector *Lutzomyia longipalpis*. Insect Mol Biol. 2003;33:279–87.

Ramalho-Ortigao J, Kamhawi S, Joshi M, Reynoso D, Lawyer P, Dwyer D, et al. Characterization of a blood activated chitinolytic system in the midgut of the sand fly vectors *Lutzomyia longipalpis* and *Phlebotomus papatasi*. Insect Mol Biol. 2005;14:703–12.

Rangel EF, Lainson R, Souza AA, Ready P, Azevedo ACR. Geographical variation in populations of *Lutzomyia* (*Nyssomyia*) *whitmani* (Antunes & Coutinho, 1939) *sensu lato* (Diptera: Psychodidae: Phlebotomine) in Brazil. Mem Inst Oswaldo Cruz. 1996;91:43–50.

Ready PD, Lainson R, Shaw JJ. 1984 (unpublished observations).

Ready PD, Arias JR, Freitas RA. A pilot study to control *Lutzomyia umbratilis* (Diptera: Psychodidae), the major vector of *Leishmania braziliensis guyanensis*, in a peri-urban rainforest of Manaus, Amazonas State, Brazil. Mem Inst Oswaldo Cruz. 1985;80:27–36.

Ready PD, Lainson R, Shaw JJ, Ward RD. The ecology of *Lutzomyia umbratilis* Ward & Fraiha (Diptera Psychodidae), the major vector to man of *Leishmania braziliensis guyanensis* Floch in northeast Amazonian Brazil. Bull Entomol Res. 1986;76:21–40.

Ready PD, Day JC, de Souza AA, Rangel EF, Davies CR. Mitochondrial DNA characterization of populations of *Lutzomyia whitmani* (Diptera: Psychodidae) incriminated in the peri-domestic and silvatic transmission of *Leishmania* species in Brazil. Bull Entomol Res. 1997;87: 187–95.

Rioux J, Killick-Kendrick R, Perieres J, Turner D, Lanotte G. Ecologie des Leishmanioses dans le Sud de la France. 13. Les sites de "flanc de coteau", biotopes de transmission privilégiés de la Leishmaniose viscérale en Cévennes. Annales de Parasitologie humaine et comparée. 1980;55:445–53.

Roberts DR, Hsi BP. An index of species abundance for use with mosquito surveillance data. Environ Entomol. 1979;8:1007–13.

Rogers M. The role of *Leishmania* proteophosphoglycans in sand fly transmission and infection of the mammalian host. Front Microbiol. 2012;3:223.

Rogers M, Bates P. *Leishmania* manipulation of sand fly feeding behavior results in enhanced transmission. PLoS Pathog. 2007;3:818–25.

Rogers K, Titus R. Immunomodulatory effects of Maxadilan and *Phlebotomus papatasi* sand fly salivary gland lysates on human primary *in vitro* immune responses. Parasite Immunol. 2003;25:127–34.

Rogers M, Chance M, Bates P. The role of promastigote secretory gel in the origin and transmission of the infective stage of *Leishmania mexicana* by the sandfly *Lutzomyia longipalpis*. Parasitology. 2002;124:495–507.

Rogers M, Ilg T, Nikolaev A, Ferguson M, Bates P. Transmission of cutaneous leishmaniasis by sand flies is enhanced by regurgitation of fPPG. Nature. 2004;430:463–7.

Rogers M, Hajmova M, Joshi M, Sadlova J, Dwyer D, Volf P, et al. *Leishmania* chitinase facilitates colonization of sand fly vectors and enhances transmission to mice. Cell Microbiol. 2008;10:1363–72.

Rogers M, Kropf P, Choi B, Dillon R, Podinovskaia M, Bates P, et al. Proteophosophoglycans regurgitated by *Leishmania*-infected sand flies target the L-arginine metabolism of host macrophages to promote parasite survival. PLoS Pathog. 2009;5:e1000555.

Rogers M, Downing T, Smith B, Imamura H, Sanders M, Svobodova M, et al. Genomic confirmation of hybridization and recent inbreeding in vector-isolated *Leishmania* population. PLoS Genet. 2014;10:e1004092.

Rohousova I, Hostomska J, Vlkova M, Kobets T, Lipoldova M, Volf P. The protective effect against *Leishmania* infection conferred by sand fly bites is limited to short-term exposure. Int J Parasitol. 2011;41:481–5.

Romano A, Inbar E, Debrabant A, Charmoy M, Lawyer P, Ribeiro-Gomes F. Cross-species genetic exchange between visceral and cutaneous strains of *Leishmania* in the sand fly vector. Proc Natl Acad Sci U S A. 2014;111:16808–13.

Rougeron V, De Meeûs T, Kako Ouraga S, Hide M, Bañuls AL. "Everything you always wanted to know about sex (but were afraid to ask)" in *Leishmania* after two decades of laboratory and field analyses. PLoS Pathog. 2010;6:e1001004.

Ryan L, Fraiha H, Lainson R, Shaw JJ. New phlebotomine sandflies of the *walkeri* group (Diptera: Psychodidae) from Pará State, Brazil, with a pictorial key 1. Mem Inst Oswaldo Cruz. 1986;84:323–31.

Ryan L, Lainson R, Shaw JJ, Wallbanks KR. The transmission of suprapylarian leishmania by the bite of experimentally infected sand flies (Diptera: Psychodidae). Mem Inst Oswaldo Cruz. 1987;82:425–30.

Sacks D, Perkins P. Development of infective stage *Leishmania* promastigotes within phlebotomine sandflies. Am J Trop Med Hyg. 1985;34:456–9.

Sacks D, Brodin T, Turco S. Developmental modification of the lipophosphoglycan from *Leishmanial major* promastigotes during metacyclogenesis. Mol Biochem Parasitol. 1990; 42:225–33.

Sadlova J, Volf P. Occurrence of *Leishmania major* in sandfly urine. Parasitology. 1999;118:455–60.

Sadlova J, Volf P. Peritrophic matrix of *Phlebotomus duboscqi* and its kinetics during *Leishmania major* development. Cell Tissue Res. 2009;337:313–25.

Sadlova J, Price H, Smith B, Votypka J, Volf P, Smith D. The stage-regulated HASPB and SHERP proteins are essential for differentiation of the protozoan parasite Leishmania major in its sand fly vector, *Phlebotomus papatasi*. Cell Microbiol. 2010;12:1765–79.

Sadlova J, Yeo M, Seblova V, Lewis M, Mauricio I, Volf P, et al. Visualisation of *Leishmania donovani* fluorescent hybrids during early stage development in the sand fly vector. PLoS One. 2011;6:e19851.

Sadlova J, Dvorak V, Seblova V, Warburg A, Votypka J, Volf P. *Sergentomyia schwetzi* is not a competent vector for *Leishmania donovani* and other *Leishmania* species pathogenic to humans. Parasit Vectors. 2013;6:186.

Saldana A, Chaves LF, Rigg CA, Wald C, Smucker E, Calzada JE. Clinical cutaneous leishmaniasis rates are associated with household *Lutzomyia gomezi, Lu. panamensis*, and *Lu. trapidoi* abundance in Trinidad de Las Minas, western Panama. Am J Trop Med Hyg. 2013;88:572–4.

Salomon OD, Quintana MG, Zaidenberg M. Urban distribution of Phlebotominae in a cutaneous leishmaniasis focus, Argentina. Mem Inst Oswaldo Cruz. 2008a;103:282–7.

Salomon OD, Rosa JR, Stein M, Quintana MG, Fernandez MS, Visintin AM, et al. Phlebotominae (Diptera: Psychodidae) fauna in the Chaco region and Cutaneous Leishmaniasis transmission patterns in Argentina. Mem Inst Oswaldo Cruz. 2008b;103:578–84.

Sant'Anna M, Diaz-Albiter H, Aguiar-Martins K, Al Salem W, Cavalcante R, Dillon V, et al. Colonisation resistance in the sand fly gut: *Leishmania* protects *Lutzomyia longipalpis* from bacterial infection. Parasit Vectors. 2014;7:329.

Savani ES, Nunes VL, Galati EA, Castilho TM, Zampieri RA, Floeter-Winter LM. The finding of *Lutzomyia almerioi* and Lutzomyia longipalpis naturally infected by *Leishmania* spp. in a cutaneous and canine visceral leishmaniases focus in Serra da Bodoquena, Brazil. Vet Parasitol. 2009;160:18–24.

Scarpassa VM, Alencar RB. *Lutzomyia umbratilis*, the main vector of *Leishmania guyanensis,* represents a novel species complex? PLoS One. 2012;7:e37341.

Schlein Y, Jacobson R. Hemoglobin inhibits the development of infective promastigotes and chitinase secretion in Leishmania major cultures. Parasitology. 1994;109:23–8.

Schlein Y, Jacobson R, Messer G. Leishmania infections damage the feeding mechanism of the sandfly vector and implement parasite transmission by bite. Proc Natl Acad Sci U S A. 1992;89:9944–8.

Seblova V, Sadlova J, Carpenter S, Volf P. Development of *Leishmania* parasites in *Culicoides nubeculosus* (Diptera: Ceratopogonidae) and implications for screening vector Competence. J Med Entomol. 2012;49:967–70.

Seblova V, Volfova V, Dvorak V, Pruzinova K, Votypka J, Kassahun A, et al. Phlebotomus orientalis sand flies from two geographically distant ethiopian localities: biology, genetic analyses and susceptibility to Leishmania donovani. PLoS Negl Trop Dis. 2013;7:e2187.

Seblova V, Sadlova J, Vojtkova B, Votypka J, Carpenter S, Bates PA, et al. The biting midge *Culicoides sonorensis* (Diptera: Ceratopogonidae) is capable of developing late stage infections of *Leishmania enriettii*. PLoS Negl Trop Dis. 2015;9:e0004060.

Seccombe A, Ready P, Huddleston LA. Catalogue of old world phlebotomine sandflies (Diptera: Psychodidae). Occas Pap Syst Entomol. 1993;8:1–57.

Secundino N, Kimblin N, Peters N, Lawyer P, Capul A, Beverley S, et al. Proteophosphoglycan confers resistance of *Leishmania major* to midgut digestive enzymes induced by blood feeding in vector sand flies. Cell Microbiol. 2010;12:906–18.

Sharma N, Mahajan V, Ranjan N, Verma G, Negi A, Mehta K. The sandflies of the Satluj river valley, Himachal Pradesh (India): some possible vectors of the parasite causing human cutaneous and visceral leishmaniases in this endemic focus. J Vector Borne Dis. 2009;46:136–40.

Shaw JJ. The behaviour of Endotrypanum schaudinni (Kinetoplastidae: Trypanosomatidae) in three species of laboratory-bred neotropical sandflies (Diptera: Psychodidae) and its influence on the classification of the genus Leishmania. In: Canning EU, editor. Parasitological topics a presentation volume to PCC Garnham FRS on the occasion of his 80th birthday. Lawrence, Kansas: Society of Protozoologists Special Publication, Allen Press Inc; 1981. p. 232–41.

Shaw JJ. The relationship of sand fly ecology to the transmission of leishmaniasis in South America with particular reference to Brazil. Mem Entomol Internat. 1999;14:503–17.

Silva F, Gomes R, Prates D, Miranda JC, Andrade B, Barral-Netto M, et al. Inflammatory cell infiltration and high antibody production in BALB/c mice caused by natural exposure to *Lutzomyia longipalpis* bites. Am J Trop Med Hyg. 2005;72:94–8.

Soares R, Barron T, McCoy-Simandle K, Svobodova M, Warburg A, Turco S. *Leishmania tropica*: intraspecific polymorphisms in lipophosphoglycan correlate with transmission by different *Phlebotomus* species. Exp Parasitol. 2004;107:105–14.

Srinivasan R, Jambulingam P, Kumar N, Selvakumar M, Edwin B, Kumar T. Temporal distribution and behaviour of sand flies (Diptera: Psychodidae) in a cutaneous leishmaniasis focus of the Kani Tribe settlements in the Western Ghats, India. Acta Trop. 2015;148:147–55.

Srinivasan R, Kumar N, Jambulingam P. Detection of natural infection of *Leishmania donovani* (Kinetoplastida: Trypanosomatidae) in *Phlebotomus argentipes* (Diptera: Psychodidae) from a forest ecosystem in the Western Ghats, India, endemic for cutaneous leishmaniasis. Acta Trop. 2016;156:95–9.

Stamper LW, Patrick RL, Fay MP, Lawyer PG, Elnaiem DE, Secundino N, et al. Infection parameters in the sand fly vector that predict transmission of *Leishmania major*. PLoS Negl Trop Dis. 2011;5:e1288.

Stierhof Y, Bates P, Jacobson R, Rogers M, Schlein Y, Handman E, et al. Filamentous proteophosphoglycan secreted by *Leishmania* promastigotes forms gel-like three-dimensional networks that obstruct the digestive tract of infected sandfly vectors. Eur J Cell Biol. 1999;10:675–89.

Strelkova M. Progress in studies on Central Asian foci of zoonotic cutaneous leishmaniasis: a review. Folia Parasitol. 1996;431:1–6.

Strelkova M, Ponirovsky E, Morozov E, Zhirenkina E, Razakov S, Kovalenko D. A narrative review of visceral leishmaniasis in Armenia, Azerbaijan, Georgia, Kazakhstan, Kyrgyzstan, Tajikistan, Turkmenistan, Uzbekistan, the Crimean Peninsula and Southern Russia. Parasit Vectors. 2015;8:330.

Svarovska A, Ant T, Seblova V, Jecna L, Beverley S, Volf P. *Leishmania major* glycosylation mutants require phosphoglycans (lpg2-) but not lipophosphoglycan (lpg1-) for survival in permissive sand fly vectors. PLoS Negl Trop Dis. 2010;4:e580.

Svobodova M, Votypka J, Peckova J, Dvorak V, Nasseredin A, Baneth G, et al. Distinct transmission cycles of *Leishmania tropica* in two adjacent foci, Northern Israel. Emerg Infect Dis. 2006;12:1860–8.

Svobodova M, Alten B, Zidkova L, Dvorak V, Hlavackova J, Myskova J, et al. Cutaneous leishmaniasis caused by *Leishmania infantum* transmitted by *Phlebotomus tobbi*. Int J Parasitol. 2009;39:251–6.

Telleria E, de Araujo A, Secundino N, d'Avila-Levy C, Traub-Cseko Y. Trypsin-like serine proteases in *Lutzomyia longipalpis* – expression, activity and possible modulation by *Leishmania infantum chagasi*. PLoS One. 2010;5:e10697.

Teodoro U, La Salvia Filho V, de Lima EM, Misuta NM, Verginassi TG, Ferreira ME. Leishmaniose tegumentar americana: flebotomíneos de area de transmissão no Norte do Parana, Brasíl. Rev Saude Publica. 1991;25:129–33.

Theodos C, Titus R. Salivary gland material from the sand fly *Lutzomyia longipalpis* has an inhibitory effect on macrophage function in vitro. Parasite Immunol. 1993;15:481–7.

Theodos CM, Ribeiro JMC, Titus RG. Analysis of enhancing effect of sand fly saliva on Leishmania infection in mice. Infect Immun. 1991;59:1592–8.

Thiakaki M, Rohousova I, Volfova V, Volf P, Chang K, Soteriadou K. Sand fly specificity of saliva-mediated protective immunity in *Leishmania amazonensis* BALB/c mouse model. Microbes Infect. 2005;7:760–6.

Titus RG, Ribeiro J. Salivary gland lysates from the sand fly *Lutzomyia longipalpis* enhance *Leishmania* infectivity. Science. 1988;239:1306–8.

Travi BL, Velez ID, Brutus L, Segura I, Jaramillo C, Montoya J. *Lutzomyia evansi* an alternate vector of *Leishmania chagasi* in a Colombian focus of visceral leishmaniasis. Trans R Soc Trop Med Hyg. 1990;84:676–7.

Valenzuela J, Belkaid Y, Rowton E, Ribeiro J. The salivary apyrase of the blood-sucking sand fly *Phlebotomus papatasi* belongs to the novel Cimex family of apyrases. J Exp Biol. 2001;204:229–37.

van Zandbergen G, Klinger M, Mueller A, Dannenberg S, Gebert A, Solbach W, et al. Cutting edge: neutrophil granulocyte serves as a vector for *Leishmania* entry into macrophages. J Immunol. 2004;173:6521–5.

Villaseca P, Llanos-Cuentas A, Perez E, Davies CR. A comparative field study of the relative importance of *Lutzomyia peruensis* and *Lutzomyia verrucarum* as vectors of cutaneous leishmaniasis in the Peruvian Andes. Am J Trop Med Hyg. 1993;49:260–9.

Volf P, Myskova J. Sand flies and *Leishmania*: specific versus permissive vectors. Trends Parasitol. 2007;23:91–2.

Volf P, Kiewegova A, Nemec A. Bacterial colonisation in the gut of *Phlebotomus duboscqi* (Diptera: Psychodidae): transtadial passage and the role of female diet. Folia Parasitol. 2002;49:73–7.

Volf P, Hajmova M, Sadlova J, Votypka J. Blocked stomodeal valve of the insect vector: similar mechanism of transmission in two trypanosomatid models. Int J Parasitol. 2004;34:1221–7.

Volf P, Benkova I, Myskova J, Sadlova J, Campino L, Ravel C. Increased transmission potential of *Leishmania major/Leishmania infantum* hybrids. Int J Parasitol. 2007;37:589–93.

Volf P, Volfova V. Establishment and maintenance of sand fly colonies. J Vector Ecol 2011;36: S1–S9.

Volf P, Nogueira P, Myskova J, Turco S, Soares R. Structural comparison of lipophosphoglycan from *Leishmania turanica* and *L. major*, two species transmitted by *Phlebotomus papatasi*. Parasitol Int. 2014;63:683–6.

Walters L, Irons K, Guzman H, Tesh R. Formation and composition of the peritrophic membrane in the sand fly, *Phlebotomus perniciosus* (Diptera, Psychodidae). J Med Entomol. 1993;30:179–98.

Wang J, Cui G, Chen H, Zhou X, Gao C, Yang Y. Current epidemiological profile and features of visceral leishmaniasis in People's Republic of China. Parasit Vectors. 2012;5:31.

Ward RD, Ribeiro AL, Ready PD, Murtagh A. Reproductive isolation between different forms of *Lutzomyia longipalpis* (Lutz & Neiva), (Diptera: Psychodidae), the vector of *Leishmania donovani chagasi* Cunha & Chagas and its significance to kala-azar distribution in South America. Mem Inst Oswaldo Cruz. 1983;78:269–80.

Weigle KA, Saravia NG, de Davalos M, Moreno LH, D'Alessandro A. *Leishmania braziliensis* from the Pacific coast region of Colombia: foci of transmission, clinical spectrum and isoenzyme phenotypes. Am J Trop Med Hyg. 1986;35:722–31.

Werneck GL. Visceral leishmaniasis in Brazil: rationale and concerns related to reservoir control. Rev Saude Publica. 2014;48:851–6.

WHO. Control of the leishmaniases. World Health Org Tech Rep Ser. 2010;949:186.

Wilson R, Bates M, Dostalova A, Jecna L, Dillon R, Volf P. Stage-specific adhesion of *Leishmania* promastigotes to sand fly midguts assessed using an improved comparative binding assay. PLoS Negl Trop Dis. 2010;4:e816.

Yared S, Gebresilassie A, Akililu E, Balkew M, Warburg A, Hailu A, et al. Habitat preference and seasonal dynamics of *Phlebotomus orientalis* in urban and semi-urban areas of kala-azar endemic district of Kafta Humera, northwest Ethiopia. Acta Trop. 2017;166:25–34.

Young DG, Duncan MA, Young DG, Duncan MA. Guide to the identification and geographic distribution of *Lutzomyia* sand flies in Mexico, the West Indies, Central and South America (Diptera: Psychodidae). Mem Am Entomol Inst. 1994;54:1–881.

Zeledón R, Murillo J, Gutierrez H. Observaciones sobre la ecologia de *Lutzomyia longipalpis* (Lutz & Neiva, 1912) y posibilidades de existencia de leishmaniasis visceral en Costa Rica. Mem Inst Oswaldo Cruz. 1984;79:455–9.

Zer R, Yaroslavski I, Rosen L, Warburg A. Effect of sand fly saliva on *Leishmania* uptake by murine macrophages. Int J Parasitol. 2001;31:810–4.

Zhang L, Ma Y, Xu J. Genetic differentiation between sandfly populations of *Phlebotomus chinensis* and *Phlebotomus sichuanensis* (Diptera: Psychodidae) in China inferred by microsatellites. Parasit Vectors. 2013;6:115.

Parasite Biology: The Reservoir Hosts

4

Carla Maia, Filipe Dantas-Torres, and Lenea Campino

4.1 Introduction

Leishmaniases are parasitic diseases caused by protozoa belonging to the genus *Leishmania* (order Kinetoplastida, family Trypanosomatidae), which infects several mammal species, including humans. These parasites are primarily transmitted by the bite of an insect vector, the phlebotomine sand fly (order Diptera, family Psychodidae; subfamily Phlebotominae) of the genera *Phlebotomus* (Old World) and *Lutzomyia* (New World) (Killick-Kendrick 1999; WHO 2010; Maroli et al. 2013). Human leishmaniases have diverse clinical manifestations. Visceral leishmaniasis (VL) caused by parasites of the *Leishmania donovani* complex (*L. donovani* in the Old World and *L. infantum* in both the Old and New Worlds) is a severe disease of humans and other mammals, which leads to death if left untreated. A number of different *Leishmania* spp. cause localized cutaneous (LCL) or diffuse cutaneous (DCL) or mucocutaneous (MCL) leishmaniasis, which are responsible for considerable morbidity of a vast number of people in endemic foci. Leishmaniases are endemic in 98 countries on 4 continents, with more than 350 million people at risk. Published figures indicate an

C. Maia (✉)
Global Health and Tropical Medicine, GHTM, Instituto de Higiene e Medicina Tropical, IHMT, Universidade Nova de Lisboa, UNL, Lisboa, Portugal
e-mail: CarlaMaia@ihmt.unl.pt

F. Dantas-Torres
Department of Immunology, Instituto Aggeu Magalhães, Fundação Oswaldo Cruz, Recife, Brazil
e-mail: filipe.dantas@cpqam.fiocruz.br

L. Campino
Global Health and Tropical Medicine, GHTM, Instituto de Higiene e Medicina Tropical, IHMT, Universidade Nova de Lisboa, UNL, Lisboa, Portugal
e-mail: Campino@ihmt.unl.pt

F. Bruschi, L. Gradoni (eds.), *The Leishmaniases: Old Neglected Tropical Diseases*, https://doi.org/10.1007/978-3-319-72386-0_4

estimated incidence of 0.2–0.4 million VL cases and 0.7–1.3 million cutaneous leishmaniasis (CL) cases (WHO 2010; Alvar et al. 2012). These figures are, most probably, underestimated as official data and are often obtained through passive case detection, and the extent of underreporting in most leishmaniasis endemic countries (even in those where the disease is of compulsory notification) is substantial (Desjeux 2004; Dujardin et al. 2008; WHO 2010; Alvar et al. 2012).

Leishmaniases are dynamic diseases, and the circumstances of transmission are continually changing in relation to environmental, demographic and human behavioural factors. In most endemic regions, leishmaniases are characterized by a patchy distribution with discrete transmission foci due to microecological conditions that affect the vector, the parasite and the reservoir host. Changes in the habitat of the natural host and vector, immunosuppressive conditions (e.g. HIV infection or organ transplantation-associated therapies) and the consequences of military conflicts, all contribute to the changing leishmaniasis landscape, which can result in either an increase or a decrease in the incidence of the disease (Gramiccia and Gradoni 2005; WHO 2010; Alvar et al. 2012; Antoniou et al. 2013).

Leishmaniases can be grouped into two broad epidemiological categories according to the source of human infection: zoonotic leishmaniases, in which the reservoir hosts are wild or domestic animals and humans play a role of an accidental host, and anthroponotic leishmaniases, in which man is the sole reservoir host and source of vector's infection (Desjeux 2004; Gramiccia and Gradoni 2007; WHO 2010).

A reservoir host of leishmaniases is an animal in which an infectious agent survives persistently in a way that the animal may serve as a source of parasites to the vectors. The mere presence of the infection in a particular mammal species, even in large numbers, does not necessarily indicate that this mammal is a reservoir host. In order to incriminate a reservoir host formally, it is necessary to demonstrate that the parasite population depends on that particular mammal for its long-term maintenance (Ashford 1996; WHO 2010). A "good" reservoir should be susceptible to the parasites, live in close contact with man, and it should be a good source of parasites to the vectors. The proportion of individuals that become infected during their lifetime should be considerable, although the incidence can vary greatly with season. A good reservoir should provide a significant food source for the phlebotomine sand fly, and both should rest and breed in the same habitat. Infection should present a chronic evolution allowing the animal to survive at least until the next transmission season (Bray 1982; Ashford 1996; WHO 2010). *Leishmania* parasites identified in reservoir hosts must be biochemically and genetically the same as those in humans.

When more than one host species can be infected, they are often divided on epidemiological grounds into primary and secondary (or minor) reservoir hosts and accidental (or incidental) hosts: primary reservoir host is a host that is responsible for maintaining the parasite indefinitely in nature. In these hosts, the infection is normally without clinical signs; secondary reservoir host is a host that can transmit infection but cannot maintain parasite transmission in the absence of the primary host(s), while accidental host is a host that although infected plays no role in the maintenance of the transmission cycle (Silva et al. 2005; WHO 2010; Quinnell and Courtenay 2009).

4.2 Anthroponotic Leishmaniases

Human beings are directly involved as a principal reservoir host in two forms of the disease: VL caused by *L. donovani* in Indian subcontinent (Bangladesh, India and Nepal) and East Africa (Djibouti, Ethiopia, Eritrea, Kenya, Somalia, South Sudan, Sudan and Uganda) and CL caused by *L. tropica* in semiarid subtropical regions from south-east Turkey to north-west of India. Small foci have also been described in Arabia, Ethiopia, Greece, Namibia, North Africa (Algeria, Egypt, Morocco and Tunisia) and in Central Asia (Gramiccia and Gradoni 2007; WHO 2010).

L. donovani-infected animals have been increasingly reported in several foci, despite the predominant anthroponotic transmission pattern, where post-kala-azar dermal leishmaniasis patients might constitute the main interepidemic reservoir host (WHO 2010). In certain districts of Sudan, rodents from the *Arvicanthis* genus (Hoogstraal and Heyneman 1969) and the Egyptian mongoose (*Herpestes ichneumon*) were suspected to be reservoir hosts of the parasite (Elnaiem et al. 2001). Anti-*Leishmania* antibodies have also been detected in donkeys, cows, goats and sheep in a kala-azar endemic region in Sudan, suggesting exposure of these animals to *L. donovani* infection (Mukhtar et al. 2000). In addition, canine leishmaniasis (CanL) seroprevalence between 42.9 and 74.3% and the same zymodemes were found in both humans and dogs in an endemic VL focus in Eastern Sudan (Dereure et al. 2000, 2003). However, in a more recent study performed in the same geographic region, a low number of dogs were found to have specific antibodies against *Leishmania* or to harbour parasites (Hassan et al. 2009).

In north-western Ethiopia, antibodies to and/or DNA of *L. donovani* complex have been detected in the blood of several domestic animals such as goats, sheep, cows, dogs and donkeys (Kalayou et al. 2011; Kenubih et al. 2015; Rohousova et al. 2015). *L. donovani* has also been molecularly amplified from the bone marrow of dogs (Bashaye et al. 2009) and from the spleen, bone marrow or liver of wild Ethiopian rodents (*Arvicanthis*, *Gerbilliscus* and *Mastomys* genera) (Kassahun et al. 2015a; Lemma et al. 2017). Finally, in the Indian subcontinent, *L. donovani* DNA has been detected in the blood of goats, cows and buffaloes in Nepal (Bhattarai et al. 2010), in goats (Singh et al. 2013) and domestic dogs (Jambulingam et al. 2017) in India and in stray dogs from Bangladesh (Akter et al. 2016).

Similarly, and despite *L. tropica* is considered to depend on humans for its survival, at least in long-established endemic foci in urban settings (WHO 2010; Antoniou et al. 2013), in foci with few or sporadic cases, the disease is known or suspected to be zoonotic (Ashford 2000; WHO 2010). CanL due to *L. tropica* have been reported in Morocco (Dereure et al. 1991; Guessous-Idrissi et al. 1997), in Iran (Hajjaran et al. 2013; Bamorovat et al. 2015), in Israel (Baneth et al. 2014) and in Crete (Ntais et al. 2014). In addition, *L. tropica* promastigotes have recently been isolated from the blood of a young stray dog from Israel admitted to a veterinarian hospital with a complaint of lethargy (Baneth et al. 2017). The isolation of the same zymodemes from dogs as those found in man in the same focus raised a potential role of dogs as reservoir hosts of this dermotropic *Leishmania* species. Nevertheless, the small number of canine cases and the short duration of the lesions in dogs make it difficult to define the precise role of this mammal in the epidemiological cycle (Dereure et al. 1991). In a broader geographical context of the Mediterranean region, several zoonotic foci have been described, with rock hyraxes (*Procavia capensis*) as reservoir hosts in Israel (Svobodova et al. 2006; Talmi-Frank et al. 2010) and the North African gundi (*Ctenodactylus gundi*) as probably serving as reservoir host of *Leishmania killicki* (synonym of *L. tropica*, Pratlong et al. 2009) in the area of Maghreb (Jaouadi et al. 2011; Bousslimi et al. 2012). In addition, *L. tropica* DNA has recently been detected in the spleen of wild rodents (*Acomys*, *Arvicanthis*, *Gerbillus* genera) (Kassahun et al. 2015a) and of one heart-nosed bat (*Cardioderma cor*) in Ethiopia (Kassahun et al. 2015b) as well as in the blood of stray cats from Izmir, Turkey (Can et al. 2016).

Despite these recent findings, more extensive studies to clarify the role of domestic animals in maintenance and transmission of *L. donovani* and *L. tropica* focusing on isolation and typing of the parasite and xenodiagnosis should be advocated.

4.3 Zoonotic Visceral Leishmaniasis

Leishmania infantum (synonymous of *L. chagasi*) is the etiological agent for zoonotic VL in several countries of Central and South America, the Mediterranean Basin, Middle East and Asia. The main vector in the New World is *Lutzomyia longipalpis*, while in the Old World, several species belonging to the subgenus *Phlebotomus* (*Larrossius*) (e.g. *Phlebotomus ariasi*, *Phlebotomus perniciosus*, *Phlebotomus tobbi*) are involved in *L. infantum* transmission (Maroli et al. 2013). Domestic dogs are the main domestic reservoir hosts for human infection.

In the Old and New Worlds, several indigenous wild mammal species have been found infected by or exposed to *L. infantum* (Table 4.1).

The role of foxes (*Vulpes* spp. and *Cerdocyon thous*), jackals (*Canis aureus*), wolves (*Canis lupus*) and raccoon dogs (*Nyctereutes procyonoides*) as sylvatic reservoir hosts has been suggested (Abranches 1989; WHO 2010). The existence of an autonomous or semi-autonomous sylvatic cycle in the Mediterranean Basin maintained by red foxes (*Vulpes vulpes*) has been proposed (Abranches et al. 1984), but the dependence level and the direction of parasite transmission (i.e. if foxes are inoculated with *L. infantum* through the bite of competent vectors that become infected after feeding on dogs harbouring parasites or vice versa) between these animal species were not evaluated. In fact, there is no strong evidence that wild carnivores are an important source of infection stressing the need of further quantitative studies to confirm their infectiousness to the vectors (Quinnell and Courtenay 2009). On the other hand, the ability to transmit infection has been confirmed by xenodiagnosis in black rats (*Rattus rattus*), hares and wild rabbits suggesting that they may represent a secondary reservoir host for *L. infantum*

Table 4.1 *Leishmania infantum* infection in wild animals in the Old and New Worlds (updated from Ashford 1996; Quinnell and Courtenay 2009; Savani et al. 2010; Millán et al. 2014; Chemkhi et al. 2015; de Oliveira et al. 2015; Ebani et al. 2016; Montoya et al. 2016; de Rezende et al. 2017; Pourmohammadi et al. 2017)

Order	Common name (scientific name)
Carnivora	Bush dog (*Speothos venaticus*)
	Common genet (*Genetta genetta*)
	Corsac fox (*Vulpes corsac*)
	Crab-eating fox (*Cerdocyon thous*)
	Egyptian mongoose (*Herpestes ichneumon*)
	European badger (*Meles meles*)
	European mink (*Mustela lutreola*)
	European pine marten (*Martes martes*)
	European wildcat (*Felis silvestris silvestris*)
	Fennec fox (*Vulpes zerda*)
	Golden jackal (*Canis aureus*)
	Grey wolf (*Canis lupus*)
	Hoary fox (*Lycalopex vetulus*)
	Iberian lynx (*Lynx pardinus*)
	Jaguar (*Panthera onca*)
	Least weasel (*Mustela nivalis*)
	Maned wolf (*Chrysocyon brachyurus*)
	Mediterranean monk seal (*Monachus monachus*)
	Polecat (*Mustela putorius*)
	Puma (*Puma concolor*)
	Raccoon dog (*Nyctereutes procyonoides*)
	Red fox (*Vulpes vulpes*)
	Stone marten (*Martes foina*)
Chiroptera	Broad-nosed bat (*Platyrrhinus helleri*)
	Common vampire bat (*Desmodus rotundus*)
	Flat-faced fruit-eating bat (*Artibeus planirostris*)
	Great fruit-eating bat (*Artibeus lituratus*)
	Pale spear-nosed bat (*Phyllostomus discolor*)
	Pallas's long-tongued bat (*Glossophaga soricina*)
	Pallas's mastiff bat (*Molossus molossus*)
	Seba's short-tailed bat (*Carollia perspicillata*)
	White-lined broad-nosed bat (*Platyrrhinus lineatus*)
Didelphimorphia	Black-eared opossum (*Didelphis aurita*)
	Common opossum (*Didelphis marsupialis*)
	Bennett's wallaby (*Macropus rufogriseus rufogriseus*)
Diprotodontia	Bennett's wallaby (*Macropus rufogriseus rufogriseus*)
Eulipotyphla	North African hedgehog (*Atelerix algirus*)
Lagomorpha	European hare (*Lepus europaeus*)
	European rabbit (*Oryctolagus cuniculus*)
	Iberian hare (*Lepus granatensis*)
Pilosa	Southern tamandua (*Tamandua tetradactyla*)

(continued)

Table 4.1 (continued)

Order	Common name (scientific name)
Primata	Black-fronted titi monkey (*Callicebus nigrifrons*)
	Black-headed night monkey (*Aotus nigriceps*)
	Brown howler (*Alouatta guariba*)
	Emperor tamarin (*Saguinus imperator*)
	Golden-bellied capuchin (*Sapajus xanthosternos*)
	Golden-headed lion tamarin (*Leontopithecus chrysomelas*)
	Vanzolini's bald-faced saki (*Pithecia vanzolinii*)
Rodentia	Algerian mouse (*Mus spretus*)
	Amazonian marsh rat (*Holochilus sciureus*)
	Azara's agouti (*Dasyprocta azarae*)
	Black rat (*Rattus rattus*)
	Brazilian porcupine (*Coendou prehensilis*)
	Broad-headed spiny rat (*Clyomys laticeps*)
	Brown rat (*Rattus norvegicus*)
	Colombian spiny rat (*Proechimys canicollis*)
	European wood mouse (*Apodemus sylvaticus*)
	Grey hamster (*Cricetulus migratorius*)
	House mouse (*Mus musculus*)
	Long-tailed climbing mouse (*Rhipidomys mastacalis*)
	Persian jird (*Meriones persicus*)
	Punaré (*Thrichomys laurentius*)
	South American water rat (*Nectomys squamipes*)
	Syrian hamster (*Mesocricetus auratus*)
	Tome's spiny rat (*Proechimys semispinosus*)

(Gradoni et al. 1983; Molina et al. 2012; Jiménez et al. 2014). The evidence that hares and, to a lesser extent, rabbits can play a role as reservoir hosts of *L. infantum* in a new focus in Fuenlabrada, Spain, linked to the urbanization of a sylvatic transmission cycle due to the creation of an urban periphery where both lagomorphs and phlebotomine sand fly vectors have the optimal conditions to increase in numbers, is an example that leishmaniasis can emerge due to environmental changes induced by man (Molina et al. 2012; Jiménez et al. 2014).

Among reports on domestic animals recurrently found infected with *L. infantum*, those regarding cats deserve attention for the potential implications to public health. *L. infantum* infection has been reported in domestic cats from several endemic countries in Europe, the Middle East and Brazil (Ozon et al. 1998; Martín-Sánchez et al. 2007; Nasereddin et al. 2008; Hatam et al. 2010; Vides et al. 2011; Pennisi et al. 2012; Chatzis et al. 2014; Maia et al. 2014; Can et al. 2016; Attipa et al. 2017). Thus, an increasing trend to regard cats as a potential domestic reservoir host of *L. infantum* exists as they seem to be:

1. Naturally susceptible to infection by this species, normally without development of clinical signs (these, when present are usually cutaneous but systemic involvement has also been recorded)
2. A blood source for some *Leishmania* vectors

3. Present parasites in an available way to infect the vector
4. Among the most popular pet animals around the world, often present in areas where the peridomestic and domestic transmission cycles of the parasite occur (Colmenares et al. 1995; Maroli et al. 2007; Martín-Sánchez et al. 2007; da Silva et al. 2010; Maia et al. 2010; Vides et al. 2011; Pennisi et al. 2012; Chatzis et al. 2014)

In addition, parasites isolated from infected cats seem to be biochemically and genetically identical to the ones obtained from humans and dogs with leishmaniases (Maroli et al. 2007; Pennisi et al. 2012; Maia et al. 2015). Despite this evidence, the epidemiological importance of cats in leishmaniasis is still poorly understood (Gramiccia and Gradoni 2007; Gramiccia 2011; Maia and Campino 2011; Pennisi et al. 2015). Therefore, from an epidemiological and control perspective it would be very important to evaluate the proportion of transmission in endemic areas attributable to cats in order to clarify if these animals are reservoir hosts sustaining and spreading *Leishmania* infection (Maia and Campino 2011). The dependence level and the direction of parasite transmission (i.e. if cats are inoculated with *L. infantum* through the bite of competent vectors that become infected after feeding on dogs harbouring parasites or vice versa) between these animal species are also important issues (Maia and Campino 2011).

Antibodies to *L. infantum* or its DNA have also been detected in horses in endemic areas from the Old and New Worlds (Solano-Gallego et al. 2003; Rolão et al. 2005; Fernández-Bellon et al. 2006; Lopes et al. 2013; Soares et al. 2013; Gama et al. 2014; Aharonson-Raz et al. 2015) and in nonendemic areas (i.e. Switzerland and Germany) close to the border of the limit of leishmaniasis distribution in Southern Europe (Koehler et al. 2002). Clinical cases of equine leishmaniasis have been described as self-limiting nodular or ulcerated skin lesions, isolated or disseminated (Koehler et al. 2002; Portús et al. 2002; Rolão et al. 2005; Gama et al. 2014; Baneth et al. 2015). Previous experimental data did not identified *Equus asinus* as a *L. infantum* reservoir host since the lesions of the experimentally infected donkeys spontaneously disappeared and xenodiagnosis performed using the vector *L. longipalpis* was negative (Cerqueira et al. 2003). Nevertheless, the dogma that domestic equines seem to display clinical and immunological responses of the resistant type (Fernández-Bellon et al. 2006) has recently been challenged as the concomitant cutaneous and visceral *L. infantum* infection was described in three horses from Belo Horizonte, Brazil (Soares et al. 2013). In addition, in northern Israel, facial lesions due to *L. infantum* in two horses progressively proliferated and needed to be treated with intralesional injections of meglumine antimoniate (Baneth et al. 2015), which together with the presence of *L. infantum* DNA in *P. perniciosus* sand flies that fed on two parasitaemic subclinically infected horses from the same stable allowed the authors to suggest that horses may serve as secondary reservoir hosts for this *Leishmania* species. Nevertheless, more research is required to elucidate the role, if any, of horses in *L. infantum* epidemiology, namely, the isolation for a more refined genetic, biological and biochemical characterization of the parasites infecting horses and the infectiousness of horses to vectors from nature and in horse populations.

Epidemiological studies conducted worldwide in endemic areas of VL caused by *L. infantum* strongly suggest that asymptomatic human infections are common (Costa et al. 2002; Michel et al. 2011). Risk factors for progression to disease include age, malnutrition, HIV coinfection and other immunosuppressive conditions (Gramiccia and Gradoni 2007; Boelaert and Sundar 2014). Parasite transmission by blood transfusion has also been reported (Michel et al. 2011). Therefore and despite the very low parasitaemia level, at least in immunocompetent asymptomatic carriers, their potential role as reservoir hosts should be addressed (Michel et al. 2011). It would also be important to screen patients from endemic areas for *Leishmania* infection before starting an immunosuppressive treatment (Basset et al. 2005).

Leishmania siamensis (*nomen nudum*) is referred in literature as the causative agent of several recent human cases of VL and CL with and without other co-immunosuppressive states in Thailand (Sukmee et al. 2008; Suankratay et al. 2010; Bualert et al. 2012; Chusri et al. 2012) and Myanmar (Noppakun et al. 2014). A putative vector, *Sergentomyia gemmea*, has recently been proposed (Kanjanopas et al. 2013). This so-called species, which belongs to the *Leishmania enrietti* complex, has not been formally named and described and therefore is not taxonomically valid (Pothirat et al. 2014; Akhoundi et al. 2016). In fact, it was recently showed that the majority of the ITS-1 and RNAPolII sequences that have been previously identified as *L. siamensis* in Thailand may actually be *Leishmania martiniquensis* (Pothirat et al. 2014). The geographical distribution of these novel *Leishmania* strains seems to be wide, as sporadic autochthonous equine and bovine CL have been reported in Germany, in Switzerland and in the USA (Müller et al. 2009; Lobsiger et al. 2010; Reuss et al. 2012). The zoonotic potential of *L. siamensis* has been suggested, as its DNA was amplified from liver and spleen samples of two black rats collected from the affected geographical area where VL in Thai patients have been reported (Chusri et al. 2014).

4.4 Zoonotic Cutaneous Leishmaniases

4.4.1 Old World

Leishmania aethiopica shows a geographical distribution limited to the highlands of East Africa (Ethiopia, Kenya and Uganda), and stable foci of low endemicity are maintained by hyraxes (*Procavia capensis* and *Heterohyrax brucei*) (Ashford et al. 1973; Saliba and Oumeish 1999; Tonui 2006; WHO 2010; Alvar et al. 2012). *Phlebotomus longipes*, *P. pedifer* and *P. sergenti* are the proven vectors (Killick-Kendrick 1999; Maroli et al. 2013; Akhoundi et al. 2016). Rock hyrax is also suspected of being the reservoir host of *L. aethiopica* in Saudi Arabia (Morsy et al. 1997; WHO 2010). Human LCL cases, and less frequently DCL or MCL, occur mostly in rural villages built on rock hills or river banks, associated with proximity to hyrax colonies. However, human cases have also been reported in and near Ethiopian urban centres, including Addis Ababa suggesting that this parasite is probably not so uncommon at lower altitudes (Negera et al. 2008; Lemma et al. 2009). *L. aethiopica* has also been isolated from a goat in Kenya (Williams et al. 1991) and from a ground squirrel (*Xerus rutilus*) in Ethiopia (Abebe et al. 1990).

Sporadic cases of LCL due to *L. infantum* are seen throughout the Mediterranean Basin (WHO 2010). This parasite is the most frequent cause of CL in Southern Europe (Gramiccia and Gradoni 2007). As mentioned before, several phlebotomine sand flies of *Larrossius* subgenus are the proven vectors of this parasite, and dogs are the main reservoir hosts for human infection. Nevertheless, in the recent focus of VL and CL in Fuenlabrada, Spain, the role of lagomorphs as potential sylvatic reservoir hosts has been raised up (Molina et al. 2012; Jiménez et al. 2014).

Leishmania major is the main cause of zoonotic CL in an area that stretches from India through Central Asia, the Middle East, to North and West Africa (WHO 2010). CL due to this *Leishmania* species, which is transmitted by several *Phlebotomus* species of *Paraphlebotomus* and *Phlebotomus* subgenera (Killick-Kendrick 1999; Maroli et al. 2013; Akhoundi et al. 2016), is widely distributed in rural arid areas with proneness to epidemic pattern with seasonal occurrence of cases (WHO 2010; Aoun and Bouratbine 2014). Several rodent species have been identified as reservoir hosts: the great gerbil (*Rhombomys opimus*) in Central Asia, Northern Afghanistan and Iran, the Indian desert jird (*Meriones hurrianae*) in India, the fat sand rat (*Psammomys obesus*) and Sundevall's jird (*Meriones crassus*) in Northern Africa and Middle East, Libyan jird (*Meriones libycus*) in the Arabian Peninsula and Central Asia, the short-tailed bandicoot rat (*Nesokia indica*) in Iran and several rodent species (*Arvicanthis*, *Tatera*, *Mastomys* or *Xerus* spp.) in sub-Saharan Africa (Ashford 2000; Gramiccia and Gradoni 2005; Pourmohammadi et al. 2008; WHO 2010; Aoun and Bouratbine 2014; Chaara et al. 2014). The Shaw's jird (*Meriones shawi*) also seems to play an important role in the transmission of *L. major* in Morocco (Rioux et al. 1982) and Tunisia (Ghawar et al. 2011a). The voles of the species *Microtus tristrami* and *Microtus guentheri* have recently been implicated as *L. major* reservoir hosts in a CL focus in northern Israel (Faiman et al. 2013). The sympatric occurrence of both vector (*Phlebotomus papatasi*) and *M. guentheri* in Turkey, Central Asia and Southern Europe suggests a threat for the spread of *L. major* into these regions (Antoniou et al. 2013; Faiman et al. 2013).

Leishmania major DNA has also been detected in internal organs of North African hedgehogs (*Atelerix algirus*) collected in Algeria (Tomás-Pérez et al. 2014) and North-Western Tunisia (Chemkhi et al. 2015), in the liver and spleen of Baluchistan gerbils (*Gerbillus nanus*) and brown rats (*Rattus norvegicus*) (Motazedian et al. 2010) and in the ears of long-eared hedgehogs (*Hemiechinus auritus*) in Iran (Azizi et al. 2011; Rouhani et al. 2014). The parasite was also detected by molecular techniques in the spleen of a hairy slit-faced bat (*Nycteris hispida*) in Ethiopia (Kassahun et al. 2015b), in the blood of domestic cats in the Ege Region of Turkey (Paşa et al. 2015) and in two dogs with dermal lesions from Israel (Baneth et al. 2016, 2017). *L. major* was isolated from an ear ulcer of a dog in Saudi Arabia (Elbihari et al. 1987) and from the spleen of an emaciated dog and from the blood of a dog with mild generalized alopecia, both from Egypt (Morsy et al. 1987). Isolation of parasites has also been made from cutaneous lesions in a vervet monkey (*Chlorocebus aethiops*) in Kenya (Binhazim et al. 1987) and in a least weasel (*Mustela nivalis*) in Tunisia (Ghawar et al. 2011b). In Kenya, specific antibodies to *L. major* have been reported in feral nonhuman primates: vervet monkeys, olive baboons (*Papio cynocephalus anubis*) and Sykes' Monkeys (*Cercopithecus albogularis*) (Gicheru et al. 2009). Nevertheless and despite the detection of this *Leishmania* species in a variety of mammals, most of them are probably accidental hosts as they are rarely infected.

4.4.2 New World

Most of *Leishmania* species responsible for CL in the Americas are native to tropical rainforests, where a variety of wild animal species and phlebotomine sand flies maintain the enzootic cycle (Table 4.2).

Leishmania amazonensis (syn. *Leishmania garnhami*) is endemic in Argentina, Bolivia, Brazil, Colombia, Costa Rica, Ecuador, French Guyana, Peru, Suriname and Venezuela (Lainson 2010; WHO 2010; Alvar et al. 2012). The main clinical human forms are localized or DCL, although this last form of the disease, an anergic variant of LCL, as well as a visceralization of infection in immunocompetent people have also been documented (WHO 2010; Boelaert and Sundar 2014). *Lutzomyia flaviscutellata*, the major vector of this dermotropic *Leishmania* species, feeds predominantly on ground-dwelling rodents, the primary reservoir hosts of *L. amazonensis*. Several other wild mammals (Table 4.2) are suspected of being secondary reservoir hosts (Ashford 2000; Gramiccia and Gradoni 2005; Lainson 2010; WHO 2010). This parasite has also been documented in dogs and cats (de Souza et al. 2005; Tolezano et al. 2007; reviewed by Dantas-Torres 2009 and Pennisi et al. 2015; Ferreira et al. 2015; Ramirez et al. 2016; Sanches et al. 2017). In addition, *L. amazonensis* DNA has recently been detected in the skin and/or spleen of different species of insectivores, frugivorous or haematophagous bats captured in non-urban and urban areas of São Paulo state, Brazil (Savani et al. 2010; de Oliveira et al. 2015).

Leishmania braziliensis is reported in almost all countries of Central and South America (WHO 2010; Alvar et al. 2012). The usual clinical form caused by the parasite is a localized CL, although diffused CL has also been reported. In addition, about 5% of the patients evolve towards a severe mucocutaneous disease (WHO 2010; Alvar et al. 2012). Visceralizing disease has also been reported for *L. braziliensis* in HIV coinfected patients (Boelaert and Sundar 2014) and in dogs coinfected with *Hepatozoon canis* (Morgado et al. 2016). Several *Lutzomyia* species of the *Lutzomyia*, *Nyssomyia*, *Psathyromyia*, *Psychodopygus* and *Verrucarum* subgenera are implicated in its transmission (Killick-Kendrick 1999; Maroli et al. 2013; Akhoundi et al. 2016). Albeit *L. braziliensis* is primarily associated to tropical forests and several bats, edentates, marsupials, opossums and wild rodents have been found infected (Table 4.2), this parasite has adapted to human-modified environments, being frequently found in the peridomestic

Table 4.2 *Leishmania* spp. causing CL in the New World and their proven/putative wild hosts (updated from Grimaldi and Tesh 1993; Lainson and Shaw 2005; Brandão-Filho et al. 2011; Marcelino et al. 2011; Shapiro et al. 2013; Roque and Jansen 2014; Kipp et al. 2016; Caldart et al. 2017; Carreira et al. 2017; de Castro Ferreira et al. 2017)

Leishmania	Order	Common name (scientific name)
L. amazonensis	Carnivora	Crab-eating fox (*Cerdocyon thous*)
		Hog-nosed skunk (*Conepatus chinga*)
		Kinkajou (*Potos flavus*)
	Chiroptera	Black bonneted bat (*Eumops auripendulus*)
		Black mastiff bat (*Molossus rufus*)
		Black myotis (*Myotis nigricans*)
		Broad-eared free-tailed bat (*Nyctinomops laticaudatus*)
		Common vampire bat (*Desmodus rotundus*)
		Flat-faced fruit-eating bat (*Artibeus planirostris*)
		Great fruit-eating bat (*Artibeus lituratus*)
		Little yellow-shouldered bat (*Sturnira lilium*)
		Pallas's long-tongued bat (*Glossophaga soricina*)
		Pallas's mastiff bat (*Molossus molossus*)
		Wagner's bonneted bat (*Eumops glaucinus*)
		White-lined broad-nosed bat (*Platyrrhinus lineatus*)
	Didelphimorphia	Brown four-eyed opossum (*Metachirus nudicaudatus*)
		Common opossum (*Didelphis marsupialis*)
		Grey four-eyed opossum (*Philander opossum*)
		Linnaeus's mouse opossum (*Marmosa murina*)
		Tate's woolly mouse opossum (*Marmosa paraguayana*)
		White-eared opossum (*Didelphis albiventris*)
		Woolly mouse opossum (*Marmosa demerarae*)
	Pilosa	Southern tamandua (*Tamandua tetradactyla*)
	Primata	Geoffroy's tamarin (*Saguinus geoffroyi*)
		Red-faced spider monkey (*Ateles paniscus*)
		Three-striped night monkey (*Aotus trivirgatus*)
	Rodentia	Black-eared rice rat (*Oryzomys melanotis*)
		Black-rumped agouti (*Dasyprocta prymnolopha*)
		Bolivian Hylaeamys (*Hylaeamys acritus*)
		Common punaré (*Thrichomys apereoides*)
		Cuvier's spiny rat (*Proechimys cuvieri*)
		Elegant rice rat (*Oryzomys nitidus*)
		Elegant-spined Atlantic spiny rat (*Trinomys setosus*)
		Hairy-tailed bolo mouse (*Necromys lasiurus*)
		Large-headed rice rat (*Hylaeamys megacephalus*)
		Red-tailed squirrel (*Sciurus granatensis*)
		Spiny rat (*Proechimys guyanensis*)
		Trinidad spiny pocket mouse (*Heteromys anomalus*)

(continued)

Table 4.2 (continued)

Leishmania	Order	Common name (scientific name)
L. braziliensis	Carnivora	Hog-nosed skunk (*Conepatus chinga*)
	Chiroptera	Flat-faced fruit-eating bat (*Artibeus planirostris*)
		Pallas's long-tongued bat (*Glossophaga soricina*)
		Pallas's mastiff bat (*Molossus molossus*)
		White-lined broad-nosed bat (*Platyrrhinus lineatus*)
	Didelphimorphia	Agile Gracile mouse opossum (*Gracilinanus agilis*)
		Common opossum (*Didelphis marsupialis*)
		Tate's woolly mouse opossum (*Marmosa paraguayana*)
		White-eared opossum (*Didelphis albiventris*)
		Woolly mouse opossum (*Marmosa demerarae*)
	Pilosa	Linnaeus's two-toed sloth (*Choloepus didactylus*)
	Primata	Three-striped night monkey (*Aotus trivirgatus*)
	Rodentia	Atlantic bamboo rat (*Kannabateomys amblyonyx*)
		Azara's agouti (*Dasyprocta azarae*)
		Black rat (*Rattus rattus*)
		Brown rat (*Rattus norvegicus*)
		Diminutive akodont (*Akodon arviculoides*)
		Dusky rice rat (*Melanomys caliginosus*)
		Common punaré (*Thrichomys apereoides*)
		Hairy-tailed bolo mouse (*Necromys lasiurus*)
		Hispid cotton rat (*Sigmodon hispidus*)
		House mouse (*Mus musculus*)
		Large-headed rice rat (*Hylaeamys megacephalus*)
		Lowland paca (*Cuniculus paca*)
		Natterer's Oecomys (*Oryzomys concolor*)
		Punaré (*Thrichomys laurentius*)
		South American water rat (*Nectomys squamipes*)
		White-footed climbing mouse (*Rhipidomys leucodactylus*)
L. colombiensis	Pilosa	Hoffman's two-toed sloth (*Choloepus hoffmanni*)
L. equatoriensis	Pilosa	Hoffman's two-toed sloth (*Choloepus hoffmanni*)
	Rodentia	Red-tailed squirrel (*Sciurus granatensis*)
L. guyanensis	Carnivora	Kinkajou (*Potos flavus*)
	Cingulata	Nine-banded armadillo (*Dasypus novemcinctus*)
	Didelphimorphia	Common opossum (*Didelphis marsupialis*)
		Grey slender mouse opossum (*Marmosops incanus*)
	Pilosa	Linnaeus's two-toed sloth (*Choloepus didactylus*)
		Southern tamandua (*Tamandua tetradactyla*)
	Rodentia	Common punaré (*Thrichomys apereoides*)
		Guyenne spiny rat (*Proechimys guyanensis*)
L. lainsoni	Rodentia	Lowland paca (*Cuniculus paca*)
L. lindenbergi		Unknown

Table 4.2 (continued)

Leishmania	Order	Common name (scientific name)
L. mexicana	Chiroptera	Commissaris's long-tongued bat (*Glossophaga commissarisi*)
		Common vampire bat (*Desmodus rotundus*)
		Godman's long-tailed bat (*Choeroniscus godmani*)
		Great fruit-eating bat (*Artibeus lituratus*)
		Highland yellow-shouldered bat (*Sturnira ludovici*)
		Jamaican fruit bat (*Artibeus jamaicensis*)
		Little yellow-shouldered bat (*Sturnira lilium*)
		Pale spear-nosed bat (*Phyllostomus discolor*)
		Pallas's long-tongued bat (*Glossophaga soricina*)
		Pygmy fruit-eating bat (*Dermanura phaeotis*)
		Southern long-nosed bat (*Leptonycteris curasoae*)
		Sowell's short-tailed bat (*Carollia sowelli*)
		Texas mouse (*Peromyscus attwateri*)
		Wagner's moustached bat (*Pteronotus personatus*)
	Didelphimorphia	Common opossum (*Didelphis marsupialis*)
		Mexican mouse opossum (*Marmosa mexicana*)
		Robinson's mouse opossum (*Marmosa robinsoni*)
	Rodentia	Big-eared climbing rat (*Ototylomys phyllotis*)
		Black-eared rice rat (*Handleyomys melanotis*)
		Black rat (*Rattus rattus*)
		Desmarest's spiny pocket mouse (*Heteromys desmarestianus*)
		Eastern woodrat (*Neotoma floridana*)
		Hispid cotton rat (*Sigmodon hispidus*)
		Slender harvest mouse (*Reithrodontomys gracilis*)
		Southern plains woodrat (*Neotoma micropus*)
		Sumichrast's vesper rat (*Nyctomys sumichrasti*)
		Yucatan deer mouse (*Peromyscus yucatanicus*)
		White-throated woodrat (*Neotoma albigula*)
L. naiffi	Cingulata	Nine-banded armadillo (*Dasypus novemcinctus*)
	Rodentia	Paraguayan punaré (*Thrichomys pachyurus*)
		Punaré (*Thrichomys laurentius*)
L. panamensis	Carnivora	Northern olingo (*Bassaricyon gabbii*)
		Kinkajou (*Potos flavus*)
		South American coati (*Nasua nasua*)
	Didelphimorphia	Brown four-eyed opossum (*Metachirus nudicaudatus*)
		Common opossum (*Didelphis marsupialis*)
	Pilosa	Brown-throated sloth (*Bradypus variegatus*)
		Brown-throated three-toed sloth (*Bradypus griseus*)
		Hoffman's two-toed sloth (*Choloepus hoffmanni*)
	Primata	Geoffroy's tamarin (*Saguinus geoffroyi*)
		Northern night monkey (*Aotus trivirgatus*)
	Rodentia	Black rat (*Rattus rattus*)
		Tome's spiny rat (*Proechimys semispinosus*)
		Desmarest's spiny pocket mouse (*Heteromys desmarestianus*)

(continued)

Table 4.2 (continued)

Leishmania	Order	Common name (scientific name)
L. peruviana	Didelphimorphia	Andean white-eared opossum (*Didelphis pernigra*)
	Rodentia	Andean pericote (*Phyllotis andium*)
L. shawi	Carnivora	South American coati (*Nasua nasua*)
	Pilosa	Linnaeus's two-toed sloth (*Choloepus didactylus*)
		Pale-throated three-toed sloth (*Bradypus tridactylus*)
	Primata	Black bearded saki (*Chiropotes satanas*)
		Tufted capuchin (*Cebus apella*)
	Rodentia	Highlands punaré (*Thrichomys inermis*)
		Punaré (*Thrichomys laurentius*)
L. venezuelensis		Unknown
L. waltoni		Unknown

environment of rural houses. In these settings, domestic animals (i.e. horses, donkeys, mules, dogs and cats) may act not only as blood sources to phlebotomine sand flies but might also participate in the transmission cycle (Bonfante-Garrido et al. 1981, 1992; Aguilar et al. 1984; Passos et al. 1996; Schubach et al. 2004; Madeira et al. 2006; Vedovello et al. 2008; Rougeron et al. 2011; Santaella et al. 2011; Truppel et al. 2014). Nevertheless, their role as reservoir hosts is still considered circumstancial (Reithinger and Davies 1999; Dantas-Torres 2007; Truppel et al. 2014; Pennisi et al. 2015). In order to prove that these animals can act as domestic reservoirs in the peridomestic environment, it will be necessary to conduct infectivity tests on phlebotomine sand flies and perform the isolation and characterization of the parasites from samples accessible to the vectors. In addition, insights derived from recent research suggest that humans might be important domestic reservoir hosts of *L. braziliensis*, at least during outbreaks (Dantas-Torres 2007; WHO 2010).

Leishmania colombiensis, which is responsible for single or multiple cutaneous lesions, is endemic in Colombia, Panama and Venezuela (WHO 2010; Alvar et al. 2012). *Lutzomyia gomezi*, *Lutzomyia hartmanni* and *Lutzomyia panamensis* are the proven or suspected vectors, and the Hoffmann's two-toed sloth (*Choloepus hoffmanni*) is the reservoir host in Panama (Killick-Kendrick 1999; Lainson 2010; WHO 2010; Maroli et al. 2013; Akhoundi et al. 2016). This parasite has also been isolated from the bone marrow of a dog in Venezuela (Delgado et al. 1993).

Leishmania guyanensis is responsible for LCL, and less frequently by DCL, being endemic in Argentina, Bolivia, Brazil (Acre, Amapá, Amazonas, Pará and Roraima states), Colombia, Ecuador, French Guiana, Guyana, Peru, Suriname and Venezuela (Lainson 2010; WHO 2010; Alvar et al. 2012). Transmission is associated with activities in forests (WHO 2010). The parasite can cause mucocutaneous lesions in a small proportion of cases (WHO 2010). The main vector is *Lutzomyia umbratilis* (Lainson 2010). Linnaeus's two-toed sloth (*Choloepus didactylus*) is a major reservoir host of *L. guyanensis* in Brazil and in French Guiana maintaining the zoonosis in the forest canopy (Table 4.2). The southern tamandua (*Tamandua tetradactyla*) has been suggested as responsible for dispersal of the parasite due to its nomadic behaviour (WHO 2010). Occasional infections in rodents and opossums have been documented (Ashford 2000; Lainson 2010; WHO 2010). The DNA of the parasite has also been detected in one dog from Colombia (Santaella et al. 2011), but the contribution of domestic dogs in the life cycle of *L. guyanensis* seems limited.

Leishmania lainsoni causes CL, usually presenting as a single ulcer. The disease is found in Bolivia (subtropical areas), Brazil (Acre, Amapa, Pará and Rondônia states), Ecuador, French Guiana, Peru (tropical areas) and Suriname (Silveira et al. 1987; WHO 2010; Alvar et al. 2012; Kato et al. 2016). The vectors are *Lutzomyia ubiquitalis* in Brazil and Peru and *Lutzomyia nuneztovari anglesi* in Bolivia (Silveira et al. 1991a; Killick-Kendrick 1999; WHO 2010; Maroli et al. 2013; Akhoundi et al. 2016). The lowland paca (*Cuniculus paca*) is said to be the reservoir host (Silveira et al. 1991b; WHO 2010).

Leishmania lindenbergi causes CL in Brazil (Pará state) (Silveira et al. 2002; Lainson 2010; WHO 2010; Alvar et al. 2012). The suspected vector is *Lutzomyia antunesi*, and the reservoir host remains unknown (Silveira et al. 2002; WHO 2010; Maroli et al. 2013; Akhoundi et al. 2016).

Leishmania mexicana (syn. *Leishmania pifanoi*) is endemic in Belize, Colombia, Costa Rica, Ecuador, Guatemala, Mexico, Southern USA and Venezuela (WHO 2010; Alvar et al. 2012). Localized CL is the most common clinical form in humans, although diffuse CL has also been reported (WHO 2010). Many species of sylvatic ground-dwelling rodents (*Heteromys*, *Neotoma*, *Nyctomys*, *Ototylomys* and *Sigmodon* spp.) and marsupials have been implicated in the transmission cycle of *L. mexicana* (Table 4.2). In Texas, USA, cases of feline CL due to parasites belonging to the *L. mexicana* complex have been reported in the same areas where human cases occurred (Craig et al. 1986; Barnes et al. 1993; Trainor et al. 2010). *L. mexicana* DNA has also been detected in skin biopsies taken from a stray dog in Texas (Kipp et al. 2016) and in different tissues (i.e. heart, liver, skin and spleen) of several species of bats collected in six states of Mexico (Berzunza-Cruz et al. 2015). This dermotropic species has also been isolated from the liver aspirate of a dog from Ecuador (Hashiguchi et al. 1991). *Lutzomyia olmeca olmeca* is the main vector, and various other species are suspected to be involved in the life cycle of the parasite (Killick-Kendrick 1999; Lainson 2010; WHO 2010; Maroli et al. 2013; Akhoundi et al. 2016).

Leishmania naiffi causes a single, small, self-limiting lesion. It is found in Brazil (Rondônia state), Ecuador, Suriname and French Guiana (WHO 2010; van Thiel et al. 2010; Alvar et al. 2012; Kato et al. 2013). The proven vector is *Lutzomyia ayrozai*, while several other species are suspected to be involved in the transmission (Killick-Kendrick 1999; Maroli et al. 2013; Akhoundi et al. 2016); the reservoir host is the nine-banded armadillo (*Dasypus novemcinctus*) (Lainson and Shaw 1989; Naiff et al. 1991; WHO 2010).

Leishmania panamensis is responsible for LCL with some patients developing diffuse or mucocutaneous disease. This species is endemic in Colombia, Costa Rica, Ecuador (Pacific littoral), Guatemala, Honduras, Nicaragua and Panama

(WHO 2010; Alvar et al. 2012). The major vector is considered to be *Lutzomyia trapidoi*, but several other species (e.g. *Lutzomyia gomezi* and *Lutzomyia panamensis*) have also been found to be naturally infected (Killick-Kendrick 1999; Maroli et al. 2013; Akhoundi et al. 2016). *Lutzomyia trapidoi* prefers to feed in the canopy, on arboreal mammals, such as sloths, which are the primary hosts of *L. panamensis*. Various wild mammalian species, including monkeys and several rodent species (Table 4.2), have been found to be infected, but their role as possible reservoir hosts is poorly known. According to WHO (2010), humans seem to play a reservoir role in some outbreaks caused by this *Leishmania* species. Dogs have also found infected with *L. panamensis*, but there is no evidence that they can play a role as reservoir hosts (Dereure et al. 1994; Vélez et al. 2012; Ramírez et al. 2016).

Leishmania peruviana distribution is limited to the Peruvian Andes, confined to areas with scant vegetation of the Western slopes between 800 and 3000 m altitude (Lainson 2010; WHO 2010). The clinical form is a localized ulcerative CL, and *Lutzomyia ayacuchensis*, *Lutzomyia peruensis* and *Lutzomyia verrucarum* are the proven vectors (Killick-Kendrick 1999; Maroli et al. 2013; Akhoundi et al. 2016). The natural reservoir hosts are probably wild marsupials and rodents (Table 4.2). Dogs are reputed to be the principal peridomestic reservoir hosts (Llanos-Cuentas et al. 1999). This assumption is based on a positive correlation observed between the risk of human CL and CanL prevalence in Huanuco, Peru. However, the scarcity of parasites in cutaneous lesions together with the high serorecovery rates suggest that dogs are able to control infection and thus may not be the main reservoir host of the parasite (Reithinger et al. 2003). Therefore, the role of dogs as reservoir hosts of *L. peruviana* should be confirmed by experimental transmission studies (Dantas-Torres 2007).

Leishmania shawi found in Brazil (Atlantic Forest of Pará state) causes localized CL (WHO 2010). In primary forest, the vector is *Lutzomyia whitmani* (Lainson et al. 1989; Killick-Kendrick 1999; Maroli et al. 2013; Akhoundi et al. 2016). The sylvatic reservoir hosts are monkeys, coatis and sloths (Lainson et al. 1989) (Table 4.2).

Leishmania venezuelensis is responsible for localized and DCL in Venezuela (Bonfante-Garrido et al. 1996; WHO 2010). *Lutzomyia olmeca bicolor* is suspected of being the vector (Killick-Kendrick 1999; Lainson 2010; Maroli et al. 2013; Akhoundi et al. 2016), while domestic cats are suspected to be the reservoir hosts (Bonfante-Garrido et al. 1991).

Leishmania waltoni is a recently described species associated with cases of DCL in humans in Dominican Republic (Shaw et al. 2015). This species belongs to the *L. mexicana* complex, and its reservoir hosts and vectors are still unknown.

4.5 Control of Reservoir Hosts

In 2010, a WHO Expert Committee defined that control strategies of leishmaniases should combine case management, integrated vector control and, in the case of zoonotic transmission, animal reservoir host control (WHO 2010).

For the control of anthroponotic leishmaniasis, an effective strategy for active case detection, surveillance and effective treatment of patients with clinical forms of leishmaniasis, accompanied by measures for preventing reinfection, should reduce or eliminate the parasite load and reduce transmission (WHO 2010). In fact, better tools have been made available to developing countries, such as: improvement of VL diagnosis (e.g. recombinant antigen (K39)-dipstick tests for in-field diagnosis), (ii) affordable VL treatment (e.g. the first oral antileishmanial drug, miltefosine; short course of therapy) and (iii) a more efficient phlebotomine sand fly control for both anthroponotic VL and CL (e.g. long-lasting insecticide-treated bed nets) (Desjeux 2004; Gramiccia and Gradoni 2005).

Control of reservoir hosts has been recommended for zoonotic VL and CL. Due to the exophilic habit of the phlebotomine vectors and the sylvatic nature of the reservoir hosts, the control of the zoonotic CL forms in both the Old and New Worlds is not easy and may even not be feasible, as it would require expensive environmental management difficult to implement and sustain (Gramiccia and Gradoni 2005; WHO 2010; Boelaert and Sundar 2014). As there is currently no vaccine for human use, the ways to protect individuals from contracting the infection include avoiding intrusion in natural zoonotic foci as well as the adoption of personal protective measures against phlebotomine sand fly bites with repellents and other devices (WHO 2010; Boelaert and Sundar 2014).

In the case of zoonotic CL caused by *L. major*, where the reservoir hosts are peridomestic rodent species, their elimination could be achieved by the destruction of the burrow systems by deep ploughing followed by planting. Another approach is by poisoning the colonies of rodents with wheat grains mixed with zinc phosphide along with the prior treatment of burrows with the anticoagulant dicoumarol (Saliba and Oumeish 1999; Ashford 2000; WHO 2010; Boelaert and Sundar 2014). This method of control may be effective against *Rhombomys* and *Meriones* rodents that feed on grains but not against *Psammomys obesus*. Because zinc phosphide is very toxic to man and other animals, care should be taken during its application (Saliba and Oumeish 1999; WHO 2010). The removal of chenopod plants, the only ones that *P. obesus* feed on, from areas close to inhabitants would also lead to the reduction of their numbers (Desjeux 1996; Saliba and Oumeish 1999; WHO 2010). Transmission of zoonotic CL due to *L. aethiopica* could also be reduced by controlling hyraxes around villages. Elimination of hyraxes within 1 km of settlements is thought to be effective in reducing transmission. As reinvasion is likely, control must be continuous. In some countries, hyraxes are protected animals, and their control is illegal and prohibited (Saliba and Oumeish 1999; Ashford 2000; WHO 2010).

In the New World, where most of *Leishmania* cycles are maintained by edentates, procyonids, arboreal or ground sylvatic rodents, an integrated environmental management approach, combining clearance of primary forest around villages and spraying of the cleared areas with insecticides to remove both the reservoir hosts and the vector, thus creating a "vector- and reservoir-free" zone around villages, might be effective for the control of zoonotic CL (WHO 2010). However, even clearing forest around villages may not reach the objective, as various *Leishmania* species (e.g. *L. braziliensis*) have proved to be remarkably adaptable to environmental degradation leading to peridomestic transmission rather to the elimination of the infections (Brandão-Filho et al. 1999; Ashford 2000; Boelaert and Sundar 2014).

Regarding zoonotic VL, infection in the canine domestic reservoir host should be monitored, and the management of infected dogs should be treatment or elimination (WHO 2010). Albeit test-and-treat strategies are performed in several Mediterranean countries, treating infected dogs alone may not be an effective control measure as relapses are frequent, and because despite clinical cure, dogs can recover infectivity weeks after treatment (Gradoni et al. 1987; Alvar et al. 1994; Miró et al. 2011); therefore, the use of repellents on dogs during and after treatment is imperative. In addition, the widespread use of the available anti-*Leishmania* drugs for both canine and human treatment might contribute to the generation and spread of drug-resistant parasites (Campino and Maia 2012). On the other hand, and despite culling dogs infected with *L. infantum* has been recommended by WHO, the implementation of this measure in countries where dogs are considered part of the family is impracticable. In Brazil, seropositive dogs are eliminated as part of a control programme, although its effectiveness in the control of infection is not clear-cut and it has not been tested in trials measuring clinical disease (González et al. 2015). Failure may occur due to several reasons (e.g. poor sensitivity of diagnostic methods, delay between diagnosis and culling and rapid replacement of culled dogs by new susceptible animals).

Leishmania life cycle can be interrupted through the use of impregnated dog collars and topical application of insecticide with repellent effect against phlebotomine sand flies (Killick-Kendrick et al. 1997; Mencke et al. 2003; Liénard et al. 2013; Dumont et al. 2015; Franc et al. 2015). In fact, a significant decrease in the incidence of zoonotic VL in children (Gavgani et al. 2002) and dogs has been observed in areas where most dogs used deltamethrin collars or have been treated with permethrin-based spot-on formulations (Maroli et al. 2001; Manzillo et al. 2006; Courtenay et al. 2009; Otranto et al. 2010). The impact of this type of control measure is dependent on the correct application and frequency of reapplication of the topical insecticides and in the loss rate of collars. In addition, the application of insecticides/repellents would have less impact on disease transmission if not integrated with stray dog control (Gramiccia and Gradoni 2005). Additional measures to control phlebotomine sand flies include reducing microhabitats favourable to them in the vicinity of the house and in other locations where dogs spend time, housing pets at dusk and indoor insecticide spraying of homes and animal shelters (Alexander and Maroli 2003; Maroli et al. 2010; Solano-Gallego et al. 2011).

Vaccination could be another strategy to reduce both CanL and the incidence in humans (Alvar et al. 2004). An effective vaccine would control both infection progression and the parasite transmissibility via the vector (Gradoni 2015). In Brazil, two canine vaccines (Leishmune® and Leishtec®) have been commercialized. Leishmune® was shown to induce a significant, long-lasting and strong protective effect against CanL in phase III of clinical trials (Silva et al. 2000; Borja-Cabrera et al. 2002). Although this vaccine was also proposed to be used as immunotherapeutic in infected dogs and as a transmission-blocking vaccine (Borja-Cabrera et al. 2004; Saraiva et al. 2006), in 2014 the Brazilian Ministry of Agriculture, Livestock and Food Supply suspended its commercialization due to non-compliance with all the requirements for phase III studies (http://www.agricultura.gov.br/assuntos/politica-agricola/arquivos/nota-tecnica-dfip-38-14-leishmune.pdf/view). Leish-Tec® conferred a significant reduction in the number of cases of CanL with an efficacy of 71.4% estimated according to parasitological results (i.e. imprinting, culture, or histopathology of dog tissues) (Regina-Silva et al. 2016). The infectiousness to reared *L. longipalpis* of vaccinated dogs presenting antibodies against the A2 antigen was 46.6% lower in comparison with non-vaccinated animals (Regina-Silva et al. 2016). In Europe, a vaccine consisting of purified excreted-secreted proteins of *L. infantum* and with QA-21 saponin as adjuvant (CaniLeish®) has provided a significant reduction in the risk of progressing to active infection or overt disease, with a clinical efficacy of 68% (Oliva et al. 2014). In vaccinated dogs that developed disease and that were exposed to the bites of reared *P. perniciosus*, the reduction in parasite transmission was found significant when compared to matched controls (Bongiorno et al. 2013). More recently (in 2017), a second vaccine (Letifend®) consisting of a recombinant Protein Q from *L. infantum* MON-1 has been commercialized in Europe. According to the product information available at the European Medicines Agency, a vaccinated dog has five times less risk to develop clinical disease than a non-vaccinated dog (https://ec.europa.eu/health/documents/communityregister/2016/20160420134483/anx_134483_en.pdf).

In last years, CanL expanded northwards in Europe, mainly due to movement of infected dogs from endemic to previously nonendemic areas (Maia and Cardoso 2015). Therefore, control of CanL should also include the compulsory certification by veterinarians of the non-infective state of animals moving from one place to another to avoid the introduction of infected dogs in areas previously nonendemic, especially in those having competent vectors which might result in the persistence of *L. infantum* (WHO 2010; Maia and Cardoso 2015).

4.6 Final Remarks

The development of efficient tools for reservoir host control depends on proper understanding of the local epidemiology of leishmaniasis (including whether transmission is anthroponotic or zoonotic). Apart from the proven reservoir hosts, *Leishmania* parasites have been found in a variety of wild and domestic animals around the world, but their role in sustaining the life cycle of the parasite is unknown. In some instances, the parasites have been isolated and formally characterized, but in many cases, the infection status and parasite species have been inferred based on the detection of DNA fragments of the parasite through PCR-based tools. Therefore, it would be crucial to isolate and formally identify *Leishmania* parasites infecting any suspected reservoir host. As in many cases, information about food sources, breeding season, movement and migration activities and longevity of the potential reservoir host(s) is lacking; further work along these lines should also be performed.

Acknowledgements The authors wish to thank A. Pereira for his work on the references. C. Maia holds a FCT Investigator Starting Grant (IF/01302/2015) from Fundação para a Ciência e a Tecnologia, Ministério da Ciência, Tecnologia e Ensino Superior, Portugal.

References

Abebe A, Evans DA, Gemetchu T. The isolation of *Leishmania aethiopica* from the ground squirrel *Xerus rutilus*. Trans R Soc Trop Med Hyg. 1990;84(5):691.

Abranches P. Reservoirs of visceral leishmaniasis. In: Hart DT, editor. Leishmaniasis: the current status and new strategies for control. New York: Plenum Press; 1989. p. 61–9.

Abranches P, et al. Kala-azar in Portugal. V. The sylvatic cycle in the enzootic endemic focus of Arrabida. J Trop Med Hyg. 1984;87(5):197–200.

Aguilar C, et al. Study of an outbreak of cutaneous leishmaniasis in Venezuela. The role of domestic animals. Mem Inst Oswaldo Cruz. 1984;79:181–95.

Aharonson-Raz K, et al. Low seroprevalence of *Leishmania infantum* and *Toxoplasma gondii* in the horse population in Israel. Vector Borne Zoonotic Dis. 2015;15(12):726–31.

Akhoundi M, et al. A historical overview of the classification, evolution, and dispersion of *Leishmania* parasites and sandflies. PLoS Negl Trop Dis. 2016;10(3):e0004349.

Akter S, et al. Molecular and serological evidence of *Leishmania* infection in stray dogs from visceral leishmaniasis-endemic areas of Bangladesh. Am J Trop Med Hyg. 2016;95(4):795–9.

Alexander B, Maroli M. Control of phlebotomine sandflies. Med Vet Entomol. 2003;17(1):1–18.

Alvar J, et al. Canine leishmaniasis: clinical, parasitological and entomological follow-up after chemotherapy. Ann Trop Med Parasitol. 1994;88(4):371–8.

Alvar J, et al. Canine leishmaniasis. Adv Parasitol. 2004;57:1–88.

Alvar J, et al. Leishmaniasis worldwide and global estimates of its incidence. PLoS One. 2012;7(5):e35671.

Antoniou M, et al. The role of indigenous phlebotomine sandflies and mammals in the spreading of leishmaniasis agents in the Mediterranean region. Euro Surveill. 2013;18(30):20540.

Aoun K, Bouratbine A. Cutaneous leishmaniasis in North Africa: a review. Parasite. 2014;21:14.

Ashford RW. Leishmaniasis reservoirs and their significance in control. Clin Dermatol. 1996;14(5):523–32.

Ashford RW. The leishmaniases as emerging and reemerging zoonoses. Int J Parasitol. 2000;30(12–13):1269–81.

Ashford RW, et al. The epidemiology of cutaneous leishmaniasis in Ethiopia. Trans R Soc Trop Med Hyg. 1973;67(4):568–601.

Attipa C, et al. Prevalence study and risk factor analysis of selected bacterial, protozoal and viral, including vector-borne, pathogens in cats from Cyprus. Parasit Vectors. 2017;10(1):130.

Azizi K, et al. *Gerbillus nanus* (Rodentia: Muridae): a new reservoir host of *Leishmania major*. Ann Trop Med Parasitol. 2011;105(6):431–7.

Bamorovat M, et al. *Leishmania tropica* in stray dogs in southeast Iran. Iran J Public Health. 2015;44(10):1359–66.

Baneth G, et al. Mucocutaneous *Leishmania tropica* infection in a dog from a human cutaneous leishmaniasis focus. Parasit Vectors. 2014;7:118.

Baneth G et al. (2015) A cluster of naturally occurring equine leishmaniosis with persistent progressive clinical manifestations, sub-clinical infection and evaluation of the horse as a reservoir host. In: 25th international conference of the world association for the advancement of veterinary parasitology. p. 100.

Baneth G, et al. *Leishmania major* infection in a dog with cutaneous manifestations. Parasit Vectors. 2016;9(1):246.

Baneth G, et al. Canine leishmaniosis caused by Leishmania major and Leishmania tropica: comparative findings and serology. Parasit Vectors. 2017;10(1):113.

Barnes JC, Stanley O, Craig TM. Diffuse cutaneous leishmaniasis in a cat. J Am Vet Med Assoc. 1993;202(3):416–8.
Bashaye S, et al. Risk factors for visceral leishmaniasis in a new epidemic site in Amhara Region, Ethiopia. Am J Trop Med Hyg. 2009;81(1):34–9.
Basset D, et al. Visceral leishmaniasis in organ transplant recipients: 11 new cases and a review of the literature. Microbes Infect. 2005;7(13):1370–5.
Berzunza-Cruz M, et al. *Leishmania* (L.) *mexicana* infected bats in Mexico: novel potential reservoirs. PLoS Negl Trop Dis. 2015;9(1):e0003438.
Bhattarai N, et al. Domestic animals and epidemiology of visceral leishmaniasis, Nepal. Emerg Infect Dis. 2010;16(2):231–7.
Binhazim AA, et al. Isolation of *Leishmania major* from a naturally infected vervet monkey (*Cercopithecus aethiops*) caught in Kiambu District, Kenya. J Parasitol. 1987;73(6):1278–9.
Boelaert M, Sundar S. Leishmaniasis. In: Farrar J, Hotez P, Junghanss T, Kang G, Lalloo D, White N, editors. Manson's tropical diseases. 23rd ed. Beijing: Elsevier Saunders; 2014. p. 631–51.
Bonfante-Garrido R, et al. Enzootic equine cutaneous leishmaniasis in Venezuela. Trans R Soc Trop Med Hyg. 1981;75(3):471.
Bonfante-Garrido R, et al. Natural infection of cats with *Leishmania* in Barquisimeto, Venezuela. Trans R Soc Trop Med Hyg. 1991;85(1):53.
Bonfante-Garrido R, et al. Cutaneous leishmaniasis in western Venezuela caused by infection with *Leishmania venezuelensis* and *L. braziliensis* variants. Trans R Soc Trop Med Hyg. 1992;86(2):141–8.
Bonfante-Garrido R, et al. Disseminated American cutaneous leishmaniasis. Int J Dermatol. 1996;35(8):561–5.
Bongiorno G, et al. Vaccination with LiESP/QA-21 (CaniLeish®) reduces the intensity of infection in *Phlebotomus perniciosus* fed on *Leishmania infantum* infected dogs—a preliminary xenodiagnosis study. Vet Parasitol. 2013;197(3–4):691–5.
Borja-Cabrera GP, et al. Long lasting protection against canine kala-azar using the FML-QuilA saponin vaccine in an endemic area of Brazil (São Gonçalo do Amarante, RN). Vaccine. 2002;20(27–28):3277–84.
Borja-Cabrera GP, et al. Effective immunotherapy against canine visceral leishmaniasis with the FML-vaccine. Vaccine. 2004;22(17–18):2234–43.
Bousslimi N, et al. Natural infection of North African gundi (*Ctenodactylus gundi*) by *Leishmania tropica* in the focus of cutaneous leishmaniasis, Southeast Tunisia. Am J Trop Med Hyg. 2012;86(6):962–5.
Brandão-Filho SP, et al. Epidemiological surveys confirm an increasing burden of cutaneous leishmaniasis in north-east Brazil. Trans R Soc Trop Med Hyg. 1999;93(5):488–94.
Brandão-Filho SP, et al. Hospedeiros reservatórios de *Leishmania* spp. associados à leishmaniose tegumentar americana com especial ênfase no Brasil. In: Barral A, Costa J, editors. Leishmanias e a leishmaniose tegumentar nas Américas. Salvador: Cyted, CNPq Fiocruz-BA; 2011. p. 65–74.
Bray R. The zoonotic potential of reservoirs of leishmaniasis in the old world. Ecol Dis. 1982;1(4):257–67.
Bualert L, et al. Autochthonous disseminated dermal and visceral leishmaniasis in an AIDS patient, southern Thailand, caused by *Leishmania siamensis*. Am J Trop Med Hyg. 2012;86(5):821–4.
Caldart ET, et al. *Leishmania* in synanthropic rodents (*Rattus rattus*): new evidence for the urbanization of *Leishmania* (*Leishmania*) *amazonensis*. Rev Bras Parasitol Vet. 2017;26(1):17–27.
Campino L, Maia C. The role of reservoirs: canine leishmaniasis. In: Ponte-Sucre A, Padron-Nieves M, Diaz E, editors. Drug resistance in Leishmania parasites—consequences, molecular mechanism and possible treatments. Wien: Springer; 2012. p. 45–64.
Can H, et al. Seroprevalence of *Leishmania* infection and molecular detection of *Leishmania tropica* and *Leishmania infantum* in stray cats of İzmir, Turkey. Exp Parasitol. 2016;167:109–14.
Carreira J, et al. *Leishmania* in marsupials—an overview of infection records in the Americas and Australia. Open J Anim Sci. 2017;7(3):315–43.

Cerqueira EJ, et al. Experimental infection of *Equus asinus* with *Leishmania chagasi* Cunha & Chagas, 1937. Rev Soc Bras Med Trop. 2003;36(6):695–701.

Chaara D, et al. Leishmaniases in Maghreb: an endemic neglected disease. Acta Trop. 2014;132:80–93.

Chatzis MK, et al. Cytological and molecular detection of *Leishmania infantum* in different tissues of clinically normal and sick cats. Vet Parasitol. 2014;202(3–4):217–25.

Chemkhi J, et al. Natural infection of Algerian hedgehog, *Atelerix algirus* (Lereboullet 1842) with *Leishmania* parasites in Tunisia. Acta Trop. 2015;150:42–51.

Chusri S, et al. Consecutive cutaneous and visceral leishmaniasis manifestations involving a novel *Leishmania* species in two HIV patients in Thailand. Am J Trop Med Hyg. 2012;87(1):76–80.

Chusri S, et al. Animal reservoirs and potential vectors of *Leishmania siamensis* in southern Thailand. Southeast Asian J Trop Med Public Health. 2014;45(1):13–9.

Colmenares M, et al. Identification of blood meals of *Phlebotomus perniciosus* (Diptera: Psychodidae) in Spain by a competitive enzyme-linked immunosorbent assay biotin/avidin method. J Med Entomol. 1995;32(3):229–33.

Costa CH, et al. Asymptomatic human carriers of *Leishmania chagasi*. Am J Trop Med Hyg. 2002;66(4):334–7.

Courtenay O, et al. A long-lasting topical deltamethrin treatment to protect dogs against visceral leishmaniasis. Med Vet Entomol. 2009;23(3):245–56.

Craig TM, et al. Dermal leishmaniasis in a Texas cat. Am J Trop Med Hyg. 1986;35(6):1100–2.

da Silva SM, et al. First report of infection of *Lutzomyia longipalpis* by *Leishmania* (*Leishmania*) *infantum* from a naturally infected cat of Brazil. Vet Parasitol. 2010;174(1–2):150–4.

Dantas-Torres F. The role of dogs as reservoirs of *Leishmania* parasites, with emphasis on *Leishmania* (*Leishmania*) *infantum* and *Leishmania* (*Viannia*) *braziliensis*. Vet Parasitol. 2007;149(3–4):139–46.

Dantas-Torres F. Canine leishmaniosis in South America. Parasit Vectors. 2009;2(Suppl 1):S1.

de Castro Ferreira E, et al. *Leishmania* (*V.*) *braziliensis* infecting bats from Pantanal wetland, Brazil: first records for *Platyrrhinus lineatus* and *Artibeus planirostris*. Acta Trop. 2017;172:217–22.

de Oliveira FM, et al. First detection of *Leishmania* spp. DNA in Brazilian bats captured strictly in urban areas. Acta Trop. 2015;150:176–81.

de Rezende MB, et al. Detection of *Leishmania* spp. in bats from an area of Brazil endemic for visceral leishmaniasis. Transbound Emerg Dis. 2017;64(6):e36–42. https://doi.org/10.1111/tbed.12597.

de Souza AI, et al. Feline leishmaniasis due to *Leishmania* (*Leishmania*) *amazonensis* in Mato Grosso do Sul state, Brazil. Vet Parasitol. 2005;128(1–2):41–5.

Delgado O, et al. *Leishmania colombiensis* in Venezuela. Am J Trop Med Hyg. 1993;48(1):145–7.

Dereure J, et al. *Leishmania tropica* in Morocco: infection in dogs. Trans R Soc Trop Med Hyg. 1991;85(5):595.

Dereure J, et al. Leishmaniose en Equateur. 4. Infestation naturelle du chien par *Leishmania panamensis*. Ann Soc Belg Med Trop. 1994;74(1):29–33.

Dereure J, et al. Visceral leishmaniasis in Sudan: first identifications of *Leishmania* from dogs. Trans R Soc Trop Med Hyg. 2000;94(2):154–5.

Dereure J, et al. Visceral leishmaniasis in eastern Sudan: parasite identification in humans and dogs; host-parasite relationships. Microbes Infect. 2003;5(12):1103–8.

Desjeux P. Leishmaniasis. Public health aspects and control. Clin Dermatol. 1996;14(5):417–23.

Desjeux P. Leishmaniasis: current situation and new perspectives. Comp Immunol Microbiol Infect Dis. 2004;27(5):305–18.

Dujardin JC, et al. Spread of vector-borne diseases and neglect of leishmaniasis, Europe. Emerg Infect Dis. 2008;14(7):1013–8.

Dumont P, et al. Repellent and insecticidal efficacy of a new combination of fipronil and permethrin against the main vector of canine leishmaniosis in Europe (*Phlebotomus perniciosus*). Parasit Vectors. 2015;8:49.

Ebani VV, et al. Serological survey on some pathogens in wild brown hares (*Lepus europaeus*) in Central Italy. Asian Pac J Trop Med. 2016;9(5):465–9.

Elbihari S, Cheema AH, el-Hassan AM. *Leishmania* infecting man and wild animals in Saudi Arabia. 4. Canine cutaneous leishmaniasis in the Eastern Province. Trans R Soc Trop Med Hyg. 1987;81(6):925–7.

Elnaiem DA, et al. The Egyptian mongoose, *Herpestes ichneumon*, is a possible reservoir host of visceral leishmaniasis in eastern Sudan. Parasitology. 2001;122(5):531–6.

Faiman R, et al. A newly emerged cutaneous leishmaniasis focus in northern Israel and two new reservoir hosts of *Leishmania major*. PLoS Negl Trop Dis. 2013;7(2):e2058.

Ferreira Ede C, et al. Mixed infection of Leishmania infantum and Leishmania braziliensis in rodents from endemic urban area of the New World. BMC Vet Res. 2015;11:71.

Fernández-Bellon H, et al. Immune response to *Leishmania infantum* in healthy horses in Spain. Vet Parasitol. 2006;135(2):181–5.

Franc M, et al. Efficacy of a new combination of fipronil and permethrin (Effitix®) against *Phlebotomus perniciosus* in dogs. Vet Parasitol. 2015;212(3–4):156–60.

Gama A, et al. Cutaneous leishmaniosis in a horse from northern Portugal. Vet Parasitol. 2014;200(1–2):189–92.

Gavgani AS, et al. Effect of insecticide-impregnated dog collars on incidence of zoonotic visceral leishmaniasis in Iranian children: a matched-cluster randomised trial. Lancet. 2002;360(9330):374–9.

Ghawar W, et al. *Leishmania major* infection among *Psammomys obesus* and *Meriones shawi*: reservoirs of zoonotic cutaneous leishmaniasis in Sidi Bouzid (central Tunisia). Vector Borne Zoonotic Dis. 2011a;11(12):1561–8.

Ghawar W, et al. First report of natural infection of least weasel (*Mustela nivalis* Linnaeus, 1776) with *Leishmania major* in Tunisia. Vector Borne Zoonotic Dis. 2011b;11(11):1507–9.

Gicheru MM, et al. Prevalence of antibodies and cell mediated immune response against *Leishmania major* in feral nonhuman primates from Kenya. Acta Trop. 2009;109(2): 136–40.

González U, et al. Vector and reservoir control for preventing leishmaniasis. Cochrane Database Syst Rev. 2015;8:CD008736.

Gradoni L, et al. Leishmaniasis in Tuscany (Italy): VII. Studies on the role of the black rat, *Rattus rattus*, in the epidemiology of visceral leishmaniasis. Trans R Soc Trop Med Hyg. 1983;77(4):427–31.

Gradoni L, et al. *Leishmania infantum* infection rates in *Phlebotomus perniciosus* fed on naturally infected dogs under antimonial treatment. Med Vet Entomol. 1987;1(4):339–42.

Gradoni L. Canine *Leishmania* vaccines: still a long way to go. Vet Parasitol. 2015;208(1–2):94–100.

Gramiccia M. Recent advances in leishmaniosis in pet animals: epidemiology, diagnostics and anti-vectorial prophylaxis. Vet Parasitol. 2011;181(1):23–30.

Gramiccia M, Gradoni L. The current status of zoonotic leishmaniases and approaches to disease control. Int J Parasitol. 2005;35(11–12):1169–80.

Gramiccia M, Gradoni L. The leishmaniases in Southern Europe. In: Takken W, Knols BGJ, editors. Emerging pests and vector-borne diseases, ecology and control of vectorborne diseases, vol. 1. Wageningen: Wageningen Academic Publishers; 2007. p. 75–95.

Grimaldi G Jr, Tesh RB. Leishmaniases of the new world: current concepts and implications for future research. Clin Microbiol Rev. 1993;6(3):230–50.

Guessous-Idrissi N, et al. Short report: *Leishmania tropica*: etiologic agent of a case of canine visceral leishmaniasis in northern Morocco. Am J Trop Med Hyg. 1997;57(2):172–3.

Hajjaran H, et al. Molecular identification and polymorphism determination of cutaneous and visceral leishmaniasis agents isolated from human and animal hosts in Iran. Biomed Res Int. 2013;2013:789326.

Hashiguchi Y, et al. Andean leishmaniasis in Ecuador caused by infection with *Leishmania mexicana* and *L. major*-like parasites. Am J Trop Med Hyg. 1991;44(2):205–17.

Hassan M, et al. Role of the domestic dog as a reservoir host of *Leishmania donovani* in eastern Sudan. Parasit Vectors. 2009;2(1):26.

Hatam GR, et al. First report of natural infection in cats with *Leishmania infantum* in Iran. Vector Borne Zoonotic Dis. 2010;10(3):313–6.

Hoogstraal H, Heyneman D. Leishmaniasis in the Sudan Republic. 30. Final epidemiological report. Am J Trop Med Hyg. 1969;18(6):1087–210.

Jambulingam P, et al. Domestic dogs as reservoir hosts for *Leishmania donovani* in the southernmost western Ghats in India. Acta Trop. 2017;171:64–7.

Jaouadi K, et al. First detection of *Leishmania killicki* (Kinetoplastida, Trypanosomatidae) in *Ctenodactylus gundi* (Rodentia, Ctenodactylidae), a possible reservoir of human cutaneous leishmaniasis in Tunisia. Parasit Vectors. 2011;4:159.

Jiménez M, et al. Could wild rabbits (*Oryctolagus cuniculus*) be reservoirs for *Leishmania infantum* in the focus of Madrid, Spain? Vet Parasitol. 2014;202(3–4):296–300.

Kalayou S, et al. Serological evidence of *Leishmania donovani* infection in apparently healthy dogs using direct agglutination test (DAT) and rk39 dipstick tests in Kafta Humera, north-west Ethiopia. Transbound Emerg Dis. 2011;58(3):255–62.

Kanjanopas K, et al. *Sergentomyia (Neophlebotomus) gemmea*, a potential vector of *Leishmania siamensis* in southern Thailand. BMC Infect Dis. 2013;13:333.

Kassahun A, et al. Detection of *Leishmania donovani* and *L. tropica* in Ethiopian wild rodents. Acta Trop. 2015a;145:39–44.

Kassahun A, et al. Natural infection of bats with *Leishmania* in Ethiopia. Acta Trop. 2015b;150:166–70.

Kato H, et al. First human cases of *Leishmania* (*Viannia*) *naiffi* infection in Ecuador and identification of its suspected vector species. Acta Trop. 2013;128(3):710–3.

Kato H, et al. First human cases of *Leishmania* (*Viannia*) *lainsoni* infection and a search for the vector sand flies in Ecuador. PLoS Negl Trop Dis. 2016;10(5):e0004728.

Kenubih A, et al. Preliminary survey of domestic animal visceral leishmaniasis and risk factors in north-west Ethiopia. Trop Med Int Health. 2015;20(2):205–10.

Killick-Kendrick R. The biology and control of phlebotomine sand flies. Clin Dermatol. 1999;17(3):279–89.

Killick-Kendrick R, et al. Protection of dogs from bites of phlebotomine sandflies by deltamethrin collars for control of canine leishmaniasis. Med Vet Entomol. 1997;11(2):105–11.

Kipp EJ, et al. Genetic evidence of enzootic leishmaniasis in a stray canine and Texas mouse from sites in west and central Texas. Mem Inst Oswaldo Cruz. 2016;111(10):652–4.

Koehler K, et al. Cutaneous leishmaniosis in a horse in southern Germany caused by *Leishmania infantum*, Vet Parasitol. 2002;109(1–2):9–17.

Lainson R. The neotropical *Leishmania* species: a brief historical review of the discovery, ecology and taxonomy. Rev Pan-Amaz Saude. 2010;1:13–32.

Lainson R, Shaw JJ. *Leishmania* (*Viannia*) *naiffi* sp. n., a parasite of the armadillo, *Dasypus novemcinctus* (L.) in Amazonian Brazil. Ann Parasitol Hum Comp. 1989;64(1):3–9.

Lainson R, et al. *Leishmania* (*Viannia*) *shawi* sp. n., a parasite of monkeys, sloths and procyonids in Amazonian Brazil. Ann Parasitol Hum Comp. 1989;64(3):200–7.

Lainson R, Shaw JJ. New world leishmaniasis. In: Cox FEG, Kreier JP, Wakelin D, editors. Topley & Wilson's microbiology and microbial infections. London: Arnold; 2005. p. 313–49.

Lemma W, et al. A zoonotic focus of cutaneous leishmaniasis in Addis Ababa, Ethiopia. Parasit Vectors. 2009;2(1):60.

Lemma W, et al. Preliminary study on investigation of zoonotic visceral leishmaniasis in endemic foci of Ethiopia by detecting *Leishmania* infections in rodents. Asian Pac J Trop Med. 2017;10(4):418–22.

Liénard E, et al. Efficacy of dinotefuran, permethrin and pyriproxyfen combination spot-on on dogs against *Phlebotomus perniciosus* and *Ctenocephalides canis*. Parasitol Res. 2013;112(11):3799–805.

Llanos-Cuentas EA, et al. Natural infections of *Leishmania peruviana* in animals in the Peruvian Andes. Trans R Soc Trop Med Hyg. 1999;93(1):15–20.

Lobsiger L, et al. An autochthonous case of cutaneous bovine leishmaniasis in Switzerland. Vet Parasitol. 2010;169(3–4):408–14.

Lopes AP, et al. Prevalence of antibodies to Leishmania infantum and *Toxoplasma gondii* in horses from the north of Portugal. Parasit Vectors. 2013;6:178.

Madeira MF, et al. Mixed infection with *Leishmania* (*Viannia*) *braziliensis* and *Leishmania* (*Leishmania*) *chagasi* in a naturally infected dog from Rio de Janeiro, Brazil. Trans R Soc Trop Med Hyg. 2006;100(5):442–5.

Maia C, Campino L. Can domestic cats be considered reservoir hosts of zoonotic leishmaniasis? Trends Parasitol. 2011;27(8):341–4.

Maia C, Cardoso L. Spread of *Leishmania infantum* in Europe with dog travelling. Vet Parasitol. 2015;213(1–2):2–11.

Maia C, et al. Feline *Leishmania* infection in a canine leishmaniasis endemic region, Portugal. Vet Parasitol. 2010;174(3–4):336–40.

Maia C, et al. Bacterial and protozoal agents of feline vector-borne diseases in domestic and stray cats from southern Portugal. Parasit Vectors. 2014;7:115.

Maia C, et al. First case of feline leishmaniosis caused by *Leishmania infantum* genotype E in a cat with concurrent nasal squamous cell carcinoma. J Feline Med Surg Open Rep. 2015;1(2):2055116915593969.

Manzillo V, et al. Deltamethrin-impregnated collars for the control of canine leishmaniasis: evaluation of the protective effect and influence on the clinical outcome of *Leishmania infection* in kennelled stray dogs. Vet Parasitol. 2006;142(1–2):142–5.

Marcelino AP, et al. Molecular detection of *Leishmania braziliensis* in *Rattus norvegicus* in an area endemic for cutaneous leishmaniasis in Brazil. Vet Parasitol. 2011;183:54–8.

Maroli M, et al. Evidence for an impact on the incidence of canine leishmaniasis by the mass use of deltamethrin-impregnated dog collars in southern Italy. Med Vet Entomol. 2001;15(4): 358–63.

Maroli M, et al. Infection of sandflies by a cat naturally infected with *Leishmania infantum*. Vet Parasitol. 2007;145(3–4):357–60.

Maroli M, et al. Guidelines for prevention of leishmaniasis in dogs. J Am Vet Med Assoc. 2010;236(11):1200–6.

Maroli M, et al. Phlebotomine sandflies and the spreading of leishmaniases and other diseases of public health concern. Med Vet Entomol. 2013;27(2):123–47.

Martín-Sánchez J, et al. Infection by *Leishmania infantum* in cats: epidemiological study in Spain. Vet Parasitol. 2007;145(3–4):267–73.

Mencke N, et al. Repellent efficacy of a combination containing imidacloprid and permethrin against sand flies (*Phlebotomus papatasi*) in dogs. Parasitol Res. 2003;90(Suppl 3):S108–11.

Michel G, et al. Importance of worldwide asymptomatic carriers of *Leishmania infantum* (*L. chagasi*) in human. Acta Trop. 2011;119(2–3):69–75.

Millán J, Ferroglio E, Solano-Gallego L. Role of wildlife in the epidemiology of *Leishmania infantum* infection in Europe. Parasitol Res. 2014;113(6.):2005–2014.

Miró G, et al. Infectivity to *Phlebotomus perniciosus* of dogs naturally parasitized with *Leishmania infantum* after different treatments. Parasit Vectors. 2011;4:52.

Molina R, et al. The hare (*Lepus granatensis*) as potential sylvatic reservoir of *Leishmania infantum* in Spain. Vet Parasitol. 2012;190(1–2):268–71.

Montoya A, et al. *Leishmania infantum* infection in Bennett's wallabies (*Macropus rufogriseus rufogriseus*) in a Spanish wildlife park. J Zoo Wildl Med. 2016;47(2):586–393.

Morgado FN, et al. *Hepatozoon canis* and *Leishmania* spp. coinfection in dogs diagnosed with visceral leishmaniasis. Rev Bras Parasitol Vet. 2016;25(4):450–8.

Morsy T, et al. Natural infections of *Leishmania major* in domestic dogs from Alexandria, Egypt. Am J Trop Med Hyg. 1987;37(1):49–52.

Morsy TA, al Dakhil MA, el Bahrawy AF. Characterization of *Leishmania aethiopica* from rock hyrax, *Procavia capensis* trapped in Najran, Saudi Arabia. J Egypt Soc Parasitol. 1997;27(2):349–53.

Motazedian MH, et al. First detection of *Leishmania major* in *Rattus norvegicus* from Fars Province, southern Iran. Vector Borne Zoonotic Dis. 2010;10(10):969–75.

Mukhtar MM, et al. Detection of antibodies to *Leishmania donovani* in animals in a kala-azar endemic region in eastern Sudan: a preliminary report. Trans R Soc Trop Med Hyg. 2000;94(1):33–6.

Müller N, et al. Occurrence of *Leishmania* sp. in cutaneous lesions of horses in Central Europe. Vet Parasitol. 2009;166(3–4):346–51.

Naiff RD, et al. Epidemiological and nosological aspects of *Leishmania naiffi* Lainson Shaw, 1989. Mem Inst Oswaldo Cruz. 1991;86(3):317–21.

Nasereddin A, Salant H, Abdeen Z. Feline leishmaniasis in Jerusalem: serological investigation. Vet Parasitol. 2008;158(4):364–9.

Negera E, et al. Outbreak of cutaneous leishmaniasis in Silti District, Ethiopia: risk factor assessment and causative agent identification. Trans R Soc Trop Med Hyg. 2008;102(9):883–90.

Noppakun N, Kraivichian K, Siriyasatien P. Disseminated dermal leishmaniasis caused by *Leishmania siamensis* in a systemic steroid therapy patient. Am J Trop Med Hyg. 2014;91(5):869–70.

Ntais P, et al. Will the introduction of *Leishmania tropica* MON-58, in the island of Crete, lead to the settlement and spread of this rare zymodeme? Acta Trop. 2014;132:125–30.

Oliva G, et al. A randomised, double-blind, controlled efficacy trial of the LiESP/QA-21 vaccine in naïve dogs exposed to two *Leishmania infantum* transmission seasons. PLoS Negl Trop Dis. 2014;8(10):e3213.

Otranto D, et al. Prevention of endemic canine vector-borne diseases using imidacloprid 10% and permethrin 50% in young dogs: a longitudinal field study. Vet Parasitol. 2010;172(3–4): 323–32.

Ozon C, et al. Disseminated feline leishmaniosis due to *Leishmania infantum* in southern France. Vet Parasitol. 1998;75(2–3):273–7.

Paşa S, et al. Detection of *Leishmania major* and *Leishmania tropica* in domestic cats in the Ege region of Turkey. Vet Parasitol. 2015;212(3–4):389–92.

Passos VM, et al. Natural infection of a domestic cat (*Felis domesticus*) with *Leishmania* (*Viannia*) in the metropolitan region of Belo Horizonte, state of Minas Gerais, Brazil. Mem Inst Oswaldo Cruz. 1996;91(1):19–20.

Pennisi MG, et al. Serological and molecular prevalence of *Leishmania infantum* infection in cats from southern Italy. J Feline Med Surg. 2012;14:656–7.

Pennisi MG, et al. LeishVet update and recommendations on feline leishmaniosis. Parasit Vectors. 2015;8:302.

Pourmohammadi B, et al. Rodent infection with *Leishmania* in a new focus of human cutaneous leishmaniasis, in northern Iran. Ann Trop Med Parasitol. 2008;102:127–33.

Pourmohammadi B, et al. Natural infection of *Nesokia indica* with *Leishmania major* and *Leishmania infantum* parasites in Damghan city, northern Iran. Acta Trop. 2017;170:134–9.

Portús M, et al. Wild and domestic mammals in the life cycle of *Leishmania infantum* in Southwest Europe. A literature review and studies performed in Catalonia (Spain). Rev Iber Parasitol. 2002;62:72–6.

Pothirat T, et al. First isolation of *Leishmania* from northern Thailand: case report, identification as *Leishmania martiniquensis* and phylogenetic position within the *Leishmania enriettii* complex. PLoS Negl Trop Dis. 2014;8(12):e3339.

Pratlong F, et al. Geographical distribution and epidemiological features of old world cutaneous leishmaniasis foci, based on the isoenzyme analysis of 1048 strains. Trop Med Int Health. 2009;14(9):1071–85.

Quinnell R, Courtenay O. Transmission, reservoir hosts and control of zoonotic visceral leishmaniasis. Parasitology. 2009;136(14):1915–34.

Ramírez J, et al. Taxonomy, diversity, temporal and geographical distribution of cutaneous leishmaniasis in Colombia: a retrospective study. Sci Rep. 2016;6:28266.

Regina-Silva S, et al. Field randomized trial to evaluate the efficacy of the Leish-Tec® vaccine against canine visceral leishmaniasis in an endemic area of Brazil. Vaccine. 2016;34(19):2233–9.

Reithinger R, Davies CR. Is the domestic dog (*Canis familiaris*) a reservoir host of American cutaneous leishmaniasis? A critical review of the current evidence. Am J Trop Med Hyg. 1999;61(4):530–41.

Reithinger R, Espinoza J, Davies C. The transmission dynamics of canine American cutaneous leishmaniasis in Huánuco, Peru. Am J Trop Med Hyg. 2003;69(5):473–80.

Reuss SM, et al. Autochthonous *Leishmania siamensis* in horse, Florida, USA. Emerg Infect Dis. 2012;18(9):1545–7.

Rioux JA, et al. *Meriones shawi* (Duvernoy, 1842) [Rodentia, Gerbillidae] a reservoir of *Leishmania major*, Yakimoff and Schokhor, 1914 [Kinetoplastida, Trypanosomatidae] in South Morocco. C R Seances Acad Sci III. 1982;294(11):515–7.

Rohousova I, et al. Exposure to *Leishmania* spp. and sand flies in domestic animals in northwestern Ethiopia. Parasit Vectors. 2015;8:360.

Rolão N, et al. Equine infection with *Leishmania* in Portugal. Parasite. 2005;12(2):183–6.

Roque AL, Jansen AM. Wild and synanthropic reservoirs of *Leishmania* species in the Americas. Int J Parasitol Parasites Wildl. 2014;3(3):251–62.

Rougeron V, et al. First clinical case of cutaneous leishmaniasis due to *Leishmania* (*Viannia*) *braziliensis* in a domestic cat from French Guiana. Vet Parasitol. 2011;181(2–4):325–8.

Rouhani S, et al. Novel identification of *Leishmania major* in *Hemiechinus auritus* and molecular detection of this parasite in *Meriones libycus* from an important foci of zoonotic cutaneous leishmaniasis in Iran. J Infect Public Health. 2014;7(3):210–7.

Saliba EK, Oumeish OY. Reservoir hosts of cutaneous leishmaniasis. Clin Dermatol. 1999;17(3):275–7.

Sanches LD, et al. Natural canine infection by *Leishmania infantum* and *Leishmania amazonensis* and their implications for disease control. Rev Bras Parasitol Vet. 2017;25(4):465–9.

Santaella J, et al. Leishmania (Viannia) infection in the domestic dog in chaparral, Colombia. Am J Trop Med Hyg. 2011;84(5):674–80.

Saraiva EM, et al. The FML-vaccine (Leishmune) against canine visceral leishmaniasis: a transmission blocking vaccine. Vaccine. 2006;24(13):2423–31.

Savani ES, et al. Detection of *Leishmania* (*Leishmania*) *amazonensis* and *Leishmania* (*Leishmania*) *infantum chagasi* in Brazilian bats. Vet Parasitol. 2010;168(1–2):5–10.

Schubach TM, et al. American cutaneous leishmaniasis in two cats from Rio de Janeiro, Brazil: first report of natural infection with *Leishmania* (*Viannia*) *braziliensis*. Trans R Soc Trop Med Hyg. 2004;98:165–7.

Shapiro JT, et al. First record of *Leishmania braziliensis* presence detected in bats, Mato Grosso do Sul, southwest Brazil. Acta Trop. 2013;128:171–4.

Shaw J, et al. Characterization of *Leishmania* (*Leishmania*) *waltoni* n.sp. (Kinetoplastida: Trypanosomatidae), the parasite responsible for diffuse cutaneous leishmaniasis in the Dominican Republic. Am J Trop Med Hyg. 2015;93(3):552–8.

Silva ES, Gontijo CM, Melo MN. Contribution of molecular techniques to the epidemiology of neotropical *Leishmania* species. Trends Parasitol. 2005;21(12):550–2.

Silva VO, et al. A phase III trial of efficacy of the FML-vaccine against canine kala-azar in an endemic area of Brazil (São Gonçalo do Amaranto, RN). Vaccine. 2000;19(9–10):1082–92.

Silveira FT, et al. Dermal leishmaniasis in the Amazon region of Brazil: *Leishmania* (*Viannaia*) *lainsoni* sp.n., a new parasite from the state of Pará. Mem Inst Oswaldo Cruz. 1987;82(2):289–91.

Silveira FT, et al. Cutaneous leishmaniasis in the Amazon region: natural infection of the sandfly *Lutzomyia ubiquitalis* (Psychodidae: Phlebotominae) by *Leishmania* (*Viannia*) *lainsoni* in Pará state, Brazil. Mem Inst Oswaldo Cruz. 1991a;86(1):127–30.

Silveira FT, et al. Leishmaniose cutânea na Amazônia: isolamento de *Leishmania* (*Viannia*) *lainsoni* do roedor *Agouti* paca (Rodentia: Dasyproctidae), no Estado do Pará, Brasil. Rev Inst Med Trop Sao Paulo. 1991b;33(1):18–22.

Silveira FT, et al. An outbreak of cutaneous leishmaniasis among soldiers in Belém, Pará state, Brazil, caused by *Leishmania* (*Viannia*) *lindenbergi* n. sp. A new leishmanial parasite of man in the Amazon region. Parasite. 2002;9(1):43–50.

Singh N, et al. Animal reservoirs of visceral leishmaniasis in India. J Parasitol. 2013;99(1):64–7.

Soares IR, et al. First evidence of autochthonous cases of *Leishmania* (*Leishmania*) *infantum* in horse (*Equus caballus*) in the Americas and mixed infection of *Leishmania infantum* and *Leishmania* (*Viannia*) *braziliensis*. Vet Parasitol. 2013;197(3–4):665–9.

Solano-Gallego L, et al. Cutaneous leishmaniosis in three horses in Spain. Equine Vet J. 2003;35(3):320–3.

Solano-Gallego L, et al. LeishVet guidelines for the practical management of canine leishmaniosis. Parasit Vectors. 2011;4:86.
Suankratay C, et al. Autochthonous visceral leishmaniasis in a human immunodeficiency virus (HIV)-infected patient: the first in thailand and review of the literature. Am J Trop Med Hyg. 2010;82(1):4–8.
Sukmee T, et al. A suspected new species of *Leishmania*, the causative agent of visceral leishmaniasis in a Thai patient. Int J Parasitol. 2008;38(6):617–22.
Svobodova M, et al. Distinct transmission cycles of *Leishmania tropica* in 2 adjacent foci, northern Israel. Emerg Infect Dis. 2006;12(12):1860–8.
Talmi-Frank D, et al. *Leishmania tropica* in rock hyraxes (*Procavia capensis*) in a focus of human cutaneous leishmaniasis. Am J Trop Med Hyg. 2010;82(5):814–8.
Tolezano JE, et al. The first records of *Leishmania* (*Leishmania*) *amazonensis* in dogs (*Canis familiaris*) diagnosed clinically as having canine visceral leishmaniasis from Araçatuba County, São Paulo state, Brazil. Vet Parasitol. 2007;149(3–4):280–4.
Tomás-Pérez M, et al. First report of natural infection in hedgehogs with *Leishmania major*, a possible reservoir of zoonotic cutaneous leishmaniasis in Algeria. Acta Trop. 2014;135:44–9.
Tonui WK. Situational analysis of leishmaniases research in Kenya. Afr J Health Sci. 2006; 13(1–2):7–21.
Trainor KE, et al. Eight cases of feline cutaneous leishmaniasis in Texas. Vet Pathol. 2010; 47(6):1076–81.
Truppel JH, et al. Can equids be a reservoir of *Leishmania braziliensis* in endemic areas? PLoS One. 2014;9(4):e93731.
van Thiel PP, et al. First cases of cutaneous leishmaniasis caused by *Leishmania* (*Viannia*) *naiffi* infection in Surinam. Am J Trop Med Hyg. 2010;82(4):588–90.
Vedovello FD, et al. American cutaneous leishmaniasis in horses from endemic areas in the north-central mesoregion of Paraná state, Brazil. Zoonoses Public Health. 2008;55(3):149–55.
Vélez ID, et al. An epidemic outbreak of canine cutaneous leishmaniasis in Colombia caused by *Leishmania braziliensis* and *Leishmania panamensis*. Am J Trop Med Hyg. 2012;86(5):807–11.
Vides J, et al. *Leishmania chagasi* infection in cats with dermatologic lesions from an endemic area of visceral leishmaniosis in Brazil. Vet Parasitol. 2011;178(1–2):22–8.
WHO. Control of the Leishmaniasis: Report of the WHO Expert Committee on the Control of Leishmaniases. World Health Organization; 2010. http://www.who.int/neglected_diseases/resources/who_trs_949/en. Accessed Sept 2016.
Williams AO, Mutinga J, Rodgers M. Leishmaniasis in a domestic goat in Kenya. Mol Cell Probes. 1991;5(5):319–25.

5 Immunological and Immunopathological Aspects

Juan José Lauthier and Masataka Korenaga

5.1 Introduction

Leishmaniases are a group of diseases caused by protozoa of the genus *Leishmania*. These parasites are classified by genetic and biological criteria that trigger diverse clinical manifestations, which ultimately are the result of interactions between the infecting species and the host's immune response. At least 20 species of *Leishmania* are infective to humans, and are often classified according to the world regions as Old World (OW) leishmaniases and New World (NW) leishmaniases (WHO 2016; Kevric et al. 2015). Some species cause only or mainly cutaneous disease (i.e., *Leishmania major*, *L. braziliensis*, *L. amazonensis*), others result in visceralization (i.e., *L. donovani*), while others can cause both symptoms (i.e., *L. infantum*). *Leishmania* infection can result in asymptomatic (le Fichoux et al. 1999; Hide et al. 2013). Some authors propose that variability in the symptoms is probably due to genetic differences acquired during evolution of *Leishmania*, making some species more adaptable to the skin while others invade visceral organs (Zhang and Matlashewski 2001). These parasites alternate between two stages: a flagellated form called promastigote that resides in the midgut of sandfly vectors and a non-flagellated form called amastigote that resides within phagolysosomal vesicles in macrophages of vertebrate hosts. The life cycle of *Leishmania* sp. initiates when an infected female sandfly injects metacyclic promastigotes while feeding on its vertebrate host. Once in the blood, promastigotes are phagocyted principally by macrophages and neutrophils. Since neutrophils have a short life span, macrophages are the ultimate host cells in which *Leishmania* parasites proliferate (Liu and Uzonna 2012). Inside the phagolysosomes, promastigotes start to differentiate into small, non-motile amastigote forms, which divide by binary fission. Amastigote

J.J. Lauthier • M. Korenaga (✉)
Department of Parasitology, Kochi Medical School, Kochi University, Nankoku, Kochi, Japan
e-mail: juanjoselauthier@yahoo.com.ar; korenaga@kochi-u.ac.jp

F. Bruschi, L. Gradoni (eds.), *The Leishmaniases: Old Neglected Tropical Diseases*, https://doi.org/10.1007/978-3-319-72386-0_5

proliferation ultimately causes macrophage disruption, liberating parasites into the extracellular milieu that in turn infect surrounding macrophages.

5.2 Immunological Responses in *Leishmania* Infections

The complexity of the immunological responses and pathogenesis of leishmaniases is a product of genetic and cellular factors that have been implicated in mediating resistance and susceptibility to infection. The disease exhibits three different forms, which are cutaneous leishmaniasis (CL), visceral leishmaniasis (VL), and mucocutaneous leishmaniasis (MCL) (Marco et al. 2015). The species of *Leishmania* and the immune status of the host are both important factors that determine the form and severity of disease (Kane and Mosser 2000). Metastatic characteristics are observed in some species, and up to 10% of CL cases progress to MCL causing destructive secondary lesions in the nasal and oral mucosa (Guerra et al. 2011).

After internalization of promastigotes into phagosomes followed by fusion with lysosomes, parasites face the hostile phagolysosomal environment (Rasmusson and Descoteaux 2004). A variety of host cell types and factors interact with *Leishmania* parasites: neutrophils, monocytes or macrophages, natural killer (NK) cells, and dendritic cells (DCs). Pattern recognition receptors (PRRs, i.e., Toll-like receptors [TLRs]) and soluble products (complement and released cytokines, including interleukin [IL]-1α, IL-12, tumor necrosis factor [TNF]) are also involved, and in conjunction they require the coordinated action of the innate and adaptive immune response in the host.

Protective immunity against leishmaniases is associated with a Th1 response, while disease progression is associated with an anti-Th2 response (McMahon-Pratt and Alexander 2004). The inflammation process in leishmaniases involves production of cytokines and other molecules like the lipid mediator leukotriene B4 (LTB4), which can also activate phagocytes leading to nitric oxide (NO) production. However, exacerbated production of these molecules can induce tissue damage. After the recognition of parasites, macrophages produce TNF-α and IL-1 that promote inflammation by inducing the expression of adhesion molecules (selectin and integrin ligands) on the endothelial surface (Kolodziej and Kiderlen 2005). In addition, the role of IL-1 is controversial, because it can either contribute to Th1 priming at early infection or aggravate disease outcome in the established infection (Kostka et al. 2006).

An important anti-inflammatory cytokine is IL-10, which is responsible for peripheral tolerance to self-antigens and prevents exacerbated immune responses to foreign antigens (Gannavaram et al. 2016). Overexpression of IL-10 in mice induced an early suppression of the innate and acquired immune responses, pathogen proliferation, and aggravation of the disease (Saraiva and O'Garra 2010). It was observed that phagocytes that produce IL-10 elicit a reduced production of Th1-type cytokines (IL-12 and interferon gamma [IFN-γ]) (Nandan et al. 2012). The innate mechanisms with the primordial secretion of IL-12 lead to the parallel induction of cell-mediated immunity which activates specific CD4$^+$ and CD8$^+$ T cells (Hsieh et al. 2008).

Since neutrophils are recruited to the infection site after the sandfly bite, they are the first line of defense against parasites (Peters et al. 2008; Peters and Sacks 2009). Neutrophils generate a potent oxidative burst and release toxic mediators, like oxygen intermediates, into the parasitophorous vacuole that leads to parasite death (de Menezes et al. 2016). In addition, neutrophils have the ability to release their nuclear DNA in structures called "neutrophil extracellular traps" (NETs). These structures are released upon neutrophil activation and consist of chromatin associated with cytosolic and granular proteins that induce neutrophil death (NETosis) in a different way from apoptosis and necrosis (Guimaraes-Costa et al. 2009; Rochael et al. 2015; de Menezes et al. 2016). In addition, typical microbicide responses are also elicited to eliminate parasites at the site of the infection. These include multi-protein NADPH oxidase complex activity, which leads to the production of reactive oxygen species, release of granule contents into intracellular microbial compartments, and release of defensins (Pham 2006; Mollinedo et al. 2010). It has also been shown that neutrophils can serve as intermediate host cells, acting as "Trojan horses" in order to allow parasites to silently enter into macrophages, thereby avoiding cell activation (van Zandbergen et al. 2004; John and Hunter 2008; Laskay et al. 2008; Cecilio et al. 2014).

Knowledge of the different strategies of neutrophil response toward *Leishmania* infection has been achieved principally by studies in murine models (Carlsen et al. 2015). A recent study suggested that after in vitro infection with *L. braziliensis*, neutrophils from CL patients produced more reactive oxygen species (ROS) and higher levels of CXCL8 and CXCL9 cytokines, chemokines associated with recruitment of neutrophils and Th1-type cells, than neutrophils from healthy control subjects. Despite of this result, there was no difference between the degree of activation of neutrophils from CL versus healthy subjects, assessed by CD66b and CD62L expression using flow cytometry (Conceicao et al. 2016).

Adhesion molecules and chemokine activation mechanisms, in conjunction with monocytes, direct local inflammatory responses, including granuloma assembly and lesion development that lead to cutaneous or visceral symptoms of the disease (McSorley et al. 1996; Roberts 2005). Transendothelial migration of leukocytes from blood to the site of inflammation is a complex process controlled by adhesion molecules, such as PECAM-1, ICAM-2, ICAM-1, CD99, ESAM, or junctional adhesion molecules (Ley et al. 2007). The expression of chemokine genes by the infected host cell induces the recruitment of more cells that paradoxically facilitates parasite infection and proliferation (Oghumu et al. 2010). It has been suggested that some *Leishmania* species can inhibit expression of pro-inflammatory cytokines and chemokine genes, with a consequent increase in virulence. In that regard, it was shown that lipophosphoglycan (LPG), the major glycoprotein on promastigote surface, inhibits the production of CCL2, thus affecting monocyte transendothelial migration (Lo et al. 1998). In CL, different chemokines produced in the infected area determine the type of infiltrating cells and eventually define disease outcome (Teixeira et al. 2006). In contrast, in VL patients the high concentration of CXCL9 and CXCL10 and IFN-γ over the course of the active infection plays an important immunopathogenic role in disease progression (Hailu et al. 2004).

Lo et al. (1998) described that lipophosphoglycan (LPG) reduces monocyte migration through IL-1β gene expression, elicited by bacterial endotoxin-activated, and cytokine-activated (IL-1β and TNF-α) endothelial cells. LPG exerts its inhibitory effect at the expression of the cell adhesion molecules E-selectin, ICAM-1, and VCAM-1. Similarly, LPG also regulates expression of the junctional adhesion proteins, CD31 and VE-cadherin, and induces release of MCP-1. All these data suggest that LPG as a single molecule is able to suppress the expression of inflammation mediators, providing clinically relevant strategies for treatment by blocking monocyte recruitment and infiltration in sites of inflammation.

5.3 Immunity of Cutaneous Leishmaniasis

In humans, symptomatic leishmaniasis is highly variable in its clinical presentation due to differences in the infecting *Leishmania* species and in the patient's immune status. Most *Leishmania* species that cause illness in humans initially induce a localized CL, which is characterized by singular or multiple well-defined skin ulcerations that correspond to sandfly bite sites. The time between parasite inoculation and development of ulcerated lesions is highly variable but typically takes weeks to months (Paniz Mondolfi et al. 2013). In many cases, localized CL is self-limiting after patients develop protective adaptive immunity, often showing typical extensive scarification at the site of cured lesions. However, lesion resolution may be hindered in immune-compromised patients and in those infected with particular parasite species (such as those belonging to the *L. mexicana* complex) (Soong 2012).

Localized forms of CL, especially the one caused by the Old World species *Leishmania major*, are self-healing diseases, usually characterized by—at least—partial immunity against reinfection (Sarkar et al. 2012). The site of *L. major* infection in the dermis usually displays a massive infiltration of macrophages and chemokines, such as monocyte chemoattractant protein-1 (MCP-1), monokine induced by IFN-γ (MIG), IFN-γ-inducible protein 10 (IP-10), and only a low amount of macrophage inflammatory protein 1α (MIP-1α) (Ritter and Korner 2002). In skin lesions of patients with self-healing CL (New World species, i.e., *L. mexicana*), the healing process is facilitated by high levels of MCP-1 expression, whereas high levels of local MIP-1α are associated with the nonhealing form of CL (Valencia-Pacheco et al. 2014). MCP-1 and IFN-γ synergistically activate monocytes to clear intracellular parasites, whereas IL-4 suppresses the effect of MCP-1 (Ritter and Korner 2002).

The mechanisms involved in the pathogenesis of *L. amazonensis* infection result from impaired activation of early immunity, such as reduced expression of inflammatory cytokines that are important for Th1-cell development and macrophage effector functions. Moreover, low levels of IFN-γ accompanied by the relative absence of IL-4 are typical of active disease associated with this *Leishmania* species, and protective immunity and healing are mediated by Th1-type cytokines (Calvopina et al. 2006a).

Predominant Th1-type cell responses are associated with IFN-γ-induced macrophage activation, indicating expression of a network of pleiotropic cytokines. In this network IL-12 produced by activated antigen presenting cells (APC)—such as macrophages and dendritic cells (DC)—shapes the basic response (Constantinescu et al. 1998). However, in patients with clinically apparent infections, Th1- and Th2-type responses are not characteristically polarized, as both activating (i.e., IL-12 and IFN-γ) and suppressive cytokines (i.e., IL-4, IL-10, IL-13, and transforming growth factor [TGF]-β) are detected (Kemp et al. 1993; Chtanova and Mackay 2001).

In order to resolve the infection, the protective immune response against CL is dependent on the development of a potent Th1-type response (Sharma and Singh 2009). Generally, IL-12 from activated APC, possibly increased by other cytokines (IL-1α, IL-18, IL-23, and IL-27, as innate mechanism), drives differentiation and proliferation of Th1 cells (Bacellar et al. 2000; Maspi et al. 2016). IFN-γ is produced in the first place by Th1 cells, and to a lower extent by $CD8^+$ T cells and IL-12-activated NK cells, mediating macrophage activation by the production of both superoxide $\left(O_{2^-}\right)$ and NO. It has been suggested that NO is cytotoxic to parasites and appears to be a common mechanism of killing *Leishmania* (Muller et al. 2013; Olekhnovitch and Bousso 2015; Pandya et al. 2016).

It has recently been proposed that $CD8^+$ T cells play three different roles in CL (Novais and Scott 2015): (1) confer protection against the parasite by the production of IFN-γ (Belkaid et al. 2002b; Uzonna et al. 2004), (2) promote an increased pathology during infection when they are cytolytic (Faria et al. 2009; Cardoso et al. 2015), and (3) in primary lesions, produce a small amount of IFN-γ that leads to an exaggerated inflammatory response, which promotes tissue damage, and an exacerbated pathology (Crosby et al. 2014).

In *L. (V.) braziliensis* infections, $CD8^+$ and other cell populations (NK and NKT cells) have an important role as cytotoxic cells leading to the development of the disease or cure. Specially, in a recent publication it was proposed that a contribution to cytotoxic activity can be held mainly by $CD4^{neg}CD8^{neg}$ double-negative (express CD3/TCRα/β or γ/δ receptor but lack CD4/CD8), NKT cells, and $CD4^+$T lymphocytes (Ferraz et al. 2017).

Another fundamental pathway for Th1-mediated resistance to *Leishmania* infection is the Fas/FasL interaction. In this scenario, activated Th1 cells induce apoptotic death in target cells expressing the Fas protein. In turn, infected macrophages upregulate Fas in response to IFN-γ and become susceptible to $CD4^+$ T cell-induced apoptotic death (Potestio et al. 2004; Eidsmo et al. 2005). It is known that the resolution or progression of the disease depends on distinct $CD4^+$ T cells (Th1 or Th2) and their production of regulatory cytokines (IFN-γ and IL-4). Nevertheless, it has been described that IL-4 is able to induce progression of the disease with an additive response in conjunction with IL-13 (Matthews et al. 2000; Noben-Trauth et al. 2003; Ahmadi et al. 2015).

In addition, IL-10 might play a key role in the pathogenesis of leishmaniasis, especially in the downregulation of Th1 response through the suppression of IFN-γ, and inhibition of NO production in infections with *L. amazonensis* (Barroso et al. 2007). It is important to remark that the role of each cytokine may change among infections produced by different *Leishmania* species. For example, *L. mexicana* presents a severe form of the disease as a result of a major production of IL-4 but specially predominance of IL-12 (Valencia-Pacheco et al. 2014). On the contrary, the role of IL-4 in *L. amazonensis* infections is not significant (Carvalho et al. 2016).

Moreover, it has been assessed the role of IL-10 in concomitant immunity in animal models. An endogenous population of $CD4^{+}CD25^{+}$ regulatory T cells has been shown to control the persistence of *L. major* in the skin after healing in resistant C57BL/6 mice. These data exhibit the possibility that more severe nonhealing forms of leishmaniasis, which in humans are associated with the overproduction of IL-10, might be a product of an imbalance in the number and activity of parasite-driven regulatory T cells (Belkaid et al. 2002a).

CL can progress to the metastatic mucocutaneous form (MCL), specifically in infections where species belonging to the *Viannia* subgenus are involved (predominantly *L. braziliensis*, *L. guyanensis*, and *L. panamensis*) (Guerra et al. 2011). In this case, the hyper-inflammatory response caused by the activation of numerous immune cells promotes swelling and destruction of local tissue (Costa et al. 2005; Ronet et al. 2010). Patients exhibit higher levels of pro-inflammatory cytokines (i.e., IFN-γ) and low levels of anti-inflammatory cytokines (i.e., IL-10) (Gomes-Silva et al. 2007). Exacerbated reactions have been observed in strains associated with the *Leishmania* double-stranded RNA (dsRNA) virus (LRV1) (Hartley et al. 2012; Zangger et al. 2014; Parmentier et al. 2016; Adaui et al. 2016; Bourreau et al. 2016; Rossi et al. 2017). This cytoplasmic dsRNA virus of the Totiviridae family has been detected in *L. braziliensis*, *L. guyanensis*, *L. major*, and *L. aethiopica* causing not only CL but also ML and disseminated leishmaniasis. In animal models, the viral dsRNA genome within *L. guyanensis* parasites is recognized by host Toll-like receptor 3 (TLR3) that induces pro-inflammatory cytokines and chemokines, typically IL-6 and TNF-α, which are related to human MCL (Zangger et al. 2014). It has been reported in murine models that a hyper-inflammatory response with increased IFN-β and inflammatory markers (commonly observed in patients with ML lesions) was produced against *L. guyanensis* parasites in exacerbated disease outcomes (Ives et al. 2011).

On the other hand, it has been described that there is a differential cytokine expression between patients in early stages of infections in contrast to elderly lesions. In patients infected with *L. (V.) braziliensis* with recent lesions, it was observed an increased in situ distribution of CD57, IL-10, TNF-α, and IFN-γ (Gomes et al. 2017).

It is also important to add that a leishmaniasis recidiva cutis (LRC) can develop under certain stimulus, such as the immune status of the patient, weak cutaneous delayed-type hypersensitivity response to leishmanin skin test, local trauma, topical corticoids, and the persistence of dormant parasites after a non-resolved infection (Calvopina et al. 2006b).

A brief explanation of immunity of cutaneous leishmaniasis can be seen in Fig. 5.1.

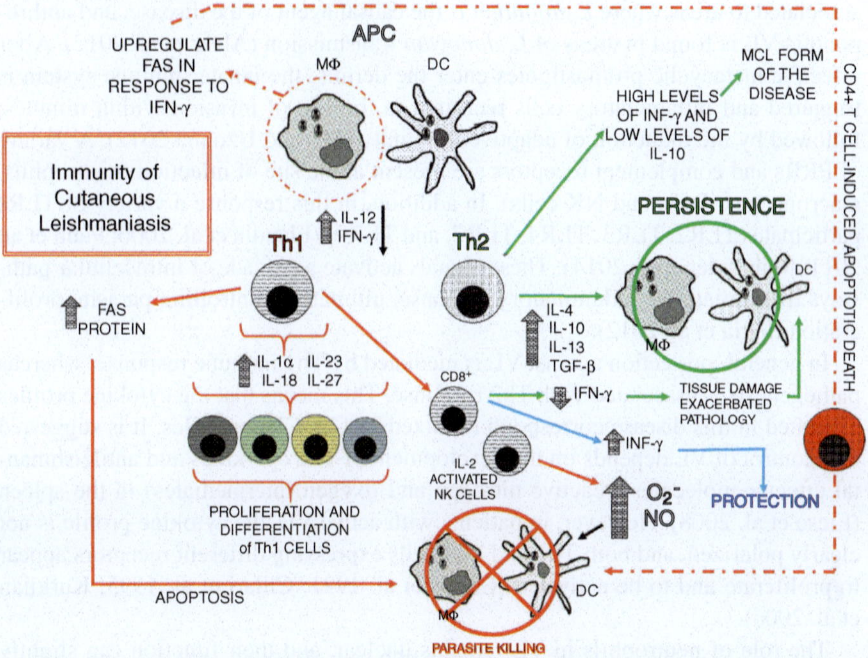

Fig. 5.1 Brief explanation of immunity of cutaneous leishmaniasis. The site infection usually displays an infiltration of activated antigen presenting cells (APC). Predominant Th1 responses are associated with IFN-γ and IL-12 is produced. Generally, IL-12 is increased by other cytokines (IL-1α, IL-18, IL-23, and IL-27, as innate mechanism) and drives differentiation and proliferation of Th1 cells. Th1 cells in the first place, and to a lower extent $CD8^+$ and IL-12-activated NK cells, produce IFN-γ. It plays an important role in mediating macrophage activation by the production of both superoxide $\left(O_{2^-}\right)$ and nitric oxide (NO). Another important process is the Fas/FasL interaction where Th1 cells induce apoptotic death in target cells expressing the Fas protein. In addition, infected macrophages upregulate Fas in response to IFN-γ and become susceptible to $CD4^+$ T cell-induced apoptotic death. $CD8^+$ T cells can play three different roles in CL: (1) confer protection against the parasite by producing IFN-γ, (2) promote an increased pathology during infection when they are cytolytic, and (3) in primary lesions, produce a small amount of IFN-γ that leads to an exaggerated inflammatory response promoting tissue damage and an exacerbated pathology. On the other hand, it has been described that IL-4, IL-10, IL-13, and TGF-β are able to induce progression of the disease. Low levels of IFN-γ accompanied by the relative absence of IL-4 are typical of active disease. CL can progress to MCL by the presence of high levels of IFN-γ and low levels of anti-inflammatory cytokines (i.e., IL-10)

5.4 Immunity of Visceral Leishmaniasis

There are two different types of VL depending on the transmission mode. Zoonotic VL occurs when the parasite is transmitted from an infected animal (reservoir) to a vector and from the vector to a human host. On the contrary, in anthroponotic VL the transmission cycle is restricted to humans, starting with an infected human (reservoir) to a vector and then to another, non-infected human. Zoonotic VL is associated to areas where *L. infantum* is the causal agent of the disease, and anthroponotic VL is found in areas of *L. donovani* transmission (Alvar et al. 2012). After infective metacyclic promastigotes enter the dermis, the innate immune system is triggered and inflammatory cells recruited to the site of invasion within minutes, followed by the induction of adaptive immunity (Liu and Uzonna 2012). A variety of PRRs and complement receptors are present at the site of infection (neutrophils, macrophages, DCs, and NK cells). In addition, in this response a variety of TLRs participate (TLR2, TLR3, TLR4, TLR7, and TLR9) (Flandin et al. 2006; Paun et al. 2011; Majumder et al. 2014). These signals activate a cascade of intracellular pathways that initiate the inflammatory response, ultimately controlling parasite proliferation (Faria et al. 2012).

In general, protection against VL is mediated by Th1 immune response, whereas pathogenesis is associated with Th2 response. This means that the cytokine profiles exhibited in this disease correspond to mixed Th1 and Th2 profiles. It is suggested that control of VL depends on the development of Th1 cytokines and antileishmanial effector molecules (reactive nitrogen and oxygen intermediates) in the spleen (Liese et al. 2008). However, in patients with active VL, the cytokine profile is not clearly polarized, and both Th1 and Th2 cells expressing different receptors appear to proliferate and to be activated (Cillari et al. 1991; Cillari et al. 1995; Kurkjian et al. 2006).

The role of neutrophils in VL remains unclear, and their function can slightly vary among different *Leishmania* species. For example, in *L. donovani* infections, neutrophil depletion at the beginning of the infection leads to an increase of parasite load in the spleen and bone marrow but not in the liver. In turn, splenomegaly delays maturation of hepatic granulomas and decreases inducible nitric oxide synthase (iNOS) expression within granulomas, followed by elevation of IL-4 and IL-10 levels (McFarlane et al. 2008). Even though promastigotes invade DCs and macrophages, TLRs also contribute to this process, having an important role in production of pro-inflammatory cytokines (Srivastava et al. 2012). In contrast, in *L. infantum* infections, it has been proposed that TLR2 signaling plays an important role in immune protection by the activation of DCs which promotes prototypal Th1 and Th17 subsets and CXCL1 to accumulate neutrophils with NO and TNF-α production, leading to the reduction of *L. infantum* uptake capacity (Sacramento et al. 2017).

In this scenario, IL-10 is the major immunosuppressive cytokine, and TGF-β also plays a significant role in disease progression (Wilson et al. 1998). The combined effect of abundant TGF-β stored at extracellular sites during infection and the ability of parasites to activate TGF-β in its local environment lead to high levels of active TGF-β near to the infected macrophage. Macrophages locally activated by TGF-β could, in turn, enhance parasite survival through its effects on innate and adaptive immune responses (Gantt et al. 2003). On the other hand, elevated concentrations of IL-10 and TGF-β in VL patients (mainly infected with *L. donovani*) could predict the development of Post-kala-azar dermal leishmaniasis (PKDL). In those cases, during the devolvement of VL, peripheral blood mononuclear cells (PBMCs) lack in production of IFN-γ. After treatment, the patients recover the systemic health, and PBMCs restart the production of IFN-γ, matching with the development of PKDL due to the persistence of the parasites in the skin (Zijlstra et al. 2003; Zijlstra 2016). Also, higher levels of promastigote surface antigen-2 (PSA-2) and glycoprotein 63 (gp63) related to dermatotropism and a decreased expression of amastigote antigen 2 (A2) which is associated with viscerotropism are observed (Mukhopadhyay et al. 2014).

Parasites of *L. donovani* are known to cause immune-depression, but this mechanism remains unclear. Both $CD4^+$ and $CD8^+$ T cells have been implicated in the resistance and healing capacity against *L. donovani*. Protection is associated with the production of IFN-γ by $CD4^+$ T cells and/or $CD8^+$ lymphocytes (Elloso and Scott 1999). Increased levels of IL-10 (restricted to $CD8^+$T and B cells) induce parasite persistence and chronicity of the disease (Nunes et al. 2005; Kariminia et al. 2005). In addition, IL-27 enhances production of IL-10 and inhibits the secretion of IL-17 by $CD4^+$ T cells (Quirino et al. 2016).

A significantly high level of apoptosis in both monocytes and T lymphocytes from acute VL patients was observed (Potestio et al. 2004). T cells expressing $CD4^+$ phenotype, in particular Th1 (evaluated by chemokine receptor 5, CCR5), were involved in this process based on a CD95-mediated mechanism (Potestio et al. 2004; Mansueto et al. 2007). Evidence that Th1 was more susceptible to apoptosis than Th2-type cells in either an autocrine or paracrine way came from the enhanced expression of CD95 in Th1-type cells from acute VL patients after activation with *Leishmania* antigens (Potestio et al. 2004). Das et al. (1999) observed an increase in apoptosis of Th1-like cells from susceptible hosts. This evidence suggests that the enhancement of Fas and FasL in $CD4^+$ $CCR5^+$ T cells can contribute to the high sensitivity to apoptosis of *Leishmania*-activated specific T cells (Potestio et al. 2004).

A summary of immunity of visceral leishmaniasis can be seen in Fig. 5.2.

5.5 Role of Sandfly Salivary Products in the Initiation of the Inflammatory Process

At the moment of feeding salivary proteins along with parasites are inoculated into the dermis. Sandfly saliva is composed of potent pharmacologically active molecules that act as anticoagulants, antiplatelets, and vasodilators (Kamhawi 2000; Mukbel et al. 2016). Furthermore, they have been shown to have immunomodulatory (Kamhawi 2000) and anti-inflammatory activities (Kamhawi et al. 2000; Kato et al. 2013).

Sandfly saliva can influence the disease outcome in two opposite ways, either by producing an exacerbated infection or by conferring protection against *Leishmania* parasites. Enhancement of infection was observed in different experimental models where saliva induced cellular recruitment to the inflammatory site, inhibited pro-inflammatory cytokines, and deactivated dendritic cells to mobilize regulatory T cells (Belkaid et al. 1998; Kamhawi et al. 2000; Andrade et al. 2005; Andrade et al. 2007). In contrast, immunogenic salivary proteins have been identified to induce a distinct Th1-delayed-type hypersensitivity (DTH) response, predictive of protection

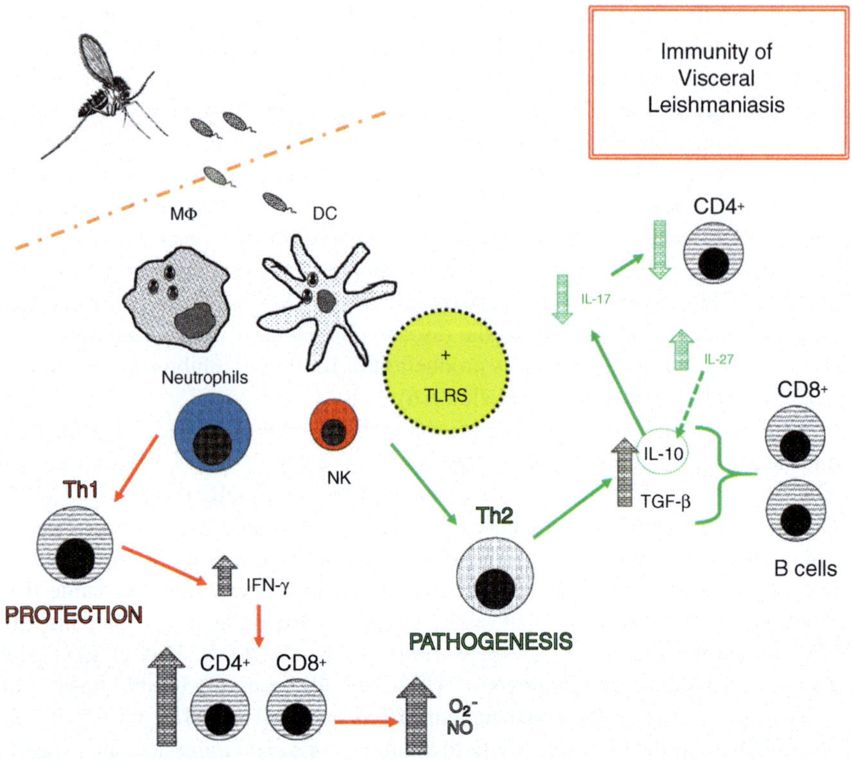

Fig. 5.2 Brief explanation of immunity in visceral leishmaniasis. After infective metacyclic promastigotes enter the dermis, the innate immune system is triggered, and inflammatory cells are recruited to the site of invasion within minutes. A variety of cells (neutrophils, macrophages, DCs, and NK cells) and TLRs (TLR2, TLR3, TLR4, TLR7, and TLR9) are present in the site of infection. Two pathways can be activated: protection against VL mediated by Th1 immune response and pathogenesis which is associated with Th2 response. Protection is associated with the production of IFN-γ by $CD4^+$ T cells and/or $CD8^+$ lymphocytes. In the last instance, control of VL depends on the development of antileishmanial effector molecules (reactive nitrogen and oxygen intermediates). Progression of the disease is enhanced by IL-10 and TGF-β production. Increased levels of IL-10 (restricted to T $CD8^+$ and B cells) induce parasite persistence and chronicity of the disease, while IL-27 enhances production of IL-10 and inhibits secretion of IL-17 by $CD4^+$ T cells

from leishmaniases (Oliveira et al. 2006; Oliveira et al. 2008; Gomes et al. 2008; Collin et al. 2009; Kamhawi et al. 2014; Gomes et al. 2016). On the other hand, it has also been demonstrated that saliva of *Lutzomyia longipalpis* could exacerbate *Leishmania major* infection (Titus and Ribeiro 1988). In addition, an experiment where salivary glands of *Lu. longipalpis* were sonicated (SGS) and exposed to mice demonstrated that SGS was able to modulate PGE2/LTB4 axis. This finding seems to point toward the importance of salivary glands products as a critical factor driving immune evasion of the parasite, which in turn represents an important mechanism for establishment of infection (Araujo-Santos et al. 2014).

Recently, Mukbel et al. (2016) compared the degree of PBMC cell proliferation after exposing to *Phlebotomus papatasi* salivary gland homogenate (SGH). They found that PBMC cells of donors from sandfly-infested areas had a higher proliferation rate in comparison to donors from noninfested areas. Individuals that exhibit humoral immunity to various salivary proteins had lower cell proliferative response. In endemic areas with leishmaniases, patients did not show a cellular proliferative response regardless of antibody reactivity or SGH concentration. On the contrary, individuals with low humoral immune response living in sandfly-infested areas without leishmaniases present a strong cell-mediated response after exposure to sandfly antigens. In this context, these results might explain the high-risk incidence of leishmaniases in endemic areas despite the presence of a strong humoral immune response against sandfly salivary proteins (Barral et al. 2000; Gomes et al. 2002; Rohousova et al. 2011).

5.6 Autoimmune Manifestations

VL is associated with clinical and laboratory autoimmune manifestations, and, for this reason, infected patients could be initially misdiagnosed as having an autoimmune disease (Nozzi et al. 2014). Some manifestations of VL are related to the host's immune response that mimics autoimmune diseases. VL infection in patients with high parasitic load stimulates Th2 lymphocytes to synthesize IL-4 and IL-10 that in turn activate B cells to produce a wide spectrum of antibodies. In addition, *Leishmania* parasites cause tissue destruction, and autoreactivity can also be stimulated by the release of self-antigens. On the other hand, polyclonal B cell activation may not be the only reason that leads to the synthesis of autoantibodies during VL infection. It is likely that a molecular mimicry process between *Leishmania* antigens and ribonucleoproteins plays an important role in autoimmune manifestations of VL (Liberopoulos et al. 2013). These aspects are significant in cases where patients are misdiagnosed with an autoimmune disease (especially systemic lupus erythematosus) and treated with immunosuppressive drugs that would exacerbate the infection leading, in the worst case, to death (Santana et al. 2015).

Conclusions

The immune response against *Leishmania* is a multifactorial process that takes several components from the bite of the sandfly to the establishment of the disease. As mentioned earlier, even the saliva of the vector can either exacerbate or confer protection against *Leishmania*. Potent anticoagulant molecules, antiplatelets, and vasodilators are released into the bite wound along with the parasites. In response, an immunomodulatory and anti-inflammatory reaction is developed in the host. Neutrophils, monocytes or macrophages, NK cells, dendritic cells, PPRs, and soluble products (complement and cytokines) arrive to the site of infection to interact with *Leishmania* parasites, unleashing the innate and adaptive immune response. If an inflammatory Th1 response develops first, it will determine a protective immunity against leishmaniases, while disease progression is associated with an anti-inflammatory Th2 response. When neutrophils reach the site of infection as part of the first line of defense, they can generate a potent oxidative burst, releasing toxic mediators into the parasitophorous vacuole that leads to parasite death. They can also serve as intermediate host cells, acting as "Trojan horses" that allow parasites to silently enter macrophages, thereby avoiding cell activation. After recognizing the parasites, macrophages produce TNF-α and IL-1 that promote inflammation by inducing the expression of adhesion molecules (selectin and integrin ligands) on the endothelial surface. In addition, the innate mechanisms with the primordial secretion of IL-12 lead to the parallel induction of cell-mediated immunity that activates specific $CD4^+$ and $CD8^+$ T cells. Finally, besides the genetic of the infecting *Leishmania* spp., the host releases a battery of adhesion molecules and chemokine activation mechanisms, in conjunction with monocytes that direct local inflammatory responses, which leads to cutaneous or visceral symptoms of the disease.

Acknowledgements We are grateful to Dr. Carolina Davies and Prof. Hirotomo Kato for their critical reading of the manuscript and valuable comments. We also thank Prof. Yoshihisa Hashiguchi for his encouragement through the research on leishmaniases.

References

Adaui V, Lye LF, Akopyants NS, Zimic M, Llanos-Cuentas A, Garcia L, Maes I, De Doncker S, Dobson DE, Arevalo J, Dujardin JC, Beverley SM. Association of the endobiont double-stranded RNA virus LRV1 with treatment failure for human leishmaniasis caused by Leishmania braziliensis in Peru and Bolivia. J Infect Dis. 2016;213(1):112–21. https://doi.org/10.1093/infdis/jiv354jiv354.

Ahmadi A, Hajilooi M, Solgi G, Abasi M, Bazmani A, Matini M, Sardarian K. Interleukin-4 receptor alpha T1432C and A1652G polymorphisms are associated with risk of visceral leishmaniasis. Adv Biomed Res. 2015;4:195. https://doi.org/10.4103/2277-9175.166131ABR-4-195.

Alvar J, Velez ID, Bern C, Herrero M, Desjeux P, Cano J, Jannin J, den Boer M. Leishmaniasis worldwide and global estimates of its incidence. PLoS One. 2012;7(5):e35671. https://doi.org/10.1371/journal.pone.0035671PONE-D-11-24894.

Andrade BB, de Oliveira CI, Brodskyn CI, Barral A, Barral-Netto M. Role of sand fly saliva in human and experimental leishmaniasis: current insights. Scand J Immunol. 2007;66(2–3):122–7. https://doi.org/10.1111/j.1365-3083.2007.01964.x.

Andrade BB, Teixeira CR, Barral A, Barral-Netto M. Haematophagous arthropod saliva and host defense system: a tale of tear and blood. An Acad Bras Cienc. 2005;77(4):665–93. https://doi.org/10.1590/S0001-37652005000400008.

Araujo-Santos T, Prates DB, Franca-Costa J, Luz NF, Andrade BB, Miranda JC, Brodskyn CI, Barral A, Bozza PT, Borges VM. Prostaglandin E2/leukotriene B4 balance induced by *Lutzomyia longipalpis* saliva favors *Leishmania infantum* infection. Parasit Vectors. 2014;7:601. https://doi.org/10.1186/s13071-014-0601-8s13071-014-0601-8.

Bacellar O, D'Oliveira A Jr, Jeronimo S, Carvalho EM. IL-10 and IL-12 are the main regulatory cytokines in visceral leishmaniasis. Cytokine. 2000;12(8):1228–31. https://doi.org/10.1006/cyto.2000.0694S1043-4666(00)90694-X.

Barral A, Honda E, Caldas A, Costa J, Vinhas V, Rowton ED, Valenzuela JG, Charlab R, Barral-Netto M, Ribeiro JM. Human immune response to sand fly salivary gland antigens: a useful epidemiological marker? Am J Trop Med Hyg. 2000;62(6):740–5.

Barroso PA, Marco JD, Calvopina M, Kato H, Korenaga M, Hashiguchi Y. A trial of immunotherapy against *Leishmania amazonensis* infection in vitro and in vivo with Z-100, a polysaccharide obtained from *Mycobacterium tuberculosis*, alone or combined with meglumine antimoniate. J Antimicrob Chemother. 2007;59(6):1123–9. https://doi.org/10.1093/jac/dkm079.

Belkaid Y, Kamhawi S, Modi G, Valenzuela J, Noben-Trauth N, Rowton E, Ribeiro J, Sacks DL. Development of a natural model of cutaneous leishmaniasis: powerful effects of vector saliva and saliva preexposure on the long-term outcome of *Leishmania major* infection in the mouse ear dermis. J Exp Med. 1998;188(10):1941–53.

Belkaid Y, Piccirillo CA, Mendez S, Shevach EM, Sacks DL. CD4+ CD25+ regulatory T cells control *Leishmania major* persistence and immunity. Nature. 2002a;420(6915):502–7. https://doi.org/10.1038/nature01152nature01152.

Belkaid Y, Von Stebut E, Mendez S, Lira R, Caler E, Bertholet S, Udey MC, Sacks D. CD8+ T cells are required for primary immunity in C57BL/6 mice following low-dose, intradermal challenge with *Leishmania major*. J Immunol. 2002b;168(8):3992–4000.

Bourreau E, Ginouves M, Prevot G, Hartley MA, Gangneux JP, Robert-Gangneux F, Dufour J, Sainte-Marie D, Bertolotti A, Pratlong F, Martin R, Schutz F, Couppie P, Fasel N, Ronet C. Presence of Leishmania RNA virus 1 in *Leishmania guyanensis* increases the risk of first-line treatment failure and symptomatic relapse. J Infect Dis. 2016;213(1):105–11. https://doi.org/10.1093/infdis/jiv355jiv355.

Calvopina M, Barroso PA, Marco JD, Korenaga M, Cooper PJ, Nonaka S, Hashiguchi Y. Efficacy of vaccination with a combination of Leishmania amastigote antigens and the lipid A-analogue ONO-4007 for immunoprophylaxis and immunotherapy against *Leishmania amazonensis* infection in a murine model of new world cutaneous leishmaniasis. Vaccine. 2006a;24(27–28):5645–52. https://doi.org/10.1016/j.vaccine.2006.03.023.

Calvopina M, Uezato H, Gomez EA, Korenaga M, Nonaka S, Hashiguchi Y. Leishmaniasis recidiva cutis due to *Leishmania (Viannia) panamensis* in subtropical Ecuador: isoenzymatic characterization. Int J Dermatol. 2006b;45(2):116–20. https://doi.org/10.1111/j.1365-4632.2004.02518.x.

Cardoso TM, Machado A, Costa DL, Carvalho LP, Queiroz A, Machado P, Scott P, Carvalho EM, Bacellar O. Protective and pathological functions of CD8+ T cells in *Leishmania braziliensis* infection. Infect Immun. 2015;83(3):898–906. https://doi.org/10.1128/IAI.02404-14IAI.02404-14.

Carlsen ED, Liang Y, Shelite TR, Walker DH, Melby PC, Soong L. Permissive and protective roles for neutrophils in leishmaniasis. Clin Exp Immunol. 2015;182(2):109–18. https://doi.org/10.1111/cei.12674.

Carvalho AK, Carvalho K, Passero LF, Sousa MG, da Matta VL, Gomes CM, Corbett CE, Kallas GE, Silveira FT, Laurenti MD. Differential recruitment of dendritic cells subsets to lymph nodes correlates with a protective or permissive T-cell response during *Leishmania (Viannia) braziliensis* or *Leishmania (Leishmania) amazonensis* infection. Mediators Inflamm. 2016;2016:7068287. https://doi.org/10.1155/2016/7068287.

Cecilio P, Perez-Cabezas B, Santarem N, Maciel J, Rodrigues V, Cordeiro da Silva A. Deception and manipulation: the arms of leishmania, a successful parasite. Front Immunol. 2014;5:480. https://doi.org/10.3389/fimmu.2014.00480.

Cillari E, Milano S, Dieli M, Maltese E, Di Rosa S, Mansueto S, Salerno A, Liew FY. Reduction in the number of UCHL-1+ cells and IL-2 production in the peripheral blood of patients with visceral leishmaniasis. J Immunol. 1991;146(3):1026–30.

Cillari E, Vitale G, Arcoleo F, D'Agostino P, Mocciaro C, Gambino G, Malta R, Stassi G, Giordano C, Milano S, et al. In vivo and in vitro cytokine profiles and mononuclear cell subsets in Sicilian patients with active visceral leishmaniasis. Cytokine. 1995;7(7):740–5. https://doi.org/10.1006/cyto.1995.0088.

Collin N, Gomes R, Teixeira C, Cheng L, Laughinghouse A, Ward JM, Elnaiem DE, Fischer L, Valenzuela JG, Kamhawi S. Sand fly salivary proteins induce strong cellular immunity in a natural reservoir of visceral leishmaniasis with adverse consequences for Leishmania. PLoS Pathog. 2009;5(5):e1000441. https://doi.org/10.1371/journal.ppat.1000441.

Conceicao J, Davis R, Carneiro PP, Giudice A, Muniz AC, Wilson ME, Carvalho EM, Bacellar O. Characterization of neutrophil function in human cutaneous leishmaniasis caused by *Leishmania braziliensis*. PLoS Negl Trop Dis. 2016;10(5):e0004715. https://doi.org/10.1371/journal.pntd.0004715PNTD-D-15-01257.

Constantinescu CS, Hondowicz BD, Elloso MM, Wysocka M, Trinchieri G, Scott P. The role of IL-12 in the maintenance of an established Th1 immune response in experimental leishmaniasis. Eur J Immunol. 1998;28(7):2227–33. https://doi.org/10.1002/(SICI)1521-4141(199807)28:07<2227::AID-IMMU2227>3.0.CO;2-N.

Costa AA, Saldanha AC, Leite BM, Ramos B, Junior IA, Noronha AL, Bafica A, Barral A, Corbett CE, Costa JM. Imaging exams of bone lesions in patients with diffuse cutaneous leishmaniasis (DCL). Acta Trop. 2005;96(1):9–15. https://doi.org/10.1016/j.actatropica.2005.06.016.

Crosby EJ, Goldschmidt MH, Wherry EJ, Scott P. Engagement of NKG2D on bystander memory CD8 T cells promotes increased immunopathology following *Leishmania major* infection. PLoS Pathog. 2014;10(2):e1003970. https://doi.org/10.1371/journal.ppat.1003970PPATHOGENS-D-13-01933.

Chtanova T, Mackay CR. T cell effector subsets: extending the Th1/Th2 paradigm. Adv Immunol. 2001;78:233–66.

Das G, Vohra H, Rao K, Saha B, Mishra GC (1999) Leishmania donovani infection of a susceptible host results in CD4+ T-cell apoptosis and decreased Th1 cytokine production Scand J Immunol 49(3):307–10. https://doi.org/10.1046/j.1365-3083.1999.00486.x.

de Menezes JP, Saraiva EM, da Rocha-Azevedo B. The site of the bite: Leishmania interaction with macrophages, neutrophils and the extracellular matrix in the dermis. Parasit Vectors. 2016;9(1):264. https://doi.org/10.1186/s13071-016-1540-310.1186/s13071-016-1540-3.

Eidsmo L, Nylen S, Khamesipour A, Hedblad MA, Chiodi F, Akuffo H. The contribution of the Fas/FasL apoptotic pathway in ulcer formation during *Leishmania major*-induced cutaneous Leishmaniasis. Am J Pathol. 2005;166(4):1099–108. https://doi.org/10.1016/S0002-9440(10)62330-9.

Elloso MM, Scott P. Expression and contribution of B7-1 (CD80) and B7-2 (CD86) in the early immune response to Leishmania major infection. J Immunol. 1999;162(11):6708–15.

Faria DR, Souza PE, Duraes FV, Carvalho EM, Gollob KJ, Machado PR, Dutra WO. Recruitment of CD8(+) T cells expressing granzyme A is associated with lesion progression in human cutaneous leishmaniasis. Parasite Immunol. 2009;31(8):432–9. https://doi.org/10.1111/j.1365-3024.2009.01125.xPIM1125.

Faria MS, Reis FC, Lima AP. Toll-like receptors in leishmania infections: guardians or promoters? J Parasitol Res. 2012;2012:930257. https://doi.org/10.1155/2012/930257.

Ferraz R, Cunha CF, Pimentel MIF, Lyra MR, Pereira-Da-Silva T, Schubach AO, Da-Cruz AM, Bertho AL. $CD3^{+}CD4^{neg}CD8^{neg}$ (double negative) T lymphocytes and NKT cells as the main cytotoxic-related-$CD107a^{+}$ cells in lesions of cutaneous leishmaniasis caused by *Leishmania (Viannia) braziliensis*. Parasites Vectors. 2017;10:219. https://doi.org/10.1186/s13071-017-2152-2.

Flandin JF, Chano F, Descoteaux A. RNA interference reveals a role for TLR2 and TLR3 in the recognition of *Leishmania donovani* promastigotes by interferon-gamma-primed macrophages. Eur J Immunol. 2006;36(2):411–20. https://doi.org/10.1002/eji.200535079.

Gannavaram S, Bhattacharya P, Ismail N, Kaul A, Singh R, Nakhasi HL. Modulation of innate immune mechanisms to enhance Leishmania vaccine-induced immunity: role of coinhibitory molecules. Front Immunol. 2016;7:187. https://doi.org/10.3389/fimmu.2016.00187.

Gantt KR, Schultz-Cherry S, Rodriguez N, Jeronimo SM, Nascimento ET, Goldman TL, Recker TJ, Miller MA, Wilson ME. Activation of TGF-beta by Leishmania chagasi: importance for parasite survival in macrophages. J Immunol. 2003;170(5):2613–20.

Gomes AHS, Martines RB, Kanamura CT, Barbo MLP, Iglezias SD, Lauletta Lindoso JA, Pereira-Chioccola VL. American cutaneous leishmaniasis: in situ immune response of patients with recent and late lesions. Parasite Immunol. 2017;39(4):e12423. https://doi.org/10.1111/pim.12423.

Gomes-Silva A, de Cassia Bittar R, Dos Santos Nogueira R, Amato VS, da Silva Mattos M, Oliveira-Neto MP, Coutinho SG, Da-Cruz AM. Can interferon-gamma and interleukin-10 balance be associated with severity of human *Leishmania (Viannia) braziliensis* infection? Clin Exp Immunol. 2007;149(3):440–4. https://doi.org/10.1111/j.1365-2249.2007.03436.x.

Gomes R, Teixeira C, Teixeira MJ, Oliveira F, Menezes MJ, Silva C, de Oliveira CI, Miranda JC, Elnaiem DE, Kamhawi S, Valenzuela JG, Brodskyn CI. Immunity to a salivary protein of a sand fly vector protects against the fatal outcome of visceral leishmaniasis in a hamster model. Proc Natl Acad Sci U S A. 2008;105(22):7845–50. https://doi.org/10.1073/pnas.0712153105pnas.0712153105.

Gomes R, Cavalcanti K, Teixeira C, Carvalho AM, Mattos PS, Cristal JR, Muniz AC, Miranda JC, de Oliveira CI, Barral A. Immunity to Lutzomyia whitmani saliva protects against experimental Leishmania braziliensis infection. PLoS Negl Trop Dis. 2016;10(11):e0005078. https://doi.org/10.1371/journal.pntd.0005078.

Gomes RB, Brodskyn C, de Oliveira CI, Costa J, Miranda JC, Caldas A, Valenzuela JG, Barral-Netto M, Barral A. Seroconversion against *Lutzomyia longipalpis* saliva concurrent with the development of anti-*Leishmania chagasi* delayed-type hypersensitivity. J Infect Dis. 2002;186(10):1530–4. https://doi.org/10.1086/344733.

Guerra JA, Prestes SR, Silveira H, Coelho LI, Gama P, Moura A, Amato V, Barbosa M, Ferreira LC. Mucosal Leishmaniasis caused by *Leishmania (Viannia) braziliensis* and *Leishmania (Viannia) guyanensis* in the Brazilian Amazon. PLoS Negl Trop Dis. 2011;5(3):e980. https://doi.org/10.1371/journal.pntd.0000980.

Guimaraes-Costa AB, Nascimento MT, Froment GS, Soares RP, Morgado FN, Conceicao-Silva F, Saraiva EM. *Leishmania amazonensis* promastigotes induce and are killed by neutrophil extracellular traps. Proc Natl Acad Sci U S A. 2009;106(16):6748–53. https://doi.org/10.1073/pnas.0900226106pnas.0900226106.

Hailu A, van der Poll T, Berhe N, Kager PA. Elevated plasma levels of interferon (IFN)-gamma, IFN-gamma inducing cytokines, and IFN-gamma inducible CXC chemokines in visceral leishmaniasis. Am J Trop Med Hyg. 2004;71(5):561–7. https://doi.org/10.4269/ajtmh.2004.71.561.

Hartley MA, Ronet C, Zangger H, Beverley SM, Fasel N. Leishmania RNA virus: when the host pays the toll. Front Cell Infect Microbiol. 2012;2:99. https://doi.org/10.3389/fcimb.2012.00099.

Hide M, Marion E, Pomares C, Fisa R, Marty P, Banuls AL. Parasitic genotypes appear to differ in leishmaniasis patients compared with asymptomatic related carriers. Int J Parasitol. 2013;43(5):389–97. https://doi.org/10.1016/j.ijpara.2012.12.008S0020-7519(13)00034-9.

Hsieh CS, Macatonia SE, Tripp CS, Wolf SF, O'Garra A, Murphy KM. Pillars article: development of TH1 CD4+ T cells through IL-12 produced by Listeria-induced macrophages. 1993. Science 260(5107): 547–549. J Immunol. 2008;181(7):4437–9.

Ives A, Ronet C, Prevel F, Ruzzante G, Fuertes-Marraco S, Schutz F, Zangger H, Revaz-Breton M, Lye LF, Hickerson SM, Beverley SM, Acha-Orbea H, Launois P, Fasel N, Masina S. Leishmania RNA virus controls the severity of mucocutaneous leishmaniasis. Science. 2011;331(6018):775–8. https://doi.org/10.1126/science.1199326331/6018/775.

John B, Hunter CA. Immunology. Neutrophil soldiers or Trojan horses? Science. 2008;321(5891):917–8. https://doi.org/10.1126/science.1162914321/5891/917.

Kamhawi S. The biological and immunomodulatory properties of sand fly saliva and its role in the establishment of Leishmania infections. Microbes Infect. 2000;2(14):1765–73. https://doi.org/10.1016/S1286-4579(00)01331-9.

Kamhawi S, Aslan H, Valenzuela JG. Vector saliva in vaccines for visceral leishmaniasis: a brief encounter of high consequence? Front Public Health. 2014;2:99. https://doi.org/10.3389/fpubh.2014.00099.

Kamhawi S, Belkaid Y, Modi G, Rowton E, Sacks D. Protection against cutaneous leishmaniasis resulting from bites of uninfected sand flies. Science. 2000;290(5495):1351–4. https://doi.org/10.1126/science.290.5495.1351.

Kane MM, Mosser DM. Leishmania parasites and their ploys to disrupt macrophage activation. Curr Opin Hematol. 2000;7(1):26–31.

Kariminia A, Bourreau E, Pascalis H, Couppie P, Sainte-Marie D, Tacchini-Cottier F, Launois P. Transforming growth factor beta 1 production by CD4+ CD25+ regulatory T cells in peripheral blood mononuclear cells from healthy subjects stimulated with *Leishmania guyanensis*. Infect Immun. 2005;73(9):5908–14. https://doi.org/10.1128/IAI.73.9.5908-5914.2005.

Kato H, Jochim RC, Gomez EA, Uezato H, Mimori T, Korenaga M, Sakurai T, Katakura K, Valenzuela JG, Hashiguchi Y. Analysis of salivary gland transcripts of the sand fly *Lutzomyia ayacuchensis*, a vector of Andean-type cutaneous leishmaniasis. Infect Genet Evol. 2013;13:56–66. https://doi.org/10.1016/j.meegid.2012.08.024S1567-1348(12)00295-X.

Kemp M, Kurtzhals JA, Bendtzen K, Poulsen LK, Hansen MB, Koech DK, Kharazmi A, Theander TG. *Leishmania donovani*-reactive Th1- and Th2-like T-cell clones from individuals who have recovered from visceral leishmaniasis. Infect Immun. 1993;61(3):1069–73.

Kevric I, Cappel MA, Keeling JH. New world and old world Leishmania infections: a practical review. Dermatol Clin. 2015;33(3):579–93. https://doi.org/10.1016/j.det.2015.03.018S0733-8635(15)00031-5.

Kolodziej H, Kiderlen AF. Antileishmanial activity and immune modulatory effects of tannins and related compounds on Leishmania parasitised RAW 264.7 cells. Phytochemistry. 2005;66(17):2056–71. https://doi.org/10.1016/j.phytochem.2005.01.011.

Kostka SL, Knop J, Konur A, Udey MC, von Stebut E. Distinct roles for IL-1 receptor type I signaling in early versus established *Leishmania major* infections. J Invest Dermatol. 2006;126(7):1582–9. https://doi.org/10.1038/sj.jid.5700309.

Kurkjian KM, Mahmutovic AJ, Kellar KL, Haque R, Bern C, Secor WE. Multiplex analysis of circulating cytokines in the sera of patients with different clinical forms of visceral leishmaniasis. Cytometry A. 2006;69(5):353–8. https://doi.org/10.1002/cyto.a.20256.

Laskay T, van Zandbergen G, Solbach W. Neutrophil granulocytes as host cells and transport vehicles for intracellular pathogens: apoptosis as infection-promoting factor. Immunobiology. 2008;213(3–4):183–91. https://doi.org/10.1016/j.imbio.2007.11.010S0171-2985(07)00156-8.

le Fichoux Y, Quaranta JF, Aufeuvre JP, Lelievre A, Marty P, Suffia I, Rousseau D, Kubar J. Occurrence of *Leishmania infantum* parasitemia in asymptomatic blood donors living in an area of endemicity in southern France. J Clin Microbiol. 1999;37(6):1953–7.

Ley K, Laudanna C, Cybulsky MI, Nourshargh S. Getting to the site of inflammation: the leukocyte adhesion cascade updated. Nat Rev Immunol. 2007;7(9):678–89. https://doi.org/10.1038/nri2156.

Liberopoulos E, Kei A, Apostolou F, Elisaf M. Autoimmune manifestations in patients with visceral leishmaniasis. J Microbiol Immunol Infect. 2013;46(4):302–5. https://doi.org/10.1016/j.jmii.2012.01.016S1684-1182(12)00033-3.

Liese J, Schleicher U, Bogdan C. The innate immune response against Leishmania parasites. Immunobiology. 2008;213(3–4):377–87. https://doi.org/10.1016/j.imbio.2007.12.005S0171-2985(08)00003-X.

Liu D, Uzonna JE. The early interaction of Leishmania with macrophages and dendritic cells and its influence on the host immune response. Front Cell Infect Microbiol. 2012;2:83. https://doi.org/10.3389/fcimb.2012.00083.

Lo SK, Bovis L, Matura R, Zhu B, He S, Lum H, Turco SJ, Ho JL. Leishmania lipophosphoglycan reduces monocyte transendothelial migration: modulation of cell adhesion molecules, intercellular junctional proteins, and chemoattractants. J Immunol. 1998;160(4):1857–65.

Majumder S, Bhattacharjee A, Paul Chowdhury B, Bhattacharyya Majumdar S, Majumdar S. Antigen-pulsed CpG-ODN-activated dendritic cells induce host-protective immune response by regulating the T regulatory cell functioning in Leishmania donovani-infected mice: critical role of CXCL10. Front Immunol. 2014;5:261. https://doi.org/10.3389/fimmu.2014.00261.

Mansueto P, Vitale G, Di Lorenzo G, Rini GB, Mansueto S, Cillari E. Immunopathology of leishmaniasis: an update. Int J Immunopathol Pharmacol. 2007;20(3):435–45. https://doi.org/10.1177/039463200702000302.

Marco JD, Barroso PA, Locatelli FM, Cajal SP, Hoyos CL, Nevot MC, Lauthier JJ, Tomasini N, Juarez M, Estevez JO, Korenaga M, Nasser JR, Hashiguchi Y, Ruybal P. Multilocus sequence typing approach for a broader range of species of Leishmania genus: describing parasite diversity in Argentina. Infect Genet Evol. 2015;30:308–17. https://doi.org/10.1016/j.meegid.2014.12.031S1567-1348(14)00487-0.

Maspi N, Ghaffarifar F, Sharifi Z, Dalimi A. Codelivery of DNA vaccination encoding LeIF gene and IL-12 increases protection against *Leishmania major* infection in BALB/c mice. Parasite Immunol. 2016;38(4):228–35. https://doi.org/10.1111/pim.12310.

Matthews DJ, Emson CL, McKenzie GJ, Jolin HE, Blackwell JM, McKenzie AN. IL-13 is a susceptibility factor for *Leishmania major* infection. J Immunol. 2000;164(3):1458–62. https://doi.org/10.4049/jimmunol.164.3.1458.

McFarlane E, Perez C, Charmoy M, Allenbach C, Carter KC, Alexander J, Tacchini-Cottier F. Neutrophils contribute to development of a protective immune response during onset of infection with *Leishmania donovani*. Infect Immun. 2008;76(2):532–41. https://doi.org/10.1128/IAI.01388-07.

McMahon-Pratt D, Alexander J. Does the *Leishmania major* paradigm of pathogenesis and protection hold for new world cutaneous leishmaniases or the visceral disease? Immunol Rev. 2004;201:206–24. https://doi.org/10.1111/j.0105-2896.2004.00190.xIMR190.

McSorley S, Proudfoot L, O'Donnell CA, Liew FY. Immunology of murine leishmaniasis. Clin Dermatol. 1996;14(5):451–64. https://doi.org/10.1016/0738-081X(96)00037-5.

Mollinedo F, Janssen H, de la Iglesia-Vicente J, Villa-Pulgarin JA, Calafat J. Selective fusion of azurophilic granules with Leishmania-containing phagosomes in human neutrophils. J Biol Chem. 2010;285(45):34528–36. https://doi.org/10.1074/jbc.M110.125302M110.125302.

Mukbel RM, Khasharmeh RH, Hijjawi NS, Khalifeh MS, Hatmal MM, McDowell MA. Human immune response to salivary proteins of wild-caught *Phlebotomus papatasi*. Parasitol Res. 2016. https://doi.org/10.1007/s00436-016-5094-210.1007/s00436-016-5094-2.

Mukhopadhyay D, Dalton JE, Kaye PM, Chatterjee M. Post kala-azar dermal leishmaniasis: an unresolved mystery. Trends Parasitol. 2014;30(2):65–74. https://doi.org/10.1016/j.pt.2013.12.004.

Muller AJ, Aeschlimann S, Olekhnovitch R, Dacher M, Spath GF, Bousso P. Photoconvertible pathogen labeling reveals nitric oxide control of *Leishmania major* infection in vivo via dampening of parasite metabolism. Cell Host Microbe. 2013;14(4):460–7. https://doi.org/10.1016/j.chom.2013.09.008S1931-3128(13)00328-4.

Nandan D, Camargo de Oliveira C, Moeenrezakhanlou A, Lopez M, Silverman JM, Subek J, Reiner NE. Myeloid cell IL-10 production in response to leishmania involves inactivation of glycogen synthase kinase-3beta downstream of phosphatidylinositol-3 kinase. J Immunol. 2012;188(1):367–78. https://doi.org/10.4049/jimmunol.1100076jimmunol.1100076.

Noben-Trauth N, Lira R, Nagase H, Paul WE, Sacks DL. The relative contribution of IL-4 receptor signaling and IL-10 to susceptibility to *Leishmania major*. J Immunol. 2003;170(10):5152–8.

Novais FO, Scott P. CD8+ T cells in cutaneous leishmaniasis: the good, the bad, and the ugly. Semin Immunopathol. 2015;37(3):251–9. https://doi.org/10.1007/s00281-015-0475-7.

Nozzi M, Del Torto M, Chiarelli F, Breda L. Leishmaniasis and autoimmune diseases in pediatric age. Cell Immunol. 2014;292(1–2):9–13. https://doi.org/10.1016/j.cellimm.2014.08.004S0008-8749(14)00129-4.

Nunes MP, Cysne-Finkelstein L, Monteiro BC, de Souza DM, Gomes NA, DosReis GA. CD40 signaling induces reciprocal outcomes in Leishmania-infected macrophages; roles of host genotype and cytokine milieu. Microbes Infect. 2005;7(1):78–85. https://doi.org/10.1016/j.micinf.2004.08.022.

Oghumu S, Lezama-Davila CM, Isaac-Marquez AP, Satoskar AR. Role of chemokines in regulation of immunity against leishmaniasis. Exp Parasitol. 2010;126(3):389–96. https://doi.org/10.1016/j.exppara.2010.02.010S0014-4894(10)00065-2.

Olekhnovitch R, Bousso P. Induction, propagation, and activity of host nitric oxide: lessons from Leishmania infection. Trends Parasitol. 2015;31(12):653–64. https://doi.org/10.1016/j.pt.2015.08.001S1471-4922(15)00172-5.

Oliveira F, Kamhawi S, Seitz AE, Pham VM, Guigal PM, Fischer L, Ward J, Valenzuela JG. From transcriptome to immunome: identification of DTH inducing proteins from a Phlebotomus ariasi salivary gland cDNA library. Vaccine. 2006;24(3):374–90. https://doi.org/10.1016/j.vaccine.2005.07.085.

Oliveira F, Lawyer PG, Kamhawi S, Valenzuela JG. Immunity to distinct sand fly salivary proteins primes the anti-Leishmania immune response towards protection or exacerbation of disease. PLoS Negl Trop Dis. 2008;2(4):e226. https://doi.org/10.1371/journal.pntd.0000226.

Pandya S, Verma RK, Khare P, Tiwari B, Srinivasarao DA, Dube A, Goyal N, Misra A. Supplementation of host response by targeting nitric oxide to the macrophage cytosol is efficacious in the hamster model of visceral leishmaniasis and adds to efficacy of amphotericin B. Int J Parasitol Drugs Drug Resist. 2016;6(2):125–32. https://doi.org/10.1016/j.ijpddr.2016.01.001.

Paniz Mondolfi AE, Duffey GB, Horton LE, Tirado M, Reyes Jaimes O, Perez-Alvarez A, Zerpa O. Intermediate/borderline disseminated cutaneous leishmaniasis. Int J Dermatol. 2013;52(4):446–55. https://doi.org/10.1111/j.1365-4632.2012.05709.x.

Parmentier L, Cusini A, Muller N, Zangger H, Hartley MA, Desponds C, Castiglioni P, Dubach P, Ronet C, Beverley SM, Fasel N. Severe cutaneous leishmaniasis in a human immunodeficiency virus patient Coinfected with *Leishmania braziliensis* and its Endosymbiotic virus. Am J Trop Med Hyg. 2016;94(4):840–3. https://doi.org/10.4269/ajtmh.15-0803ajtmh.15-0803.

Paun A, Bankoti R, Joshi T, Pitha PM, Stager S. Critical role of IRF-5 in the development of T helper 1 responses to *Leishmania donovani* infection. PLoS Pathog. 2011;7(1):e1001246. https://doi.org/10.1371/journal.ppat.1001246.

Peters NC, Egen JG, Secundino N, Debrabant A, Kimblin N, Kamhawi S, Lawyer P, Fay MP, Germain RN, Sacks D. In vivo imaging reveals an essential role for neutrophils in leishmaniasis transmitted by sand flies. Science. 2008;321(5891):970–4. https://doi.org/10.1126/science.1159194321/5891/970.

Peters NC, Sacks DL. The impact of vector-mediated neutrophil recruitment on cutaneous leishmaniasis. Cell Microbiol. 2009;11(9):1290–6. https://doi.org/10.1111/j.1462-5822.2009.01348.xCMI1348.

Pham CT. Neutrophil serine proteases: specific regulators of inflammation. Nat Rev Immunol. 2006;6(7):541–50. https://doi.org/10.1038/nri1841.

Potestio M, D'Agostino P, Romano GC, Milano S, Ferlazzo V, Aquino A, Di Bella G, Caruso R, Gambino G, Vitale G, Mansueto S, Cillari E. CD4+ CCR5+ and CD4+ CCR3+ lymphocyte subset and monocyte apoptosis in patients with acute visceral leishmaniasis. Immunology. 2004;113(2):260–8. https://doi.org/10.1111/j.1365-2567.2004.01948.xIMM1948.

Quirino GF, Nascimento MS, Davoli-Ferreira M, Sacramento LA, Lima MH, Almeida RP, Carregaro V, Silva JS. Interleukin-27 mediates susceptibility to visceral leishmaniasis by suppressing the IL-17-neutrophil response. Infect Immun. 2016;84(8):2289–98. https://doi.org/10.1128/IAI.00283-16.

Rasmusson B, Descoteaux A. Contribution of electron and confocal microscopy in the study of Leishmania-macrophage interactions. Microsc Microanal. 2004;10(5):656–61. https://doi.org/10.1017/S1431927604040851.

Ritter U, Korner H. Divergent expression of inflammatory dermal chemokines in cutaneous leishmaniasis. Parasite Immunol. 2002;24(6):295–301. https://doi.org/10.1046/j.1365-3024.2002.00467.x.

Roberts MT. Current understandings on the immunology of leishmaniasis and recent developments in prevention and treatment. Br Med Bull. 2005;75–76:115–30. https://doi.org/10.1093/bmb/ldl003.

Rochael NC, Guimaraes-Costa AB, Nascimento MT, DeSouza-Vieira TS, Oliveira MP, Garcia e Souza LF, Oliveira MF, Saraiva EM. Classical ROS-dependent and early/rapid ROS-independent release of neutrophil extracellular traps triggered by Leishmania parasites. Sci Rep. 2015;5:18302. https://doi.org/10.1038/srep18302srep18302.

Rohousova I, Hostomska J, Vlkova M, Kobets T, Lipoldova M, Volf P. The protective effect against Leishmania infection conferred by sand fly bites is limited to short-term exposure. Int J Parasitol. 2011;41(5):481–5. https://doi.org/10.1016/j.ijpara.2011.01.003S0020-7519(11)00033-6.

Ronet C, Hauyon-La Torre Y, Revaz-Breton M, Mastelic B, Tacchini-Cottier F, Louis J, Launois P. Regulatory B cells shape the development of Th2 immune responses in BALB/c mice infected with *Leishmania major* through IL-10 production. J Immunol. 2010;184(2):886–94. https://doi.org/10.4049/jimmunol.0901114jimmunol.0901114.

Rossi M, Castiglioni P, Hartley MA, Eren RO, Prével F, Desponds C, Utzschneider DT, Zehn D, Cusi MG, Kuhlmann FM, Beverley SM, Ronet C, Fasel N. Type I interferons induced by endogenous or exogenous viral infections promote metastasis and relapse of leishmaniasis. PNAS. 2017;114(19):4987–92. https://doi.org/10.1073/pnas/1621447114.

Sacramento LA, da Costa JL, de Lima MH, Sampaio PA, Almeida RP, Cunha FQ, Silva JS, Carregaro V. Toll-like receptor 2 is required for inflammatory process development during Leishmania infantum infection. Front Microbiol. 2017;8:262. https://doi.org/10.3389/fmicb.2017.00262.

Santana IU, Dias B, Nunes EA, Rocha FA, Silva FS Jr, Santiago MB. Visceral leishmaniasis mimicking systemic lupus erythematosus: case series and a systematic literature review. Semin Arthritis Rheum. 2015;44(6):658–65. https://doi.org/10.1016/j.semarthrit.2014.12.004S0049-0172(14)00327-8.

Saraiva M, O'Garra A. The regulation of IL-10 production by immune cells. Nat Rev Immunol. 2010;10(3):170–81. https://doi.org/10.1038/nri2711nri2711.

Sarkar A, Ghosh S, Pakrashi S, Roy D, Sen S, Chatterjee M. Leishmania strains causing self-healing cutaneous leishmaniasis have greater susceptibility towards oxidative stress. Free Radic Res. 2012;46(5):665–73. https://doi.org/10.3109/10715762.2012.668186.

Sharma U, Singh S. Immunobiology of leishmaniasis. Indian J Exp Biol. 2009;47(6):412–23.

Soong L. Subversion and utilization of host innate defense by *Leishmania amazonensis*. Front Immunol. 2012;3:58. https://doi.org/10.3389/fimmu.2012.00058.

Srivastava A, Singh N, Mishra M, Kumar V, Gour JK, Bajpai S, Singh S, Pandey HP, Singh RK. Identification of TLR inducing Th1-responsive *Leishmania donovani* amastigote-specific antigens. Mol Cell Biochem. 2012;359(1–2):359–68. https://doi.org/10.1007/s11010-011-1029-5.

Teixeira MJ, Teixeira CR, Andrade BB, Barral-Netto M, Barral A. Chemokines in host-parasite interactions in leishmaniasis. Trends Parasitol. 2006;22(1):32–40. https://doi.org/10.1016/j.pt.2005.11.010.

Titus RG, Ribeiro JM. Salivary gland lysates from the sand fly *Lutzomyia longipalpis* enhance Leishmania infectivity. Science. 1988;239(4845):1306–8.

Uzonna JE, Joyce KL, Scott P. Low dose *Leishmania major* promotes a transient T helper cell type 2 response that is down-regulated by interferon gamma-producing CD8+ T cells. J Exp Med. 2004;199(11):1559–66. https://doi.org/10.1084/jem.20040172jem.20040172.

Valencia-Pacheco G, Loria-Cervera EN, Sosa-Bibiano EI, Canche-Pool EB, Vargas-Gonzalez A, Melby PC, Andrade-Narvaez FJ. In situ cytokines (IL-4, IL-10, IL-12, IFN-gamma) and chemokines (MCP-1, MIP-1alpha) gene expression in human *Leishmania (Leishmania) mexicana* infection. Cytokine. 2014;69(1):56–61. https://doi.org/10.1016/j.cyto.2014.05.016S1043-4666(14)00149-5.

van Zandbergen G, Klinger M, Mueller A, Dannenberg S, Gebert A, Solbach W, Laskay T. Cutting edge: neutrophil granulocyte serves as a vector for Leishmania entry into macrophages. J Immunol. 2004;173(11):6521–5. https://doi.org/10.4049/jimmunol.173.11.6521.

WHO. Leishmaniasis. Fact sheet. 2016. http://www.whoint/mediacentre/factsheets/fs375/en/.

Wilson ME, Young BM, Davidson BL, Mente KA, McGowan SE. The importance of TGF-beta in murine visceral leishmaniasis. J Immunol. 1998;161(11):6148–55.

Zangger H, Hailu A, Desponds C, Lye LF, Akopyants NS, Dobson DE, Ronet C, Ghalib H, Beverley SM, Fasel N. *Leishmania aethiopica* field isolates bearing an endosymbiontic dsRNA virus induce pro-inflammatory cytokine response. PLoS Negl Trop Dis. 2014;8(4):e2836. https://doi.org/10.1371/journal.pntd.0002836PNTD-D-13-01835.

Zijlstra EE. The immunology of post-kala-azar dermal leishmaniasis (PKDL). Parasit Vectors. 2016;9:464. https://doi.org/10.1186/s13071-016-1721-0.

Zijlstra EE, Musa AM, Khalil EAG, El-Hassan IM, El-Hassan AM. Post-kala-azar dermal leishmaniasis. Lancet Infect Dis. 2003;3(2):87–98.

Zhang WW, Matlashewski G. Characterization of the A2-A2rel gene cluster in *Leishmania donovani*: involvement of A2 in visceralization during infection. Mol Microbiol. 2001;39(4):935–48. https://doi.org/10.1046/j.1365-2958.2001.02286.x.

Clinical Aspects in Immunocompetent and Immunocompromised Patients

6

Rojelio Mejia

6.1 Introduction

Leishmaniasis is a disease caused by several protozoan species of genus *Leishmania*. Clinical signs and symptoms vary with the *Leishmania* species that generally follow a geographical distribution (Bennett et al. 2015). The *Leishmania* species vary with region, but also with clinical presentation (Table 6.1). This chapter will divide its scope first into the different syndrome associated with leishmaniasis including cutaneous leishmaniasis (CL), mucocutaneous leishmaniasis (MCL), visceral leishmaniasis (VL), and post-kala-azar dermal leishmaniasis (PKDL). It will further divide into immunocompromised patients that can have diffuse cutaneous leishmaniasis (Develoux et al. 1996).

Signs and symptoms of all forms of leishmaniasis will be discussed, along with progression of disease. Lesions and progression of disease can be dependent on the *Leishmania* species (Guerrant et al. 2011; Tyring et al. 2006). Although many authors agree that there is considerable overlap in the clinical presentation of infections caused by different *Leishmania* species, diagnoses should not be made on appearance only (Guerrant et al. 2011; Tyring et al. 2006). It is important to speciate the *Leishmania*, since certain species can progress to VL (kala-azar) or MCL (espundia). In the clinical setting, a patient with chronic skin lesions that is from or has traveled to an endemic area needs a biopsy and tissue examination to evaluate for leishmaniasis (Guerrant et al. 2011; Tyring et al. 2006). Clinical description of the dermal lesions and using laboratory and molecular means is important for diagnosing *Leishmania* species (Boggild et al. 2010; Reithinger and Dujardin 2007). Knowing the *Leishmania* species is essential for treatment, follow-up, prognosis, and further workup for clinical pathology.

R. Mejia
National School of Tropical Medicine, Baylor College of Medicine, Houston, TX, USA
e-mail: Rojelio.Mejia@bcm.edu

F. Bruschi, L. Gradoni (eds.), *The Leishmaniases: Old Neglected Tropical Diseases*, https://doi.org/10.1007/978-3-319-72386-0_6

Table 6.1 Main *Leishmania* agents and corresponding clinical syndromes

Leishmania species	Clinical leishmaniasis syndromes	Unique presentations
New World		
L. infantum/chagasi	Visceral	
L. amazonensis	Cutaneous	Diffuse cutaneous leishmaniasis in immunocompromised patients
		Visceral
L. braziliensis	Cutaneous	Mucocutaneous
L. guyanensis	Cutaneous	Mucocutaneous
L. panamensis	Cutaneous	Mucocutaneous
L. mexicana	Cutaneous	Diffuse cutaneous leishmaniasis in immunocompromised patients
L. venezuelensis	Cutaneous	
L. peruviana	Cutaneous	
Old World		
L. donovani	Visceral	Post-kala-azar dermal leishmaniasis (PKDL)
L. infantum	Visceral	Cutaneous
L. major	Cutaneous	
L. tropica	Cutaneous	Leishmaniasis recidivans
		Mild visceral disease
L. aethiopica	Cutaneous	Diffuse cutaneous leishmaniasis in immunocompromised patients

6.2 Cutaneous Leishmaniasis

The most common form of leishmaniasis is cutaneous (Jones et al. 1987). The early common manifestation of CL is a solitary papule on exposed skin that progresses over weeks to months, evolving initially into a partial ulcerative lesion early in the clinical manifestation (Fig. 6.1a). Lesions have a wide range of presentations including papules, nodules, and plaques that resemble lepromatous leprosy or lupus vulgaris.

Differential diagnoses of CL should be categorized into acute or chronic presentation of lesions:

Acute: includes lesions less than 30 days from presentation and can include insect bites, myiasis, keloids, and furuncles/carbuncles

Chronic: are lesions more than 30 days since appearing and include limited infectious etiologies such as Buruli ulcers (*Mycobacterium ulcerans*), lepromatous leprosy, and blastomycosis. Also, noninfectious entities such as autoimmune, inflammatory conditions, and malignancies such as lupus vulgaris, sarcoidosis, discoid lupus erythematosus, and malignant melanoma, adult T-cell leukemia/lymphoma

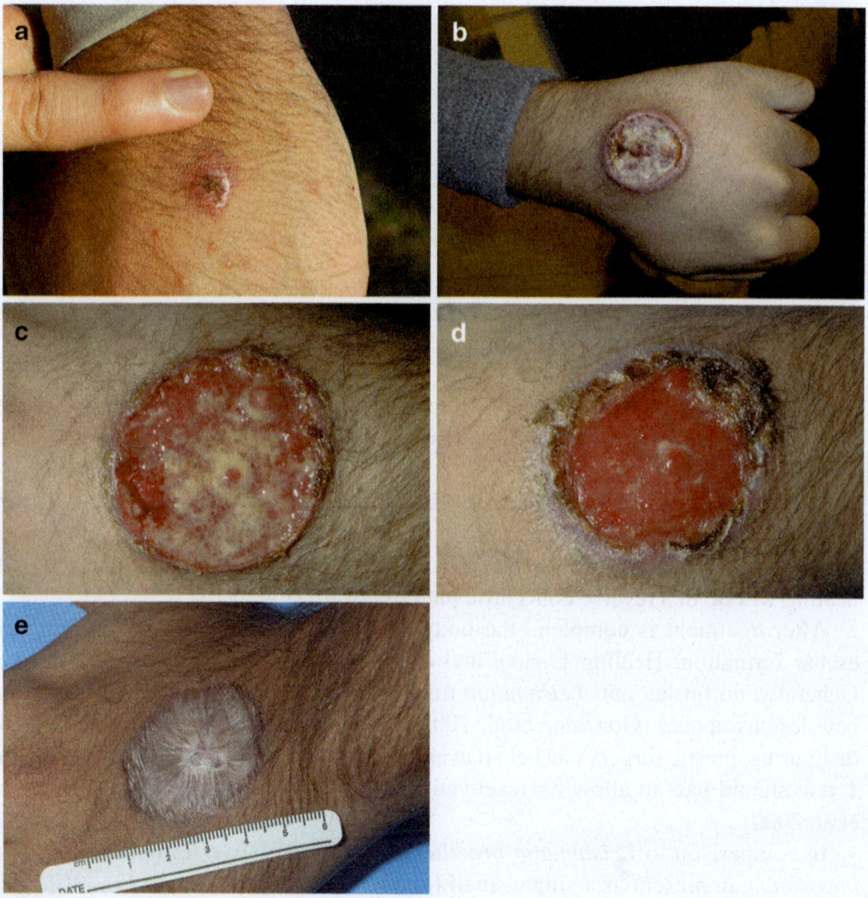

Fig. 6.1 (**a**) Cutaneous leishmaniasis, *Leishmania braziliensis* contracted in Tambopata, Peru. Painless ulcerating lesion with satellites forming distally. No erythema or discharge from the lesion. (**b**) Progression of ulcerative lesion, 3 months after initial formation. Classic rolled borders and thick white-yellow fibrinous material. (**c**) Natural progression of *Leishmania braziliensis* lesion, 6 months after initial presentation. No treatment administered to date, the lesion continues to be painless. (**d**) Treatment after eight doses of liposomal amphotericin B. Lesion has begun to decrease in size; rolled border is absent. Granulation tissue can be seen around the periphery. (**e**) Healed lesion 5 months posttreatment. No further reactivation of lesions. No edema or pain associated with site

Discrete satellites may form near the main lesion and progress/coalesce into ulcerative lesions depending on species (Table 6.1). Progression of ulcer may take months to years depending on species, burden of infection, and immunocompetency of the patient. When travelers, immigrants, or ex-patriots are infected and return to their non-endemic countries, often inexperienced healthcare workers will misdiagnose these lesions. The natural history of these lesions, generally take several months to begin healing without appropriate treatment.

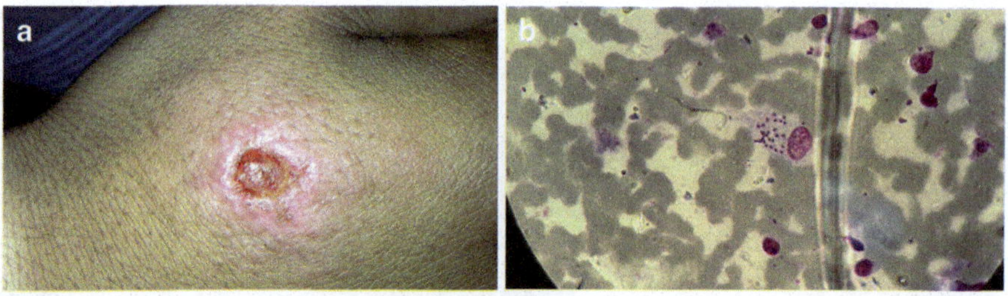

Fig. 6.2 (**a**) Solitary ulcer from a farmer in Guatemala. No rolled borders, but relative deep ulceration. (**b**) Touch prep stain of *Leishmania mexicana*. Amastigotes engulfed in a macrophage (center)

As the lesions continue to progress, changes in lesions can follow natural healing patterns with loss of rolled borders and flattening of overall lesion (Fig. 6.1c; Guerrant et al. 2011; Tyring et al. 2006).

Once diagnosis is appropriately made and treatment started, resolution of the thick fibrinous material with granulation tissue and eschar formation signals the body's healing by secondary intent. Surgical sutures would be contraindicated and healing will be in a reverse concentric pattern inward (Fig. 6.1d).

After treatment is complete, the body will cover a large open lesion with an eschar formation. Healing is slow and can take several months (Murray 2012). Generally, no further anti-*Leishmania* treatment is needed during this time, unless new lesions appear (Gonzalez et al. 2008, 2009). The final scar can be large and disfiguring, plastic surgery can help if available with cosmetic appearances, although 1 year should pass to allow for reactivation of leishmaniasis (Fig. 6.1e; Wortmann et al. 2000).

In comparison to *Leishmania braziliensis* large ulcerative lesion, *Leishmania mexicana* can present as a single small lesion. The unique feature is that they can quickly self-resolve without treatment in a span of a few months (Fig. 6.2a; Herwaldt et al. 1992). A helpful clinical correlate is that with most ulcerative lesions, a small probe can be used at the base of the borders to take a small sample for diagnoses. The sample can be smeared on a glass slide and stained with Giemsa to make a touch prep stain (Garcia 2016). This low-cost method can be performed in resource-limited settings and provide a quick answer in less than 1 h (Fig. 6.2b).

The Old World *Leishmania major* is another species that has a short incubation time and can be self-limiting. Unlike *L. mexicana*, multiple lesions are more common, classically *L. major* begins with a red furunculoid nodule (Fig. 6.3; Tyring et al. 2006). *L. major* will not cause VL (Guerrant et al. 2011).

An uncommon presentation of *L. tropica* is leishmaniasis recidivans, a syndrome associated with long incubation period from months to years, recurrent papules around the scar or trauma site, and a dry appearance of the lesions. Biopsy will show amastigotes and treatment can effectively heal lesions (Guerrant et al. 2011; Tyring et al. 2006; Marovich et al. 2001).

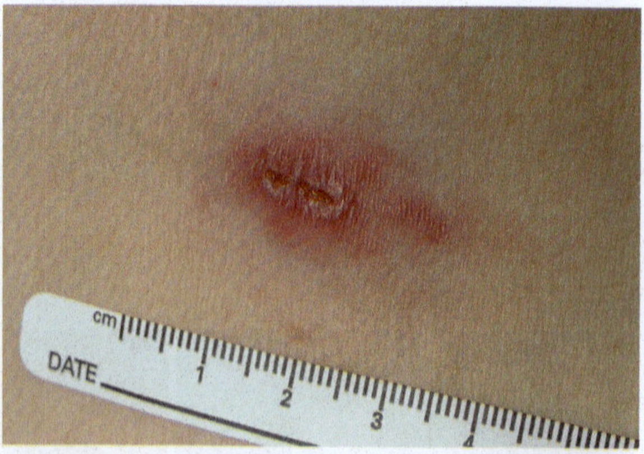

Fig. 6.3 *Leishmania major* lesion in a visitor to Morocco, he stayed overnight in the Sahara Desert in a tent. Thirty days later presented with these painless lesions on his flank

6.3 Mucocutaneous Leishmaniasis

MCL is a severe form caused by the dissemination to facial mucosae of some New World *Leishmania* species. These belong to the subgenus *Viannia* which includes *L. braziliensis*, *L. panamensis*, and less commonly *L. guyanensis*. The *Leishmania* subgenus *L. amazonensis* (Table 6.1) can also progress to MCL (Guerra et al. 2011). Old World species progressing to mucosal leishmaniasis usually occur in immunocompromised hosts, the species most associated are *L. tropica*, *L. major*, and *L. infantum* and may not include cutaneous lesions (Shirian et al. 2013). The importance of speciation of the *Leishmania* protozoa during diagnosis is because of this risk for progression to MCL. Any CL patient found infected with the above species should have a thorough mucosal examination including endoscopy if available with biopsies for erosions or abnormal tissue in the mouth, nose, and pharynx. Less commonly, genital, anal, and ocular mucosa can be involved and should be monitored (Tyring et al. 2006).

Presentation of MCL generally begins when active cutaneous lesions are present (Fig. 6.4a; Magill 2013), although there are reports of delayed presentation of mucosal involvement beyond 10 years of partially treated or untreated CL (Fig. 6.5a; Marsden 1986). Classic symptoms include nasal stuffiness with discharge or blockage that can be appreciated on imaging including magnetic resonance imaging (MRI) and computed tomography (CT) scans (Fig. 6.4b). All MCL *Leishmania* agents can be aggressive causing friable mucosal tissue with extensive bleeding (Fig. 6.5b). Because it can destroy cartilage and impact nerve ending, MCL can be painful and cause severe deformities. In Latin America, chronic mucocutaneous disease is called espundia (or sponge-like). Lesions can spread throughout the mucosal layers reaching the larynx; symptoms of hoarseness, dysphonia, and respiratory distress are concerning for severe disease (Fig. 6.4c).

The differential for mucosal involvement in MCL is limited to aphthous ulcers and lesions from herpes, cytomegalovirus, and malignancies. Clinically mucosal involvement should be monitored closely by otolaryngology and radiographic

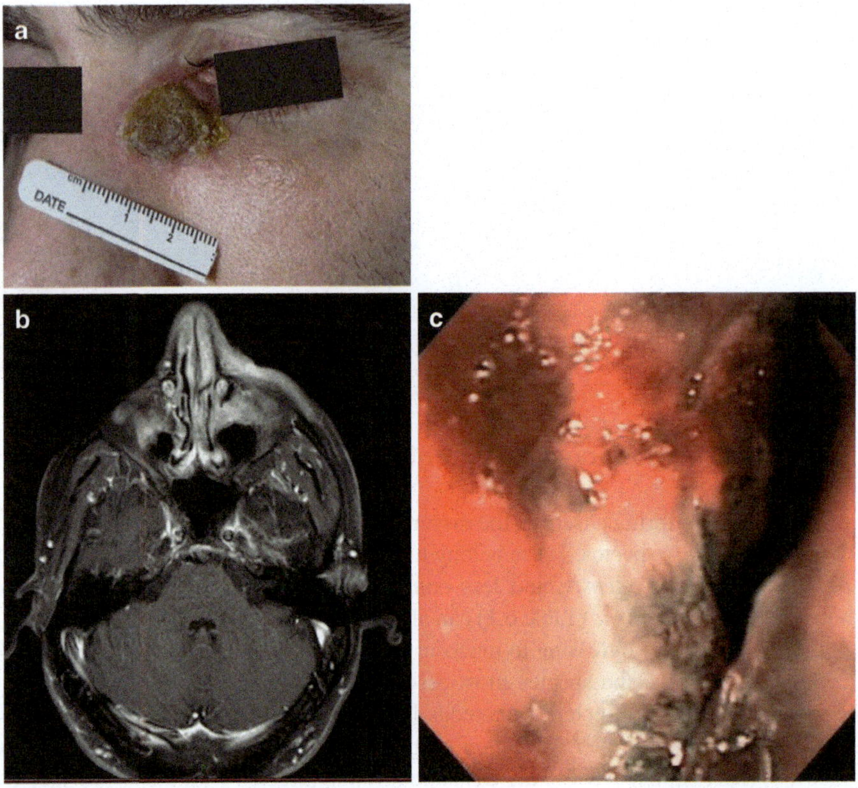

Fig. 6.4 (**a**) Traveler to Amazon region of Ecuador infected with *Leishmania braziliensis*. Face lesion presented 1 week after returning from trip. (**b**) MRI with extensive enhancement showing ethmoid inflammation. Clinically could not breathe through the nose and suffered loss of smell (anosmia). (**c**) Friable mucosal tissue seen on nasal endoscopy, up to the level of the nasopharynx. Biopsy showed amastigotes in macrophage and PCR was positive for *Leishmania braziliensis*

imaging (CT/MRI) especially if symptoms continue posttreatment; close follow-up should continue to a minimum of 1 year until the absence of symptoms (Amato et al. 2007, 2011).

6.4 Visceral Leishmaniasis

Also known as kala-azar or black fever in the Indian subcontinent, VL has the highest burden in the Old World, where it is caused by members of the *Leishmania donovani* complex (*L. donovani* and *L. infantum*). In the New World, the disease is caused by *L. infantum*/*chagasi* in children with increasing incidences in Brazil, and both species, *L. infantum*/*chagasi* and *L. amazonensis*, have been associated with VL in the immunocompromised host (Table 6.1). Visceral leishmaniasis is the syndrome associated with the progression and systemic infection of *Leishmania*

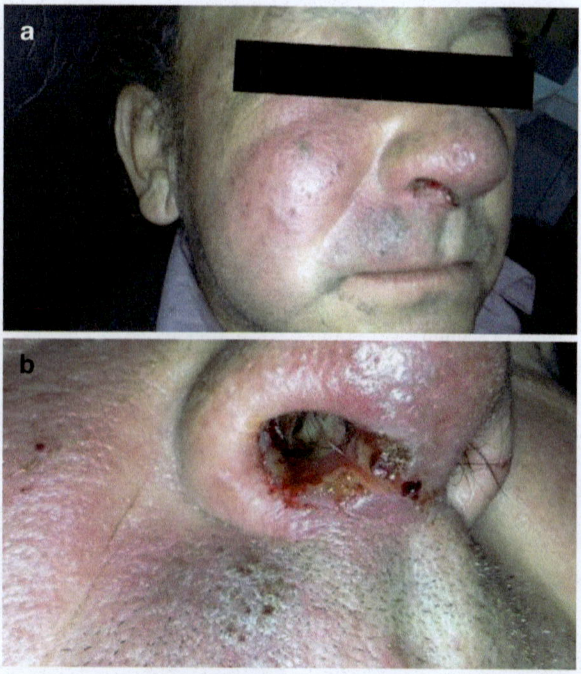

Fig. 6.5 (**a**) Sugarcane farmer presenting 15 years after primary lesion (right temporal region). Was treated with sodium stibogluconate, but could not tolerate treatment course and was partially treated. Current symptoms include nasal stuffiness, bleeding from nares, and bilateral cheek involvement. (**b**) Relapse *Leishmania braziliensis* with mucosal and subdermal involvement. Bleeding occurs frequently due to friability of mucosal layer

protozoa. Presentation includes the presence or lack of cutaneous papules at the site of inoculation (Tyring et al. 2006). Clinical symptoms present with fevers, loss of appetite, weight loss, and malaise (Guerrant et al. 2011; Tyring et al. 2006; Magill 2013). Onset can be insidious and painful progression over several months to visceral organ involvement including splenomegaly and/or hepatomegaly (Bern et al. 2000). CT/MRI or other imaging (ultrasound) can be useful to appreciate early organ enlargement and possible areas to biopsy for diagnoses. Because of visceral involvement, serum laboratory analysis can assist with disease management. Impact on liver function tests, hypoalbuminemia, thrombocytopenia, anemia, eosinopenia, jaundice, and ascites are manifestation of bone marrow, spleen, and liver involvement of the parasite (Baba et al. 2006). Prognosis is poor without treatment for VL (Seaman et al. 1996). Patients should be followed up closely for 6–12 months monitoring normalization of serum laboratory values, spleen and liver size reduction, and improvement in weight and nutritional status. Importantly coinfection with human immunodeficiency virus (HIV) can hasten progression of VL and is worsened with lower CD4 counts (<50) (Pintado et al. 2001). Coinfection should be followed up by a HIV and Tropical Medicine specialist for optimal medical management.

The differential diagnosis of VL has a vast array of infections and illnesses that have systemic manifestations including typhoid fever, malaria, disseminated tuberculosis, brucellosis, histoplasmosis, hepatosplenic schistosomiasis, hyperreactive malarial splenomegaly, subacute bacterial endocarditis, lymphoma, myeloproliferative diseases, hemophagocytic syndrome, or cirrhosis with portal hypertension.

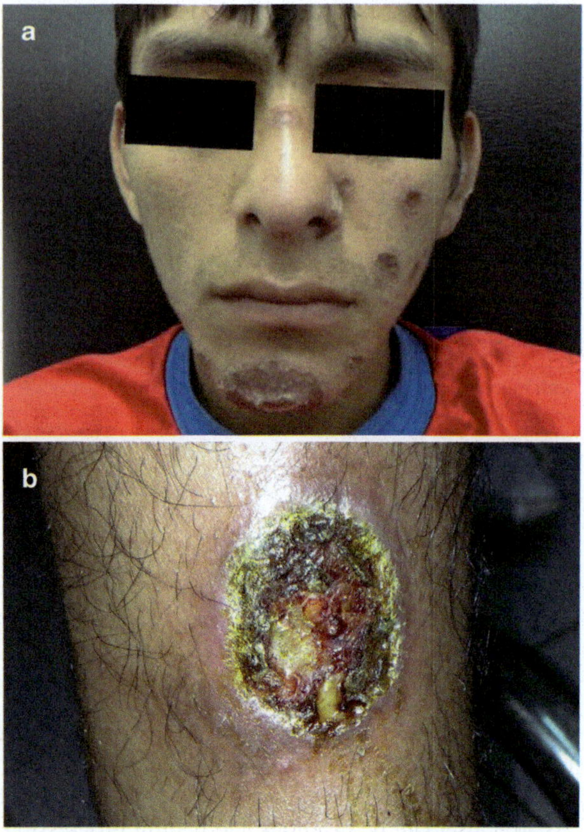

Fig. 6.6 (**a**) Diffuse cutaneous leishmaniasis in an HIV patient with CD4 count less than 50 cells/ml. Biopsy proven *Leishmania amazonensis* causing diffuse facial plaques without ulceration. (**b**) Leg lesion with slow healing and starting to ulcerate. Patient had already received full course of sodium stibogluconate, but remained with low CD4 counts

A unique phenomenon to VL is known as the post-kala-azar dermal leishmaniasis (PKDL). This cutaneous eruption of papules and nodules can occur during or after treatment for VL caused by *L. donovani* in the Indian subcontinent and East Africa (Table 6.1). Amastigotes can be seen in biopsy samples, and treatment is indicated for PKDL (Guerrant et al. 2011; Tyring et al. 2006; Magill 2013; Ansari et al. 2006; Ramesh and Mukherjee 1995).

Another systemic syndrome is viscerotropic leishmaniasis associated with *L. tropica* (Table 6.1). Signs and symptoms are similar to kala-azar; although milder, treatment is similar and has better prognosis (Magill et al. 1993).

6.5 Diffuse Cutaneous Leishmaniasis

Diffuse CL is a disseminated disease involving skin area not at the site of inoculation by sandflies. It is limited exclusively to immunocompromised patients that have a Th1-mediated decrease response (Develoux et al. 1996). The most common immunocompromised state is patients coinfected with HIV and have low CD4 counts defined as acquired immune deficiency syndrome (AIDS). Clinical presentation are several non-ulcerative lesions that are nodules or plaques on the face or limbs initially with involvement of the entire body if not treated or underlying immunosuppression not addressed (Fig. 6.6a; Guerrant et al. 2011; Tyring et al. 2006; Magill 2013). Progression of disease can have lesions ulcerate and become difficult to cure (Fig. 6.6b). Resolution of diffuse CL can be difficult and prolonged, it is dependent on addressing the underlying immunodeficiency (Motta et al. 2003).

References

Amato VS, et al. Treatment of mucosal leishmaniasis in Latin America: systematic review. Am J Trop Med Hyg. 2007;77(2):266–74.

Amato VS, et al. Can we use a lower dose of liposomal amphotericin B for the treatment of mucosal American leishmaniasis? Am J Trop Med Hyg. 2011;85(5):818–9.

Ansari NA, Ramesh V, Salotra P. Interferon (IFN)-gamma, tumor necrosis factor-alpha, interleukin-6, and IFN-gamma receptor 1 are the major immunological determinants associated with post-kala azar dermal leishmaniasis. J Infect Dis. 2006;194(7):958–65.

Baba CS, et al. Chronic diarrhea and malabsorption caused by *Leishmania donovani*. Indian J Gastroenterol. 2006;25(6):309–10.

Bennett JE, Dolin R, Blaser MJ. Mandell, Douglas, and Bennett's principles and practice of infectious diseases, vol. 2. 8th ed. Philadelphia: Elsevier; 2015.

Bern C, et al. Factors associated with visceral leishmaniasis in Nepal: bed-net use is strongly protective. Am J Trop Med Hyg. 2000;63(3–4):184–8.

Boggild AK, et al. Clinical and demographic stratification of test performance: a pooled analysis of five laboratory diagnostic methods for American cutaneous leishmaniasis. Am J Trop Med Hyg. 2010;83(2):345–50.

Develoux M, et al. Diffuse cutaneous leishmaniasis due to *Leishmania major* in Senegal. Trans R Soc Trop Med Hyg. 1996;90(4):396–7.

Dowlati Y. Cutaneous leishmaniasis. Clinical aspect. Clin Dermatol. 1996;14(5):425–31.

Garcia LS. Diagnostic medical parasitology, vol. xxii. 6th ed. Washington: ASM Press; 2016. p. 1388.

Gonzalez U, et al. Interventions for old world cutaneous leishmaniasis. Cochrane Database Syst Rev. 2008;4:CD005067.

Gonzalez U, et al. Interventions for American cutaneous and mucocutaneous leishmaniasis. Cochrane Database Syst Rev. 2009;2:CD004834.

Guerra JA, et al. Mucosal Leishmaniasis caused by Leishmania (Viannia) braziliensis and Leishmania (Viannia) guyanensis in the Brazilian Amazon. PLoS Negl Trop Dis. 2011;5(3):e980.

Guerrant RL, Walker DH, Weller PF. Tropical infectious diseases: principles, pathogens and practice, vol. xxiv. 3rd ed. Edinburgh: Elsevier; 2011. p. 1130.

Herwaldt BL, Arana BA, Navin TR. The natural history of cutaneous leishmaniasis in Guatemala. J Infect Dis. 1992;165(3):518–27.

Jones TC, et al. Epidemiology of American cutaneous leishmaniasis due to Leishmania braziliensis braziliensis. J Infect Dis. 1987;156(1):73–83.

Magill AJ. Hunter's tropical medicine and emerging infectious diseases, vol. xxiv. 9th ed. London: Elsevier; 2013. p. 1190.

Magill AJ, et al. Visceral infection caused by Leishmania tropica in veterans of operation desert storm. N Engl J Med. 1993;328(19):1383–7.

Marovich MA, et al. Leishmaniasis recidivans recurrence after 43 years: a clinical and immunologic report after successful treatment. Clin Infect Dis. 2001;33(7):1076–9.

Marsden PD. Mucosal leishmaniasis ("espundia" Escomel, 1911). Trans R Soc Trop Med Hyg. 1986;80(6):859–76.

Motta AC, et al. Disseminated mucocutaneous leishmaniasis resulting from chronic use of corticosteroid. Int J Dermatol. 2003;42(9):703–6.
Murray HW. Leishmaniasis in the United States: treatment in 2012. Am J Trop Med Hyg. 2012;86(3):434–40.
Pintado V, et al. Visceral leishmaniasis in human immunodeficiency virus (HIV)-infected and non-HIV-infected patients. A comparative study. Medicine (Baltimore). 2001;80(1):54–73.
Ramesh V, Mukherjee A. Post-kala-azar dermal leishmaniasis. Int J Dermatol. 1995;34(2):85–91.
Reithinger R, Dujardin JC. Molecular diagnosis of leishmaniasis: current status and future applications. J Clin Microbiol. 2007;45(1):21–5.
Seaman J, Mercer AJ, Sondorp E. The epidemic of visceral leishmaniasis in western upper Nile, southern Sudan: course and impact from 1984 to 1994. Int J Epidemiol. 1996;25(4):862–71.
Shirian S, et al. Three *Leishmania/L. species—L. infantum, L. major, L. tropica*—as causative agents of mucosal leishmaniasis in Iran. Pathog Glob Health. 2013;107(5):267–72.
Tyring SK, Lupi O, Hengge UR. Tropical dermatology, vol. xi. New York: Elsevier; 2006. p. 515.
Wortmann GW, et al. Cutaneous leishmaniasis following local trauma: a clinical pearl. Clin Infect Dis. 2000;31(1):199–201.

Diagnosis 7

Marina Gramiccia and Trentina Di Muccio

7.1 Introduction

When we say "leishmaniasis" or "leishmaniases", we are referring to a group of human diseases caused by protozoan parasites belonging to *Leishmania* genus (see Chap. 2). The course of the disease is variable ranging from spontaneous healing to chronicity, including severe and fatal disease, but most infected individuals remain asymptomatic or subclinical (see Chap. 6). Asymptomatic infections are widespread and contribute to maintaining the long-term presence of the parasite in endemic regions. Therefore, there is a wide infection spectrum as a result of the parasite inoculation and thus different approaches for diagnosis (Loría-Cervera and Andrade-Narváez 2014).

Up to the moment, the leishmaniasis diagnosis is performed by an association of clinical, epidemiological, and laboratory data. Particularly in relation to laboratory methods, the lack of a gold standard for human patients is a limitation for the disease control, because the achievement of accurate epidemiological data is associated with the guidance of control measures, thus helping to increase their efficiency (de Paiva-Cavalcanti et al. 2015). Parasitological, molecular, and serological assays are methods available for the diagnosis of leishmaniasis. However, the direct demonstration of parasite is the only way to confirm the disease conclusively. Diagnostic parameters have been redefined in the last decade with the development of new immunological and molecular tests (Srividya et al. 2012). However, in many areas a definitive diagnosis still relies on the century-old parasitological methods.

In this chapter we have been trying to include different diseases as a single pathological entity but grouping them in four basic pathologies: visceral (VL), cutaneous (CL), mucocutaneous (MCL), and post-kala-azar dermal leishmaniasis (PKDL). In addition a particular attention was given to special situations, as *Leishmania*-HIV

M. Gramiccia (✉) • T. Di Muccio
Vector-Borne Diseases, Department of Infectious Diseases,
Istituto Superiore di Sanità, Rome, Italy
e-mail: marina.gramiccia@iss.it

F. Bruschi, L. Gradoni (eds.), *The Leishmaniases: Old Neglected Tropical Diseases*, https://doi.org/10.1007/978-3-319-72386-0_7

coinfection or forms of immunosuppression. The diagnostic approaches have been articulated in dedicated sections (parasitological, molecular, and immunological diagnostic pathway) where different clinical samples and techniques were discussed. Moreover, for each pathology the diagnostic pathway was presented and the best validated protocols suggested. Furthermore, diagnostic methods were discussed on the basis of their cost and if directly applied on the field or strictly at high laboratory level. Basic similarities among different diseases were pointed out (Murray et al. 2005).

7.2 Diagnostic Approaches

In the USA National Institutes of Health guidelines 2016 (http://osp.od.nih.gov), all *Leishmania* spp. are classed as Risk Group 2 of biohazardous agents for human infection. Direct contact with infected hosts or handling of biological samples and parasite cultures from these hosts does not require special precautions because of the sand fly-borne nature of the infections and the lack of resistant forms in the environment (Gradoni and Gramiccia 2014).

The diagnostic approaches pivotal for a conclusive diagnosis of leishmaniasis are summarized in Fig. 7.1. In the present chapter we deeply discuss the diagnosis of leishmaniasis at laboratory level that can be made by both direct and indirect assays by the following: (a) demonstration of parasite in tissues of relevance, (b) detection of parasite nucleic acids in tissue samples, and (c) immunodiagnosis by detection of leishmania-specific antibodies or parasite antigens and cell-mediated immunity.

7.2.1 Parasitological Diagnostic Pathway

The parasitological approach is basic for the diagnostic assurance in the all pathological entities. It remains the gold standard in diagnosis for its high specificity (WHO 2010). The performance of different parasitological techniques is directly

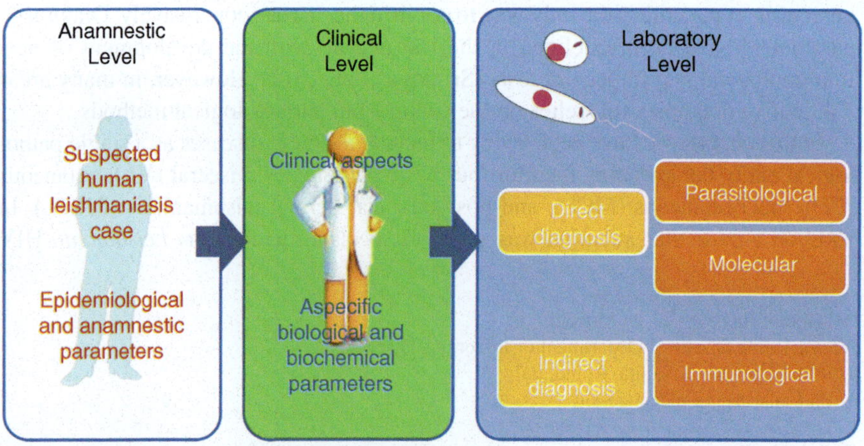

Fig. 7.1 General pathway for leishmaniasis diagnosis: from patient's history to the laboratory

influenced by the choice of biopsy samples that can be simply summarized as (a) different tissue samples (bone marrow, spleen, lymph node, liver, peripheral blood), often invasive and characterized by different sensitivity degrees, for VL and (b) skin samples useful for CL and MCL.

Classical parasitological procedures are microscopic examination and in vitro cultures used individually or in combination. In vivo isolation of *Leishmania* in susceptible animals (e.g., the Syrian hamster *Mesocricetus auratus* or BALB/c mice) is no longer recommended for routine diagnosis.

7.2.1.1 Microscopy

The visualization of the parasite amastigote form by light microscopic examination of aspirates from bone marrow, spleen, lymph nodes, or cutaneous lesion is the classical confirmatory test. The technique is advantageous as it is directly able to detect parasite load and is inexpensive to perform. The specificity of this technique is high, although the sensitivity varies depending on the tissue used (WHO 2010). Results are dependent on technical expertise and quality of prepared slides and reagents. Hence, a well-trained and competent microscopist and an effective quality control are needed for accurate diagnosis. The sample is smeared or impressed directly on the slide. Examination of Giemsa-stained slides is the technique most commonly used to visualize the amastigote form, the parasite tissue form. After fixation with methanol, the slide is stained with Giemsa stain for direct examination by light microscopy under oil immersion (600–1000×). Amastigotes appear as round or oval bodies, 2–4 μm in diameter, with characteristic nucleus and kinetoplast; they can be seen within macrophages or mononuclear cells. The amastigote's cytoplasm appears pale blue, with a relatively large nucleus that stains in deep red. In the same plane as the nucleus, but at a various angle to it, a deep red or violet road-like body, the kinetoplast, is visible (Gradoni and Gramiccia 2014; Sakkas et al. 2016; Srivastava et al. 2011). The efficiency of the method depends on the clinical forms and, in the case of CL, on the causing *Leishmania* spp. and the oldness of the cutaneous lesion. Histologic paraffin slides can be used but with a lower sensitivity. When positive, the microscopic examination allows to obtain a very rapid final diagnosis of leishmaniasis. However the taxonomical level is genus specific not allowing the species characterization.

7.2.1.2 In Vitro Culture

The detection of parasites in biopsy samples by culture is more sensitive than microscopic examination, but these techniques remain restricted to referral hospitals and research centers, despite efforts to simplify them (Gradoni and Gramiccia 2014). Culture allows to detect the promastigote form of the parasite, the form occurring in the midgut of the phlebotomine vectors. The in vitro culture must permit efficient transformation from amastigote (present in the biopsy samples) to viable promastigote and support the promastigote division. Promastigotes are elongated extracellular organisms, body size 15–20 × 1.5–3.5 μm with a single flagellum 15–28 μm long, arising close to the kinetoplast at the anterior; the nucleus is situated centrally. The promastigotes observation is carried out by light or phase contrast microscope (400×) on a drop of liquid phase directly placed on a slide with a coverslip. When liquid media were used, the observation could be done directly across inoculated flasks by an inverted microscope. Promastigotes observation is easy due to the size and motility of the parasite. It is frequent the observation of "rosette", procycles in logarithmic phase or metacycles in stationary phase. After incubation at 21–23 °C, at which temperature most leishmanias grow well, cultures are examined weekly for 4 weeks, with one subculture, until amastigotes turn into promastigotes with a diagnosis time ranging from 5 to 30 days. The method suffers of high costs and the time-consuming nature of the technique. It requires expertise and expensive equipment. In addition, the biopsy samples have to be maintained at room temperature and cultured no more than 72 h from the time of their collection. The choice of the isolation and culture methods depends on the immediate circumstances and on the technical capability and experience of the laboratory staff (WHO 2010). However, for in vitro isolation, the techniques used should be carried out under strictly sterile conditions, which is rarely feasible in the field. Unfortunately, there is still no "universal" culture medium in which all the different leishmanias will grow easily, and it is almost impossible to predict which medium will be best suited to the growth of a particular isolate of *Leishmania*. Individual laboratories have to find the most suitable medium among biphasic blood agar media and tissue culture media supplemented with fetal calf serum (Evans 1987). When attempting primary isolation of unknown organisms, a blood agar-based medium, preferably Novy, MacNeil, and Nicolle medium (NNN) otherwise Evans' modified Tobie's medium (EMTM) or brain-heart infusion (BHI) agar medium, should be used.

Clinical samples may also be cultured into monophasic liquid media as Schneider's insect, M199, HO-MEM, or Grace's medium. Liquid media, as Schneider's Drosophila or RPMI 1640, are commercially available but have less efficiency compared to the blood agar media for *Leishmania* culture isolation, then having lower diagnostic sensitivity (Evans 1987). Progress achieved in the culture method includes now the development of a more sensitive microculture method (MCM) for *Leishmania* isolation (Allahverdiyev et al. 2005). Morphological identification enables *Leishmania* identity at genus level. However it is an essential element for the parasitological diagnosis being basic for the *Leishmania* species identification pathway.

7.2.2 Molecular Pathway

The polymerase chain reaction (PCR) methods currently constitute the main molecular diagnostic culture-independent approach for *Leishmania* detection. Molecular diagnosis has overcome some drawbacks of traditional methods, increasing the speed and sensitivity and allowing using different and less invasive clinical specimens. The choice of appropriate diagnostic approach and following PCR assay depends on different purposes (diagnosis of disease, clinical management, outbreak investigations, species identification, detection of asymptomatic infection), according to the local infrastructure, training, and budget.

In the last years, a myriad of PCR protocols have been developed to detect *Leishmania* DNA in clinical samples. However, there are not defined protocols and almost each laboratory applies its own in-house method. Depending on the kind of clinical samples, sample storage, the DNA extraction protocol, the choice of sequences target of *Leishmania* genome, the use of different primers pairs, and the PCR methodology, protocols can vary considerably in sensitivity and specificity (Reithinger and Dujardin 2007). Some methods present high concordance when they applied on a given biological sample, while they differ on other kind of samples; moreover, concordance can also be affected by the *Leishmania* species targeted (Antinori et al. 2007; Bensoussan et al. 2006; Lachaud et al. 2002; Montalvo et al. 2014). Indeed, in many studies describing a new technique or marker, validation is performed on only a few reference strains, often from a region or country different from that where the test is used. Given the considerable variability of *Leishmania* parasites, the tests not validated with adequate reference strains must not be regarded as reliable (Van der Auwera and Dujardin 2015).

To the best of our knowledge, very few studies have proposed a diagnostic or *Leishmania* typing protocol for interlaboratory comparison of PCR methods, involving laboratories from different endemic regions (Cruz et al. 2013; Van der Auwera et al. 2016). Therefore PCR-based protocols urgently need standardization and optimization. In view of this, we advise investigators to choose one diagnostic protocol, preferably the most widely used, with particular attention to the epidemiological context.

7.2.2.1 Sample and Nucleic Acid Extraction

A large variety of biological samples can be used to detect *Leishmania* DNA or RNA. Usually, these samples are the same ones used for parasitological methods; however, due to the higher sensitivity of the procedures, additional low invasive or unusual samples could be investigated. Samples can be used fresh or alternatively stored, usually at 4 °C or −20 °C; long-term storage should be done at −40 °C or below. If not used fresh, the samples collected in the field or hospital can be stored in tubes, on filter paper, or slides (stained or not), sometimes requiring the addition of reagents (buffers, formalin, paraffin) to stabilize the materials. Standard extraction protocols include the use of silica, Chelex resin, or phenol-chloroform. Several commercial kits have been used successfully to extract *Leishmania* DNA or RNA from all kind of clinical samples.

Each standardized homemade PCR protocol should consider: standardized sample volume, DNA or RNA extraction controls, internal amplification controls, negative and positive PCR controls. Furthermore replicate assays should be carefully scheduled.

7.2.2.2 PCR-Based Methods

PCR methods allow a specific DNA amplification of *Leishmania* sequences. Several distinct PCR formats are available and include either conventional or real-time PCRs. Recently, the innovative methods have also been assayed in order to adjust molecular tools to field settings by simplifying PCR formats.

Conventional PCR Assays In the conventional PCR assays or end-point PCR, and its modifications (nested (n)-PCR and semi-n-PCR), PCR amplicons are resolved by electrophoresis and visualized after ethidium bromide staining. These assays require high laboratory equipment (a PCR cabinet, a thermocycler, a power supply, an electrophoresis tank, a UV transilluminator, and a camera) available in any standard molecular laboratory and are generally expensive.

Simplification of PCR products detection has been attempted by PCR-enzyme-linked immunosorbent assay (ELISA), a "reverse hybridization" method based on the capture of PCR amplicons by specific probes immobilized in ELISA microtiter wells and colorimetric visualization.

Real-Time PCR The real-time PCR is considered the best technology for quantification of *Leishmania* parasites (Quantitative real-time PCR, qPCR) (Cruz et al. 2002; Mary et al. 2004; Van der Meide et al. 2005). In the qPCR the products are analyzed during their amplification after staining with SYBR-green I dye or hybridization with fluorogenic probes (TaqMan or fluorescence resonance energy transfer). It offers many objective advantages over conventional end-point PCR, particularly speed, broad dynamic range of target DNA quantitation, and reduction of contamination; however it is a very expensive technology. This technique may be useful for different purposes as (a) diagnosis when the parasitemia is very low; (b) quantification of parasitemia at low levels, such as during post-therapy follow-up, especially in the immunocompromised host; (c) efficacy assessment of anti-leishmanial drugs; and (d) kinetic study of parasitemia. Moreover, to detect viable parasites, RNA rather than DNA quantification is obtained by reverse transcription (RT) real-time PCR (Van der Meide et al. 2008).

Innovative Methods Recently, nucleic acid sequence-based amplification (NASBA) with its variations, the quantitative (QT-NASBA) and coupled to oligochromatography (NASBA-OC), *Leishmania*-specific PCR oligochromatographic test (*Leishmania* OligoC-test), and loop-mediated isothermal amplification (LAMP) were introduced as standardized low-tech molecular diagnostics for leishmaniasis (Adams et al. 2010; Basiye et al. 2010; Deborggraeve et al. 2008a; Khan et al. 2012; Mugasa et al. 2010; Saad et al. 2010; Takagi et al. 2009; Van der Meide et al. 2005).

A part of the 18S ribosomal DNA or RNA is amplified by PCR (OligoC-TesT) or NASBA-OC after the amplification products are detected by a simple and rapid dipstick method based on oligochromatography. These assays show high sensitivity and specificity for *Leishmania* detection from VL, CL, and MCL patients; however, they are not yet an option for routine diagnosis of leishmaniasis, and they are still restricted for use in reference centers with basic molecular biology facilities (Basiye et al. 2010).

LAMP represents a promising assay developed for use in low resource settings, and it is based on dried immobilized reagents, so it does not need cold chain for transport and storage. The reaction has several advantages: (a) the amplification takes place in a simple water bath at 60–65 °C; (b) the specificity of the reaction is high because of the design of six primers; and (c) the product can be visualized directly by using SYBR-green I dye, which turns green in the presence of amplified products and remains orange in its absence. LAMP uses only one enzyme (*Bst* DNA polymerase) and is able to amplify large amounts of DNA within 30–60 min by the intricate design of primers and auto-strand displacement DNA synthesis (Notomi et al. 2000). Recently, several studies have included a reverse transcriptase step to specifically amplify RNA (RT-LAMP) (Adams et al. 2010; Curtis et al. 2008). Different studies have demonstrated the high diagnostic sensitivity and excellent specificity of LAMP ranging from 80 to 100% and 100%, respectively (Adams et al. 2010; Khan et al. 2012; Takagi et al. 2009). Although LAMP sensitivity and specificity compare favorably with other molecular diagnostics, before being introduced in the field, further evaluation will be necessary.

7.2.2.3 *Leishmania* Identification

Identification of the causative *Leishmania* species is recommended for the correct diagnosis and prognosis of the disease as well as for case management, treatment, and disease control (WHO 2010). The need to determine the infecting *Leishmania* species depends largely on the specific context, especially useful in areas with various coexisting *Leishmania* species or in context of human mobility. Furthermore, the traditional geographical connotation of clinical forms should be taken with caution, due to expansion of transmission cycles or people movement that may cause unexpected species to circulate in unexpected regions (Di Muccio et al. 2015).

PCR-based methods combine high sensitivity for direct detection of the infecting parasites with species specificity; currently, no commercial standard tests are available for *Leishmania* species typing. The detection of the PCR product can be achieved either on a conventional or in a real-time PCR. In case of a genus-specific PCR, downstream analysis is required in order to determine the parasite species on the basis of size or sequence information in the PCR amplicon. The widely accessible and most used sequence-dependent technique is restriction fragment length polymorphism (RFLP) analysis, whereby the PCR product is digested with one or several restriction endonucleases. Consequently, gel-based analysis of the resulting DNA fragment mixture allows species classification of the parasite. A more informative and equally straightforward method is sequence analysis of the PCR amplicon. The species can be determined by identification of single-nucleotide polymorphisms (SNPs) or comparison of the obtained sequence with available reference sequences (Van der Auwera and Dujardin 2015).

Target Sequences

Over the years, several molecular markers resolving genetic differences between *Leishmania* parasites at different levels have been developed to address key epidemiological and population genetic questions: (a) minicircles of kinetoplast DNA (kDNA); (b) cytochrome *b* gene (cyt*b*) on kDNA maxicircles; (c) ribosomal RNA (rRNA) genes, such as small subunit rRNA (SSU rRNA), and rRNA gene-internal transcribed spacers (ITS, in detail ITS1 and ITS2); (d) antigen genes, such as glycoprotein 63 (gp63), heat-shock protein70 (hsp70), cysteine proteinase B (cpB); (e) mini-exon (ME) genes; (f) 7SL-RNA; (g) carbohydrate metabolism enzymes: glucose-6-phosphate dehydrogenase (G6PDH), mannose-phosphate isomerase (MPI), glucose phosphate isomerase (GPI); and (h) repetitive nuclear sequences (Buitrago et al. 2011; Cruz et al. 2002; Cupolillo et al. 1995; el Tai et al. 2000; Khatri et al. 2009; Lachaud et al. 2002; Marfurt et al. 2003; Mary et al. 2004; Mauricio et al. 2004; Minodier et al. 1997; Montalvo et al. 2014; Rodgers et al. 1990; Salotra et al. 2001; Schönian et al. 2003; Sreenivas et al. 2004; Van Eys et al. 1992; Zelazny et al. 2005). The targets and primers can be genus, subgenus, species, or strain *Leishmania* specific. Depending on molecular approach, purpose of the diagnosis, and epidemiological context, the choice of appropriate genetic targets could vary (Tables 7.1, 7.2, and 7.3), usually a globally applicable typing strategy is desirable rather than those validated in limited areas.

In general, high-copy-number targets have been shown to be most sensitive; consequently they are chosen for parasite primary diagnosis, quantification, and viability studies in which sensitivity must be maximized; however they are not species-specific. In contrast, for *Leishmania* species typing, species-specific sequences are targeted, which, however, show serious limitation of sensitivity; consequently they should be combined with a genus-specific assay to rule out false negatives (Van der Auwera and Dujardin 2015). In view of this problem, a

Table 7.1 DNA target organism, primer sequences, and PCR product sizes of conventional PCR protocols for the detection of *Leishmania* genus/subgenus

Target region/*Leishmania* sp.	Primers sequence (5′–3′)	Product (bp)	Reference
SSUrRNA/*Leishmania* spp.	R221-F: GGTTCCTTTCCTGATTTACG R332-R: GGCCGGTAAAGGCCGAATAG	603	Van Eys et al. (1992)
	R223-Fn: TCCCATCGCAACCTCGGTT R333-Rn: AAAGCGGGCGCGGTGCTG	358	Cruz et al. (2002)
SSUrRNA/*L.(V.) braziliensis*	R251-F: TGACTAAAGCAGTCATTC R798-R: GGCCGGTAAAGGCCGAATAG	650	Meredith et al. (1993)
kDNA/*Leishmania* spp.	Mary-F: CTTTTCTGGTCCTCCGGGTAGG Mary-R: CCACCCGGCCCTATTTTACACCAA	120	Mary et al. (2004)
	13A: TGGGGGAGGGGCGTTCT 13B: ATTTTACACCAACCCCCAGTT	120	Rodgers et al. (1990) and Lachaud et al. (2002)
kDNA/*L. donovani cx*	LdI-F: AAATCGGCTCCGAGGCGGGAAAC LdI-R: GGTACACTCTATCAGTAGCAC	592	Salotra et al. (2001)
	LdIn-F: TCGGACGTGTGTGGATATGGC LdIn-R: CCGATAATATAGTATCTCCCG	385	Sreenivas et al. (2004)
kDNA/*Leishmania (Viannia) braziliensis*	LEIB1: GGGGTTGGTGTAATATAGTGG LEIB2: CTAATTGTGCACGGGGAGG	750	De Bruijn and Barker (1992)

bp base pairs, *F* forward, *R* reverse, *Fn* forward in nested PCR, *Rn* reverse in nested PCR

Table 7.2 DNA target, primer sequences, and PCR product sizes of conventional PCR protocols for the detection of *Leishmania* spp

Target region	Primers sequence (5′–3′)	Product (bp)	Reference
ITS	IR1: GCTGTAGGTGAACCTGCAGCAGCTGGATCATT IR2: GCGGGTAGTCCTGCCAAACACTCAGGTCTG	1000–1200[a]	Cupolillo et al. (1995) and Buitrago et al. (2011)
ITS1	LITSR: CTGGATCATTTTCCGATG L5.8S: TGATACCACTTATCGCACTT	300–350[a]	el Tai et al. (2000), Schönian et al. (2003), Mauricio et al. (2004) and Khatri et al. (2009)
hsp70	Hsp70sen: GACGGTGCCTGCCTACTTCAA Hsp70ant: CCGCCCATGCTCTGGTACATC	1422	Montalvo et al. (2014)
ME	ME-F: TATTGGTATGCGAAACTTCCG ME-R: ACAGAAACTGATACTTATATAGCG	450	Marfurt et al. (2003)
7SL-RNA	TRY7SL.F1.M13: GTAAAACGACGGCCAGTGCTCTGTAACCTCGGGGCT TRY7SL.R1.M13: CAGGAAACAGCTATGACGGCTGCTCCGTYNCCGGCCTGACCC	137–139[a]	Zelazny et al. (2005)
Repetitive nuclear sequence	T2: CGGCTTCGCACCATGCGGTG B4: ACATCCCTGCCCACATACGC	250	Minodier et al. (1997)

bp base pairs, *F* forward, *R* reverse
[a]According to *Leishmania* species amplified

Table 7.3 DNA target, primer sequences, and PCR product sizes of conventional PCR protocols for the detection of *Leishmania* spp. and/or strain

Target region	Primers sequence (5′–3′)	Product (bp)	Reference
gp63	MUS: GTGGGTGTCATCAACATCCC MUSA3: CTGCTGCCGTACACCTGGAC	870	Victoir et al. (2003) and Guerbouj et al. (2001)
ME	ME-*F*: TATTGGTATGCGAAACTTCCG ME-*R*: ACAGAAACTGATACTTATATAGCG	450	Marfurt et al. (2003)
7SL-RNA	TRY7SL.*F*1.M13: GTAAAACGACGGCCAGTG CTC TGTAACCTCGGGGCT TRY7SL.*R*1.M13: CAGGAAACAGCTATGACGGCTGCTCCGTYN CCG GCC TGACCC	137–139[a]	Zelazny et al. (2005)
cpB	PIGS1A:-CCT CAT TGCTTTGGTCCTGG PIGS2B: GGCGTGCCCACGTATATCGC	1400	Quispe Tintaya et al. (2004)
	CPB*F*: CGAACTTCGAGCGCAACCT CPB*R*: CAGCCCAGGACCAAAGCAA	1079	Quispe Tintaya et al. (2004)
Minicircles of kDNA	DRJ: CGATTTTTGAACGGGATTTCTGCAC KLK2: CTCCGGGGCGGGAAACTGG	800	Morales et al. (2002) and Laurent et al. (2007) Cortes et al. (2006)
ITS	IR1: GCTGTAGGTGAACCTGCAGCAGCTGGATCATT IR2: GCGGGTAGTCCTGCCAAACACTCAGGTCTG		Cupolillo et al. (1995)
ITS1	L5.8S: TGATACCACTTATCGCACTT LITSR: CTGGATCATTTTCCGATG	300	Schönian et al. (2003)
ITS2	L5.8SR: AAGTGCGATAAGTGGTA LITSV: ACACTCAGGTCTGTAAAC	650	Chicharro et al. (2002)
SSU rDNA	SSU-12103-D: GGGAATATCCTCAGCACGT 5.8S-13,333-R: CGACACTGAGAATATGGCATG	1200	Rotureau et al. (2006)

bp base pairs, *F* forward, *R* reverse
[a]According to *Leishmania* species amplified

diagnostic workflow based on the initial amplification of genus/subgenus-specific sequences (Table 7.1) followed by subsequent differentiation of *Leishmania* species by RFLP or sequencing of the specific amplified sequences (Tables 7.2 and 7.3) is considered a useful approach. The first diagnostic step usually makes use of targets of the *Leishmania* genus or subgenus: high-copy-number targets such as kDNA minicircles (about 10,000 copies per cell), SSU rDNA (40–200 copies per cell), or ME (100–200 copies per cell) genes have been shown to be most sensitive. Despite

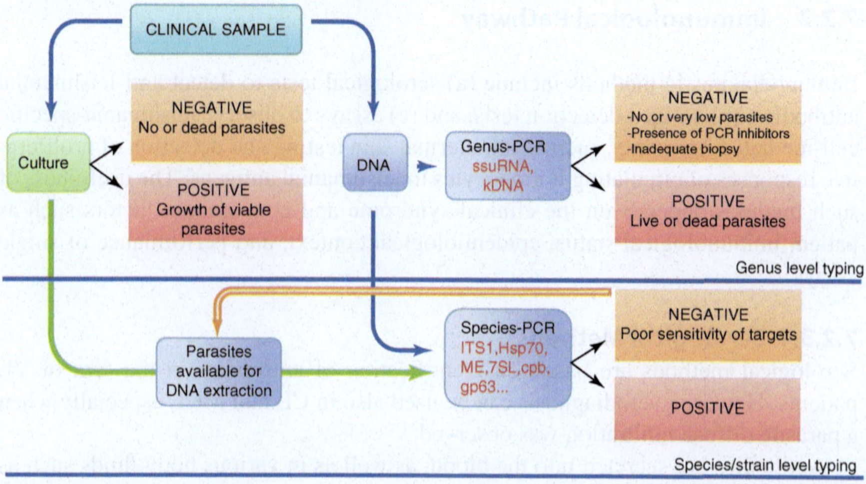

Fig. 7.2 Molecular workflow for leishmaniasis diagnosis

the difficulty of comparison of molecular methods, it is believed that the highest sensitivity is achieved when kDNA minicircles are targeted in real-time PCR (Cruz et al. 2013; Mary et al. 2004).

After *Leishmania* infection has been confirmed, the second step expects the identification below the genus/subgenus level species complex or species level. In this case, *Leishmania* species typing methods should allow to discriminate many species with less emphasis on sensitivity. Both ITS1 and hsp70 are considered the best targets for *Leishmania* species identification, in Old and New World, respectively (Table 7.2) (Van der Auwera et al. 2014, 2016).

A genotyping assay to strain level could be interesting to resolve clinical (e.g., distinguish between recrudescence and re-infection) or epidemiological questions. Many studies have used the sequencing or PCR-RFLP assays of both coding and noncoding sequences of multigene families (Table 7.3): minicircles kDNA (Chicharro et al. 2002; Cortes et al. 2006; Laurent et al. 2007; Morales et al. 2002), cpB (Quispe Tintaya et al. 2004), gp63 (Guerbouj et al. 2001; Victoir et al. 2003), ITS (Cupolillo et al. 1995; Schönian et al. 2011; Rotureau et al. 2006), and ME (Marfurt et al. 2003; Mauricio et al. 2004). However these assays show some limitations because they are not fully reproducible and comparable between laboratories due to the lack of validation and problem of wide variability within single strains; furthermore, they, having been standardized on not all *Leishmania* species or on a single strain for each species, are not usable on a global scale, but should be recommended for comparative studies between few strains (Schönian et al. 2011).

Considering the different sensitivity degrees of targets, sometimes it is not possible to obtain results of *Leishmania* typing at species and strain level from biological samples that resulted positive at the first step, unless the load of the parasite is previously amplified in culture (Fig. 7.2).

7.2.3 Immunological Pathway

Immunodiagnostic methods include (a) serological tests to detect anti-leishmanial antibodies, (b) antigen detection tests, and (c) assays to observe leishmania-specific cell-mediated immunity, such as intradermal skin testing and detection of proliferative responses of circulating lymphocytes to leishmanial antigens. The usefulness of such methods depends on the clinical syndrome and concomitant factors such as patient immunological status, epidemiological context, and performance of single assays.

7.2.3.1 Serological Methods

Serological methods are based on identification of antibodies in the sera of VL patients. However, serodiagnosis can be used also in CL and MCL especially when a parasite diffuse infiltration was observed.

Antibodies are secreted into the blood, as well as in various body fluids such as urine and saliva which have been shown to be useful in the diagnosis with various degrees of sensitivity. The development of recombinant antigens has considerably improved the sensitivity and specificity of immunological diagnosis over crude antigens. The currently available serodiagnostic tests have been based on five major formats: indirect immunofluorescence test (IFAT), enzyme-linked immunosorbent assay (ELISA), immunoblotting (or western blotting), agglutination test (DAT and LAT), and immunochromatographic tests (ICT), commonly referred to as rapid diagnostic tests (RDTs).

All serological tests are lacking in some features. Because specific antibodies remain detectable up to several months or years after cure in VL, the methods may not reliably differentiate remote from recent or current infection. For the same reason, relapse cannot faithfully be diagnosed by serology. Therefore, a significant proportion of healthy people living in endemic areas may have asymptomatic infections; thus it may result positive for anti-leishmania antibodies although to a low titer. Moreover, serological tests are generally unhelpful for diagnostic purpose in immunocompromised individuals. With regard to these limitations, antibody-based tests must always be used in combination with a standardized clinical case definition for VL diagnosis (WHO 2010).

IFAT The test is based on detecting of anti-leishmanial antibodies by employing specific antigens often homemade (cultured promastigotes represent the commonest antigen source) and secondary antibodies (anti-immunoglobulin antibody) conjugated with a fluorescent dye. The anti-leishmanial antibodies are demonstrated in the very early stages of VL infection but are undetectable 6–9 months after cure (WHO 2010; Chappuis et al. 2006). On the contrary, the persistence of antibodies in low titers is an indication of a probable relapse. The test is genus specific, although significant cross-reactivity has been reported with trypanosomiasis, malaria, tuberculosis, typhoid fever, brucellosis, and fungal infections (Kohanteb and Ardehali 2005). In Chagas' disease-free areas, IFAT showed for VL diagnosis a sensitivity of 96% and specificity of 98%, values similar to ELISA ones (Gradoni and Gramiccia

2014). The IFAT remains a routine serological assay in some endemic regions because it is easy to perform, although it is time-consuming, not fully automated and it requires relatively well-equipped laboratories.

ELISA The majority of the immunological techniques for detection of anti-leishmanial antibodies have been based on ELISA, in which sensitivity and specificity depends on the antigen used. Although several antigens have been so far identified, the better results are obtained by antigen k39 coded by kinesin-related gene that contains a repetitive epitope of 39 amino acid residues. The recombinant product rK39 was shown to be an early marker for disease progression in VL with sensitivity and specificity of 100% and 96%, respectively. Furthermore, the anti-K39 antibody titers directly correlate with 98% of cases of active VL (Badaró et al. 1996) and have potential in monitoring the chemotherapy and in predicting the clinical relapse (Kumar et al. 2001). In addition, an interesting feature of this antigen is that it can be used in HIV-coinfected patients in which anti-K39 antibody levels decline rapidly with the treatment success. ELISA is not used in the poor endemic regions due to the requirement of sophisticated equipment (de Paiva-Cavalcanti et al. 2015).

Immunoblotting It provides detailed antibody responses to various leishmanial antigens. This test is more sensitive than IFAT and ELISA, but expensive, time-consuming, and technically cumbersome. Specific antibodies can also be detected by immunoblotting where the soluble proteins obtained by logarithmic phase lysed promastigotes are run on sodium dodecyl sulfate-polyacrylamide gel electrophoresis gels. The separated proteins are electroblotted onto a nitrocellulose membrane and probed with patient's serum. The sensitivity of this technique can be enhanced using the chemiluminescent antibody probes (Mary et al. 1992; Santos-Gomes et al. 2000).

Agglutination Tests Serum-based direct agglutination tests (DAT) or urine-based latex agglutination tests (LAT) are routinely used for determining anti-leishmanial antibodies or antigens in VL patients using serum, plasma, or urine samples. These tests are suitable for both field and laboratory application (Chappuis et al. 2006; Srivastava et al. 2011; Srividya et al. 2012; Sundar et al. 2006).

DAT is simple, cost-effective, reliable, highly specific, and sensitive, safe, and adaptable to microtiter plates. This test has been validated in several endemic areas (Boelaert et al. 2004) and used for VL diagnosis in different regions for the past two decades (de Paiva-Cavalcanti et al. 2015). The antigen usually consists of cultured promastigotes, and the test can be carried out on serum, plasma, and whole blood (Moody and Chiodini 2000). DAT is a semi-quantitative method and uses microplates with V-shaped wells in which Coomassie brilliant blue stained killed *L. donovani* promastigotes are mixed with increasing dilutions of patient's specimen. After an overnight incubation, if specific antibodies are present, agglutination is visible to the naked eye (Gradoni and Gramiccia 2014). The DAT titer is indicated as the highest dilution at which agglutination is still visible. Attention is needed on determining the cutoff values for a positive test (threshold titers ranged from 1: 800 to 1: 640,033) that depend on the epidemiological context. DAT has an important limitation based on the long incubation period (12–18 h). According to a meta-analysis study, the sensitivity and specificity of DAT were estimated at 94.8% and 97.1%, respectively. Nevertheless, when DAT was performed in patients from East Africa, it showed a lower sensitivity in comparison to the Indian subcontinent immunocompetent patients (Chappuis et al. 2006). Cross-reactivity has been reported among patients with Chagas disease, malaria, leprosy and schistosomiasis. Moreover, DAT remains positive for a long time after recovery, thus cannot be used to assess cure or for diagnosis of relapses. Furthermore, as ICT, its use in endemic areas is limited because about 20–30% of healthy individuals are DAT positive (Banu et al. 2016; Sundar et al. 2006).

Antigenic detection tests are considered more rapid, simple, noninvasive, and reliable for VL diagnosis than antibody detection ones. Furthermore, these tests offer a viable alternative to antibody detection methods in immunocompromised patients with reduced antibody production, such as in *Leishmania*-HIV-coinfected patients. They are theoretically considered more specific in avoiding cross-reactivity. Among the several antigen detection methods, LAT test (KAtex, Kalon Biological) is the most used. KAtex assay detects a low molecular weight (5–20 kD) glycoconjugate, a heat-stable *Leishmania* antigen in urine samples employing latex beads sensitized with antibodies raised against *L. donovani* antigen. This antigen is present in both promastigote and amastigote forms of the parasite. The method is easy, rapid (2 min for the final agglutination), and appropriate for the primary VL diagnosis, for the detection of subclinical infection, and for monitoring the efficacy of treatment. KAtex displays positive predictive value and high specificity (82–100%) but variable sensitivity (47–95%). Higher sensitivity and specificity have been reported in immunocompromised patients, 85–100% and 96–100%, respectively (Sakkas et al. 2016; Srividya et al. 2012).

ICT Strips Immunochromatographic strips using rK39 antigen have become popular in recent years. It has been developed as a commercially available rapid test for use in difficult field conditions. The rK39 is fixed on a nitrocellulose membrane, and colloidal gold-protein A is used for detection. A drop of serum or blood is smeared over the tip of the strips and soaked in 4–5 drops of phosphate-buffered saline. If the antibody is present, it will react with the conjugate (colloidal gold-protein A) reacting with rK39 antigen on the strip, yielding a pink band. The result will be readable within 15 min. A meta-analysis (Chappuis et al. 2006) and a multicenter evaluation (Boelaert et al. 2008) confirmed the findings of high diagnostic accuracy of the rK39 ICT. However, ICT suffers to be positive in a significant proportion of healthy individuals in endemic regions and for long periods after VL cure.

Recently, a rK39 rapid test using saliva demonstrated low sensitivity; therefore, saliva is considered an unsuitable sample for the VL diagnosis (Vaish et al. 2012).

TDR/WHO carried out a worldwide multicenter evaluation of commercially available RDTs for VL, and it concluded that diagnostic accuracy varies between the major endemic regions. In Brazilian and East African territories, several RDTs, used alone, are inadequate for VL diagnosis (Cunningham et al. 2012).

7.2.3.2 Test for Cellular Immunity

Delayed hypersensitivity is an important feature of all forms of human leishmaniasis and can be measured by the leishmanin skin test (LST), also known as the Montenegro reaction (Manson-Bahr 1987). Leishmanin is a killed suspension of whole (0.5–1 × 10^6/ml) or disrupted (250 μg protein/ml) promastigotes in pyrogen-free saline containing phenol. A delayed reaction develops (induration ≥5 mm) and is read at 48–72 h. The false-positive reaction rate in otherwise healthy people is approximately 1%, but this can be higher in areas where there is a background of leishmaniasis, as many of the healthy population may show quite high rates of leishmanin sensitivity. Although there is complete cross-reactivity among all strains of *Leishmania*, heterologous antigens often give smaller reactions, which may be caused by difficulty in standardization. In VL it only measures past infections because during active disease, a complete anergy is found (Gradoni and Gramiccia 2014); in the opposite LST is used in the CL and MCL clinical diagnosis. Leishmanins are not available commercially worldwide.

Patients who recover from leishmaniasis usually show development of strong immunity and induction of gamma interferon. More recently, the measurement of gamma interferon (IFN-γ) as a surrogate marker of cellular immune responses has been assayed in human leishmaniasis, using a format similar to that used for tuberculosis (e.g., QuantiFERON-TB Gold), by using ELISA to measure the cytokine secreted by leukocytes in response to *Leishmania* antigen stimulation (Turgay et al. 2010). Different antigens and *Leishmania* species were evaluated, but results, although encouraging, were contradictory and need further research (Alimohammadian et al. 2012; Gidwani et al. 2011; Singh and Sundar 2015; Turgay et al. 2010). The standardization of these procedures should be emphasized as an alternative to the LST presently unavailable as good manufacturing practice product in many countries.

7.3 Visceral Leishmaniasis

VL implies a potentially fatal parasitic disease caused by parasites belonging to *L. donovani* complex (*L. donovani*, *L. infantum*). Although VL in patients from known endemic areas is usually diagnosed clinically, more often the signs and symptoms are inconclusive and mistaken with other co-endemic diseases (Sakkas et al. 2016).

The complete VL clinical aspects in immunocompetent and immunocompromised patients were reported in Chap. 6. Taking into account the different performances of the diagnostic tests for VL in immunocompetent and immunocompromised individuals, the variability of disease presentation, and the different conditions of the patients, the best approach for VL diagnosis is the combination of a direct (parasitological or molecular) and immunological methods, possibly to be repeated if initial results are negative but clinical suspicion remains (Van Griensven et al. 2014)

Table 7.4 Available methods for the diagnosis of human VL and their purpose

Method	Confirmation of primary clinical case	Contribution to treatment follow-up	Confirmation of clinical relapse	Prevalence of leishmanial exposure	HIV-*Leishmania* coinfection
Agent detection[a]					
Microscopy	+++	–	+++	–	+++
In vitro culture	++	+	++	+	++
PCR	+++	+++	+++	++	+++
Real-time PCR	+++	+++	+++	++	+++
Immune response or antigen detection[b]					
IFAT	++	+	++	–	+
ELISA	+++	++	+++	+	+
DAT	++	–	–	+	++
LAT	++	++	++	+	++
ICT rk39	++	–	++	+	+
Tests for cellular immunity	–	+++	–	+++	–

[a]A combination of agent identification methods applied on the same clinical sample is recommended

[b]One of the listed serological tests is sufficient, +++ recommended method, ++ suitable method, + may be used in some situations, but cost, reliability, or other factors severely limit its application, – not appropriate for this purpose, *PCR* polymerase chain reaction, *IFAT* indirect fluorescent antibody test, *ELISA* enzyme-linked immunosorbent assay, *DAT* direct agglutination test, *LAT* latex agglutination test, *ICT* immunochromatographic test

(Table 7.4). In particular, one positive direct method plus one immunological test are requested for a standardized VL diagnostic pathway (WHO 2010). No tests are currently available that can detect asymptomatic *L. donovani* complex infection or predict progression of infection to clinical VL disease (Singh and Sundar 2015).

7.3.1 Direct Diagnosis

7.3.1.1 Sample Collection

Isolation and identification of the parasite from tissue aspirates or biopsy have been used for the direct visualization of parasites by parasitological and molecular procedures. VL biopsy samples are aspirate of bone marrow (the most common), spleen (the most sensitive but feared for bleeding complications), lymph nodes (only in foci where lymphadenopathy is frequent, e.g., Sudan), liver biopsy (not frequently used), and buffy coat of peripheral blood (noninvasive method but less sensitive). Occasionally the use of samples collected from unusual sites as cerebrospinal fluid, digestive mucosa, bronchoalveolar and pleural liquid has been described in HIV or immunosuppressed patients (Van Griensven et al. 2014).

7.3.1.2 Parasitological Assays

Ideally, all cases of leishmaniasis should be confirmed by demonstration of the parasite, which is straightforward (except for needing an invasive procedure) if parasites are plentiful (e.g., in kala-azar) but otherwise can be difficult (e.g., VL by *L. infantum*).

The visualization of the parasite amastigote form by microscopic examination of aspirates from bone marrow, spleen, or lymph nodes is regarded as the most suitable diagnostic instrument for VL control. The technique is advantageous as it is directly able to detect parasite load and is inexpensive to perform. The specificity of this technique is high, although the sensitivity varies depending on the tissue used, being higher for spleen (93–99%) than for bone marrow (53–86%) or lymph node (53–65%) aspirates (Chappuis et al. 2007; Siddig et al. 1988). However, the procedure for splenic aspiration carries the risk of fatal internal bleeding or complicated by life-threatening hemorrhages in ~0.1% of individuals and therefore requires considerable technical expertise (Chappuis et al. 2007). Bone marrow aspirate is the commonest method employed being safer than splenic aspirate and liver biopsy. Its sensitivity is higher in HIV-coinfected patients. The lowest sensitivity of parasitological diagnosis is observed in peripheral blood smears, as parasitemia in immunocompetent individuals with VL is very low.

The use of microscopy in the diagnosis of VL offers the benefits of high specificity and the possibility of grading the parasite on a logarithmic scale (0–6+). Nevertheless, a negative bone marrow (or spleen) aspiration finding does not completely rule out VL, and there are several reports in which VL was diagnosed only after repeated bone marrow examinations (WHO 2010).

The detection of parasites in biopsy samples by culture is more sensitive than microscopic examination, but these techniques remain restricted to referral hospitals and research centers, despite efforts to simplify them. When attempting primary isolation of unknown organisms, a blood agar-based medium, preferably EMTM or NNN medium, should be used. Liquid media commercially available have less efficiency in VL and therefore a lower diagnostic sensitivity (Gradoni and Gramiccia 2014). Advancements in the culture method were obtained by sensitive MCM for *Leishmania* isolation (Allahverdiyev et al. 2005; Serin et al. 2005).

7.3.1.3 Molecular Assays

n-PCR and qPCR have been largely employed for the optimization of new VL diagnostic assays. In Table 7.5 is reported an overview of advantages and limitations of principal molecular methods. Given their high sensitivity, they have the capacity to estimate the parasitological burden in several less invasive specimens, thus bringing security to the patient.

The follow-up of the treatment, the evaluation of drug efficacy are common approaches performed by the qPCR (Antinori et al. 2007; de Vries et al. 2006; Van der Meide et al. 2005, 2008). Different protocols have been optimized to obtain high sensitivity. The range of sensitivity can vary depending on DNA *Leishmania* targets. SSU rDNA real-time PCR showed a sensitivity of 0.625 parasites/ml of peripheral blood (Bossolasco et al. 2003), whereas *L. infantum* kDNA real-time PCR reached

Table 7.5 Advantages and limitations of molecular methods used in human VL diagnosis

Assay	Requested skill level	Advantage	Limitation
PCR	Medium	High sensitivity, specificity, and accurate results. Easy diagnostic interpretation. Possibility of species differentiation by RFLP	Qualitative test. Unable to quantify the target DNA. Time-/cost-consuming. Limited detection range of some assays
Nested PCR	Medium	Higher specificity and sensitivity. Possibility of species differentiation by RFLP	Time-/cost-consuming. Qualitative test. Unable to quantify the target DNA
Real-time PCR	High	Higher sensitivity, specificity, security, and speedy results. Quantitative capacity. Useful in VL follow-up and/or in HIV coinfections, evaluation of drug efficacy	High cost. Difficulty in interpreting the results
QT-NASBA	Medium	High specificity. Quantitative capacity. Indicated to detect active diseases; RNA detection. Useful in evaluation of drug efficacy	Handling steps and time-consuming procedure. Assays developed only for RNA detection. Few studies yet
NASBA-OC	Low	High specificity. Speedy results. Simple dipstick format for the detection of amplification products. Applicable in the field. RNA detection	Unable to quantify the target RNA. Assays developed only for RNA detection. Few studies yet
LAMP	Low	High sensitivity. Low cost. Applicable in the field. Test of cure, marker for active disease	Unable to quantify the target DNA. Qualitative test. Few studies yet

PCR polymerase chain reaction, *QT-NASBA* quantitative nucleic acid sequence-based amplification, *NASBA-OC* nucleic acid sequence-based amplification oligochromatography, *LAMP* loop-mediated isothermal amplification, *RFLP* restriction fragment length polymorphism

a sensitivity of 0.0125 parasites/ml of peripheral blood from Mediterranean VL patients (Mary et al. 2004). An additional interest topic seems to be the kinetic study of parasitemia in the immunocompromised host, allowing biological diagnosis of relapses. Indeed, different studies showed that *Leishmania* DNA levels correlated with the clinical course of VL, and their measurement at diagnosis and during and after treatment seemed to be useful in the clinical management of HIV-infected patients. Moreover, the range of variation of *Leishmania* parasites showed a good correlation with the clinical status of the VL subjects, thus allowing to discriminate between symptomatic patients, cured patients, and asymptomatic carriers (Mary et al. 2004; Bossolasco et al. 2003).

Nevertheless, the PCR-based diagnostic tests are considered a marker of infection rather than disease, and PCR alone is of less value for VL diagnosis in endemic regions due to the fact that only diseased persons are treated. Hence, molecular tests should always be interpreted in combination with a standardized clinical case definition (WHO 2010). Moreover, molecular diagnosis should not replace the existing immunodiagnostic assays but give an added value as a marker of infection.

Innovative procedures as NASBA and LAMP show high sensitivity and specificity. In particular NASBA showed a sensitivity of 80–97% (according to the samples) and specificity of 89–100% for *Leishmania* detection in patients from different epidemiological areas (Basiye et al. 2010; Deborggraeve et al. 2008b; Mugasa et al. 2010; Saad et al. 2010). Analogously, LAMP showed sensitivity of 80–96% and specificity of 98–100%; moreover it was also considered a good marker for active disease and as a test of cure (Adams et al. 2010; Khan et al. 2012). However, both procedures are not yet an option for routine VL diagnosis.

7.3.2 Immunological Diagnosis

7.3.2.1 Serological and Antigen Detection Assays

VL serological diagnosis can be performed by many assays, such as IFAT, ELISA, immunoblotting, DAT, rK39 ICT, and KAtex. Usefulness, sensitivity, and specificity of these tests were reported in Table 7.6. Selection of the serological test during VL diagnosis should be based on different parameters such as region, cost, sensitivity, specificity, feasibility, sustainability, and field applicability, especially in problematic endemic areas. Regarding *Leishmania*-HIV coinfection, serological methods show variable low sensitivity as the determination of antibodies is difficult due to the immunosuppressive action of the virus. However, DAT, immunoblotting, and KAtex have been proven to be superior in diagnosis of these coinfection cases. The combination of different techniques is a strategy that may improve the global specificity and sensitivity.

7.3.2.2 Cellular Immunity Assays

LST is a marker of cell-mediated immunity and remains positive for many years after the initial infection. LST is known to become negative during the anergic phase of acute VL and to return to positive once the cell-mediated immune response is restored (Zijlstra et al. 1994). Due to these limitations, LST is usually used during VL epidemiological surveys furnishing over the decades numerous data on VL prevalence and disease distribution (Bettini et al. 1983; Schaefer et al. 1994; Schenkel et al. 2006; Zijlstra et al. 1994). It represents a valid epidemiologic tool to characterize the *Leishmania* prevalence exposure of populations in endemic areas (Bern et al. 2007). In vitro IFN-γ release assays have been recently developed to document latent infections. In Turkey the use of a modified QuantiFERON assay to detect the cell-mediated immune response against *L. infantum* was explored using three different *Leishmania*-specific antigens (Turgay et al. 2010) with encouraging but not conclusive results. In India five different antigens were used, and the assay based on soluble *Leishmania* antigen showed a sensitivity of 80% and specificity of 100%, even if IFN-γ response in patients with active VL needs further research (Gidwani et al. 2011). In Iran similar assays showed a higher IFN-γ-positive

Table 7.6 Serological methods suggested for VL diagnosis

Assay	Requested skill level	Sensitivity	Specificity	Comments	Reference
IFAT	Expert	96%	98%	Positive in the early stages of infection and undetectable 6–9 months after cure. Requires equipped laboratory setup, time- and/or cost-consuming	Boelaert et al. (2008) and Badaró et al. (1983)
DAT	Medium	95%	97%	Rapid test, applicable in the field, useful for epidemiological studies, useful in HIV coinfections	Chappuis et al. (2006)
ICT rK39	Low	94–100%	81–96%	Inexpensive, rapid, simple, and can be performed by untrained person. Useful for screening, not useful in treated patients	Chappuis et al. (2006) and Boelaert et al. (2008)
ELISA rK39	Medium	100%	96%	Useful for monitoring of chemotherapy, predicting clinical relapses, follow-up of HIV coinfections	Kumar et al. (2001) and Sundar and Rai (2002)

IFAT indirect fluorescent antibody test, *DAT* direct agglutination test, *ICT* immunochromatographic test, *ELISA* enzyme-linked immunosorbent assay

response in patients from zoonotic *L. infantum* VL endemic areas than patients from CL areas. In addition data indicate that LST is significantly more sensitive than IFN-γ levels in CL than VL-cured persons (Alimohammadian et al. 2012).

7.4 Cutaneous Leishmaniasis

CL implies a broad spectrum of diseases where the patient generally shows one or several ulcer(s) or nodule(s) in the skin (Chappuis et al. 2007) (for clinical details see Chap. 6). They are caused by different *Leishmania* species, and two different groups of pathologies are distinguished: (a) localized CL caused in the Old World by *L. major*, *L. tropica*, and *L. infantum* and in the New World by *L. guyanensis*, *L. peruviana*, *L. braziliensis*, and *L. infantum* and (b) diffuse CL caused by *L. aethiopica* in the Old World and *L. amazonensis* and *L. mexicana* in the New World. However one *Leishmania* species can show different clinical complaints (e.g., *L. infantum*). As the clinical presentation lacks specificity and treatment is costly, diagnostic confirmation is basic (WHO 2010) and direct parasitological diagnosis of CL lesions is necessary as neither clinical examination nor serology is adequate (Table 7.7).

Table 7.7 Methods available for diagnosis of human CL and their purpose

Method	Confirmation of primary clinical case	Prevalence of *Leishmania* exposure
Agent detection[a]		
Microscopy	+++	–
In vitro culture	++	–
Conventional PCR	+++	–
Real-time PCR	+++	–
Immune response detection[b]		
IFAT	+	–
ELISA	++	–
Tests for cellular immunity	+++	+++

[a]A combination of agent identification methods applied on the same clinical sample is recommended

[b]One of the listed serological tests is sufficient, +++ recommended method, ++ suitable method, + may be used in some situations, but cost, reliability, or other factors severely limit its application, – not appropriate for this purpose, *PCR* polymerase chain reaction, *IFAT* indirect fluorescent antibody test, *ELISA* enzyme-linked immunosorbent assay

7.4.1 Direct Diagnosis

7.4.1.1 Sample Collection

Direct diagnosis by parasitological or molecular procedures remains the reference standard for diagnosis due its high specificity. The sensitivities vary over a broad range depending on *Leishmania* species and stage of the lesion. Skin material is obtained by superficial scraping with a scarifier, a needle aspiration, or a standardized CL punch biopsy. A 2–4 mm skin punch biopsy provides abundant material useful when the parasite load is low. The sample collection site is the determining factor in the detection of parasite and depends on the clinical type of the lesion. Usually the inflammatory border of the lesion is the elected place for parasite detection.

7.4.1.2 Parasitological Assays

Parasitological diagnosis appears the gold standard in the CL diagnosis for its high specificity both as microscopic examination and in vitro cultures, used individually or in combination. Direct impression smears stained with Giemsa is the best procedure and often is the only available method in endemic areas. Characteristics, advantages, and disadvantages of the procedure are the same reported in VL. The sensitivity of the direct examination is low, at approximately 50–70% in the Old World and even lower, at approximately 15–30%, in the New World where chronic cases and MCL are frequent (Goto and Lindoso 2010).

The parasite culture on simple media requires the homogenization of the skin sample performed in phosphate-buffered saline or culture medium (liquid phase for biphasic media) under sterile conditions. Due to the microbiologically dirty site of skin lesion, a careful addition of antibiotics, preferably gentamicin (100 μg/ml) or penicillin 100 U/ml + streptomycin (100 μg/ml) as alternative, and antifungal compounds such as 5-fluorocytosine (100 μg/ml) are suggested; they should be used initially to avoid bacterial, yeast, or fungi contamination or their overgrowing. The organisms from patients with chronic CL can be very difficult to cultivate. The parasites sometimes die when subcultured, even when the initial isolation is successful. This seems especially common when the initial isolation has been into a rich medium. Often this can be overcome if subcultures are made into less nutritionally rich media, such as NNN, or one of the semisolid media such as "Sloppy Evans" or semisolid Locke blood agar (Evans 1987; Gramiccia and Gradoni 1989). The detection level is higher, reaching 44–58% by culturing the samples. Advancements in the culture method were obtained by MCM (Serin et al. 2005).

7.4.1.3 Molecular Assays

Classical molecular techniques have been increasingly recommended for CL diagnosis due to their accuracy and speed. Different biological samples were used: impression smear (stained or not) (Khatri et al. 2009), scarification of the edge lesion (Neitzke-Abreu et al. 2013), biopsy skin (Pita-Pereira et al. 2012; El-Beshbishy et al. 2013; Di Muccio et al. 2015), skin aspiration fluid (Ozensoy Toz et al. 2013), and peripheral blood (diffuse CL), confirming the higher sensitivity of the molecular method. PCR for etiological diagnosis reaches sensitivity close to 100% in CL (Nasereddin et al. 2008) and 97.1% in MCL (Oliveira et al. 2005) employing the most widely used targets, *Leishmania* kDNA or SSU rRNA. Van der Meide et al. (2008), comparing in the New World the performance among qPCR, RT-qPCR, and QT-NASBA, concluded that RT-qPCR and QT-NASBA are the most sensitive assays.

A common diagnostic approach in CL is the application of *Leishmania* typing in regions where several *Leishmania* species coexist (see New World, Middle East, or Asia). For example, n-PCR-RFLP is performed by using scraped off slides with impression smears from suspected CL cases (Khatri et al. 2009, 2016). A further interesting topic is the application of molecular methods to the quantitative assessment of the parasite load in CL and MCL patients. QT-NASBA assay has been used to monitor parasite load in skin biopsies of CL patients after treatment predicting the clinical outcome. Jara et al. (2013) and Van der Meide et al. (2008) standardized a kDNA qPCR assay highly sensitive and accurate for detection and quantification of *Leishmania* (*Viannia*) spp. showing that CL lesions had tenfold-higher parasite loads than MCL lesions. Moreover, among CL patients, the parasite load was inversely correlated with disease duration, but there was no difference in parasite load according to the parasite species, the patient's age, and the number or size of lesions. They suggest that parasite loads may differ between CL and MCL lesions and between acute and chronic CL, according to histopathology studies.

7.4.2 Immunological Diagnosis

7.4.2.1 Serological Assays

Serodiagnosis is not the usual methods for CL diagnosis due to low sensitivity in the Old World, but is routine in some centers in the New World (WHO 2010). More commonly used assays are IFAT and ELISA. In agreement with the *Leishmania* species, epidemiological context, and clinical manifestation, they show different sensitivity and specificity but anyway low. In CL, the anti-*Leishmania* antibody level does not remain high after treatment, and therefore positive results generally indicate current infection (Goto and Lindoso 2010).

7.4.2.2 Cellular Immunity Assays

LST is used in clinical diagnosis of CL and prevalence studies (Arbaji et al. 1993). LST and QuantiFERON were compared in Iran in a randomized trial in persons residing in areas endemic for zoonotic or anthroponotic CL. The results suggest that LST is significantly more sensitive than IFN-γ levels in persons who have been cured of CL (Alimohammadian et al. 2012). An absence of cutaneous delayed hypersensitivity to LST and a low or absent IFN-γ production by peripheral blood mononuclear cells characterize diffuse CL presentation.

7.5 Mucocutaneous Leishmaniasis

In MCL, patients suffer from progressively destructive ulcerations of the mucosa, extending from the nose and mouth to the pharynx and larynx. These lesions are not self-healing and are usually seen months or years after a first episode of CL, when the macrophages of the naso-oropharyngeal mucosa become colonized (Chappuis et al. 2007) (for clinical details see Chap. 6). Whereas MCL is typically associated with *L. braziliensis*, there are increasing reports of mucosal lesions (with or without concurrent cutaneous or visceral disease) caused by *L. infantum, L. donovani*, *L. major*, and *L. tropica* (Gradoni and Gramiccia 1994; Strazzulla et al. 2013). Mucosal leishmaniasis and MCL appear to be more common in immunosuppressed patients. Visceralization of dermatotropic species has been reported, besides skin dissemination of viscerotropic species (Van Griensven et al. 2014). Methods suggested for MCL diagnosis were reported in Table 7.8. Clinical monitoring is basic for the patients treatment follow-up.

7.5.1 Direct Diagnosis

7.5.1.1 Sample Collection

Skin material is obtained by the same procedures described in Sect. 7.4.1.1. However, due to the fact that mucosa tissue is involved, a cutaneous exudate could be used as biopsy sample.

7.5.1.2 Parasitological Assays

Parasitological diagnosis is made on the basis of the observation of amastigotes in mucosal samples, smeared or impressed onto slides and stained with Giemsa. MCL is difficult to diagnose, even when clinically active, because amastigotes are generally scarce. However, it seems more effective in Old World than in New World

Table 7.8 Methods available for diagnosis of human MCL and their purpose

Method	Confirmation of primary clinical case	Contribution to treatment follow-up	Confirmation of clinical relapse	Prevalence of *Leishmania* exposure
Agent detection[a]				
Microscopy	+	–	+	–
In vitro culture	+	–	++	–
Conventional PCR	+++	+++	+++	+
Real-time PCR	+++	+++	+++	+
Immune response detection[b]				
IFAT	+	–	–	–
ELISA	++	++	++	–
Tests for cellular immunity	+++	+++	+++	+++

[a]A combination of agent identification methods applied on the same clinical sample is recommended

[b]One of the listed serological tests is sufficient, +++ recommended method, ++ suitable method, + may be used in some situations, but cost, reliability, or other factors severely limit its application, – not appropriate for this purpose, *PCR* polymerase chain reaction, *IFAT* indirect fluorescent antibody test, *ELISA* enzyme-linked immunosorbent assay

MCL. Indeed, in *L. infantum* MCL histology has good sensitivity (from 50–70% to almost 100%), whereas it decreases to 35–70% in *L. braziliensis* MCL. Specificity is equally high in both infections (>95%) (Goto and Lindoso 2010; Strazzulla et al. 2013). The different sensitivity is related to the abundance of *L. infantum* bodies in mucosal lesions, whereas *L. braziliensis* bodies are moderate or scarce (Herwaldt 1999; Strazzulla et al. 2013). Samples of the lesion should be cultured although the organisms from patients with MCL can be very difficult to cultivate (Gradoni and Gramiccia 2014). By analogy with procedures described in Sect. 7.4.1.2, the skin sample should be homogenized in saline or culture medium (liquid phase for biphasic media) under sterile conditions and with the addition of gentamicin and 5-fluorocytosine. Liquid media (Schneider's Drosophila Medium, HO-MEM), Sloppy Evans, EMTM, and NNN are the media preferred.

7.5.1.3 Molecular Assays

From a general point of view, the molecular methods used for MCL were the same as described in Sect. 7.4.1.3. Furthermore, because the disease may progress to an invasive pathology, peripheral blood has been used as tissue sample (Shahbazi et al. 2008). Especially in the New World innovative procedures as QT-NASBA and LAMP were applied (Espir et al. 2016; Jara et al. 2013).

7.5.2 Immunological Diagnosis

The methods used for MCL are usually IFAT and ELISA and they show higher sensitivities than for CL. As CL, the anti-*Leishmania* antibody level does not remain high after treatment, and therefore positive results generally indicate current infection (Goto and Lindoso 2010; Herwaldt 1999).

LST is often used in the MCL clinical diagnosis (Espir et al. 2016).

7.6 Post-Kala-Azar Dermal Leishmaniasis

PKDL, characterized by a macular, maculo-papular, or nodular skin rash, is a complication of VL that is frequently observed after treatment in Sudan and more rarely in other East African countries and in the Indian subcontinent (Desjeux et al. 2013; Zijlstra et al. 2003) (for clinical details see Chap. 6). In PKDL, parasites persist in the skin after VL while they can no longer be demonstrated in the viscera. It can also occur in immunosuppressed individuals in *L. infantum*-endemic areas (Chappuis et al. 2007; Srivastava et al. 2011; Zijlstra 2014), in some cases being attributed to the immune reconstitution inflammatory syndrome and non-HIV-related immunosuppression (Zijlstra 2014). It occurs within weeks to a few months following treatment, in up to 50–60% of people who have recovered from VL. In the Indian subcontinent, 10% of VL patients go on to develop PKDL after an interval of 6 months to 4 years. Although mortality from PKDL is low, PKDL patients represent a largely neglected reservoir of infection that perpetuates anthroponotic *L. donovani* disease in India. Though PKDL in Sudan and in India is both due to *L. donovani*, Sudanese PKDL frequently self-heals (84% in 1 year), whereas Indian PKDL takes longer time to self-heal (Espir et al. 2016). The tools for diagnosis of PKDL are inadequate. In endemic areas, clinical signs and symptoms, along with a previous kala-azar episode and positive antibody tests (e.g., rK39 RDT), are being used to diagnose PKDL without any parasitological confirmation. However, this approach may not be accurate enough as ~10% of PKDL patients have no history of VL and positivity of serological tests up to several years after treatment. Skin slit smear microscopy is the only confirmatory test but is very painful and impractical with macular lesions. LST has low sensitivity, and it is hard to culture parasite due to contamination risk. n-PCR is highly sensitive, and recent development of kDNA qPCR has shown to be efficacious in diagnosis of PKDL (Verma et al. 2010). A rapid noninvasive diagnostic test is urgently required as focal VL outbreaks have been linked to an index case of PKDL.

7.7 *Leishmania*-HIV Coinfections and Non-HIV-Related Immunosuppressive Diseases

The emergence of HIV and its association with VL poses challenges as how best to diagnose and treat patients presenting with HIV-VL coinfection. Patients with HIV-VL coinfection represent an important but largely neglected reservoir of parasites, and focal reemergence of VL has been linked to an index case of HIV (Alvar et al. 2008). Sensitivity and specificity of various diagnostic methods for HIV-VL have been reviewed by Cota et al. (2012) and Monge-Maillo et al. (2014). Serological assays are considered not accurate, since the majority of these patients often do not exhibit detectable levels of antibodies. Parasitemia is higher in HIV coinfection; thus direct detection of parasite or its component in blood by PCR or qPCR is increasingly used not only for diagnosis but also for the follow-up of the patients during and after treatment, but these tests are often not readily available in poor healthcare settings. One more option for VL diagnosis is the KAtex; in fact in HIV-infected patients, sensitivity up to 85.7% was found during a primary episode of VL (Riera et al. 2004). At present, there is not any clear evidence to support recommendations on serological or molecular diagnosis of HIV-VL. Consequently, diagnosis often relies on invasive spleen or bone marrow aspiration or with a combination of parasitological and immunological pathways (Table 7.4).

VL has also been found in a wide range of non-HIV-related patients with immunosuppressive states, mainly falling under the topics of transplantation medicine, rheumatology, hematology, and oncology. Clinical presentation can be atypical in immunosuppressed individuals, being easily misdiagnosed or mistaken as a flare-up of the underlying disease. Most often, symptoms consist of detection of parasites in abnormal tissues such as the intestine, oral cavity, skin, and lung tissue. Skin manifestations appear to be more frequent and can occur before, during, or after the VL episode. CL and MCL can have a number of particular features in individuals with immunosuppression, especially if severe, including parasite dissemination, clinical polymorphism with atypical and often more severe clinical forms, and even viscerализation (Van Griensven et al. 2014). As not all cases are reported or notified, the current burden of leishmaniasis in immunosuppressed individuals is probably underestimated, being rising worldwide. The best diagnostic approach in non-HIV-related immunosuppressive diseases is again the combination of parasitological, molecular, and serological methods. In addition, KAtex method might also be useful for diagnosing VL in transplanted patients (Sakkas et al. 2016).

7.8 Agent Identification

The current classification of *Leishmania* is still based on isoenzyme typing by using the multilocus enzyme electrophoresis (MLEE). MLEE has been the most widely used technique during the last years and is still considered the gold standard method for the strain identification, separating it into groups, so-called zymodemes, according to the identification of their enzymatic patterns (WHO 2010; Rioux et al. 1990). However, the method, basic for epidemiological studies, is *Leishmania* isolation and culture-dependent, time-consuming, and limited to specialized centers (WHO 2010).

Since the advent of PCR, numerous molecular tools have been published that distinguish species and strains of *Leishmania* parasites. The tools range from the amplification and subsequent DNA sequence analysis or restriction fragment length polymorphism (RFLP) of multicopy targets or multigene families, including both coding and noncoding regions, to the recently developed multilocus sequence typing (MLST), which is now considered to be one of the highest resolution methods, and multilocus microsatellite typing (MLMT), a highly discriminatory and reproducible method used for *Leishmania* population genetic studies worldwide (Schönian et al. 2011). These approaches were deeply discussed in Chap. 2.

Each of these molecular markers has its specific discriminatory power, advantages, and drawbacks. The kind of marker most suitable depends on the appropriate level of resolution required. Details were previously summarized and discussed in Sect. 7.2.2.3.

References

Adams ER, Schoone GJ, Ageed AF, Safi SE, Schallig HD. Development of a reverse transcriptase loop-mediated isothermal amplification (LAMP) assay for the sensitive detection of *Leishmania* parasites in clinical samples. Am J Trop Med Hyg. 2010;82:591–6.

Alimohammadian MH, Jones SL, Darabi H, Riazirad F, Ajdary S, Shabani A, et al. Assessment of interferon-γ levels and Leishmanin skin test results in persons recovered for Leishmaniasis. Am J Trop Med Hyg. 2012;87:70–5.

Allahverdiyev AM, Bagirova M, Uzun S, Alabaz D, Aksaray N, Kocabas E, et al. The value of a new microculture method for diagnosis of visceral leishmaniasis by using bone marrow and peripheral blood. Am J Trop Med Hyg. 2005;73:276–80.

Alvar J, Aparicio P, Aseffa A, Den Boer M, Canãvate C, Dedet J-P, et al. The relationship between leishmaniasis and AIDS: the second 10 years. Clin Microbiol Rev. 2008;21:334–59.

Antinori S, Calattini S, Longhi E, Bestetti G, Piolini R, Magni C, et al. Clinical use of polymerase chain reaction performed on peripheral blood and bone marrow samples for the diagnosis and monitoring of visceral leishmaniasis in HIV-infected and HIV-uninfected patients: a single-center, 8-year experience in Italy and review of the literature. Clin Infect Dis. 2007;44:1602–10.

Arbaji AK, Gradoni L, Gramiccia M. Leishmanin skin test survey in a focus of high endemicity of Leishmania major in Jordan. Acta Trop. 1993;54:77–9.

Badaró R, Reed SG, Carvalho EM. Immunofluorescent antibody test in American visceral leishmaniasis: sensitivity and specificity of different morphological forms of two *Leishmania* species. Am J Trop Med Hyg. 1983;32:480–4.

Badaró R, Benson D, Eulálio MC, Freire M, Cunha S, Netto EM, et al. rK39: a cloned antigen of *Leishmania chagasi* that predicts active visceral leishmaniasis. J Infect Dis. 1996;173:758–61.

Banu SS, Meyer W, Ahmed BN, Kim R, Lee R. Detection of *Leishmania donovani* in peripheral blood of asymptomatic individuals in contact with patients with visceral leishmaniasis. Trans R Soc Trop Med Hyg. 2016;110:286–93.

Basiye FL, Mbuchi M, Magiri C, Kirigi G, Deborggraeve S, Schoone GJ, et al. Sensitivity and specificity of the *Leishmania* OligoC-TesT and NASBA-oligochromatography for diagnosis of visceral leishmaniasis in Kenya. Trop Med Int Health. 2010;15:806–10.

Bensoussan E, Nasereddin A, Jonas F, Schnur LF, Jaffe CL. Comparison of PCR assays for diagnosis of cutaneous leishmaniasis. J Clin Microbiol. 2006;44:1435–9.

Bern C, Haque R, Chowdhury R, Ali M, Kurkjian KM, Vaz L, et al. The epidemiology of visceral leishmaniasis and asymptomatic leishmanial infection in a highly endemic Bangladeshi village. Am J Trop Med Hyg. 2007;76:909–14.

Bettini S, Gramiccia M, Gradoni L, Pozio E, Mugnai S, Maroli M. Leishmaniasis in Tuscany (Italy): VIII. Human population response to leishmanin in the focus of Monte Argentario (Grosseto) and epidemiological evaluation. Ann Parasitol Hum Comp. 1983;58:539–47.

Boelaert M, Rijal S, Regmi S, Singh R, Karki B, Jacquet D, et al. A comparative study of the effectiveness of diagnostic tests for visceral leishmaniasis. Am J Trop Med Hyg. 2004;70:72–7.

Boelaert M, El-Safi S, Hailu A, Mukhtar M, Rijal S, Sundar S, et al. Diagnostic tests for kala-azar: a multi-centre study of the freeze-dried DAT, rK39 strip test and KAtex in East Africa and the Indian subcontinent. Trans R Soc Trop Med Hyg. 2008;102:32–40.

Bossolasco S, Gaiera G, Olchini D, Gulletta M, Martello L, Bestetti A, et al. Real-time PCR assay for clinical management of human immunodeficiency virus-infected patients with visceral leishmaniasis. J Clin Microbiol. 2003;41:5080–4.

Buitrago R, Cupolillo E, Bastrenta B, Le Pont F, Martinez E, Barnabe C, et al. PCR-RFLP of ribosomal internal transcribed spacers highlights inter and intra species variation among *Leishmania* strains native to La Paz, Bolívia. Infect Genet Evol. 2011;11:557–63.

Chappuis F, Rijal S, Soto A, Menten J, Boelaert M. A meta-analysis of the diagnostic performance of the direct agglutination test and rK39 dipstick for visceral leishmaniasis. BMJ. 2006;333:723.

Chappuis F, Sundar S, Hailu A, Ghalib H, Rijal S, Peeling RW, et al. Visceral leishmaniasis: what are the needs for diagnosis, treatment and control? Nat Rev Microbiol. 2007;5:873–82.

Chicharro C, Morales MA, Serra T, Ares M, Salas A, Alvar J. Molecular epidemiology of *Leishmania infantum* on the island of Majorca: a comparison of phenotypic and genotypic tools. Trans R Soc Trop Med Hyg. 2002;96:S93–9.

Cortes S, Mauricio I, Almeida A, Cristovão JM, Pratlong F, Dedet JP, et al. Application of kDNA as a molecular marker to analyse *Leishmania infantum* diversity in Portugal. Parasitol Int. 2006;55:277–83.

Cota GF, de Sousa MR, Demarqui FN, Rabello A. The diagnostic accuracy of serologic and molecular methods for detecting visceral Leishmaniasis in HIV infected patients: meta-analysis. PLoS Neg Trop Dis. 2012;6:e1665.

Cruz I, Canavate C, Rubio JM, Morales MA, Chicharro C, Laguna F, et al. A nested polymerase chain reaction (Ln-PCR) for diagnosing and monitoring *Leishmania infantum* infection in patients co-infected with human immunodeficiency virus. Trans R Soc Trop Med Hyg. 2002;96:185–9.

Cruz I, Millet A, Carrillo E, Chenik M, Salotra P, Verma S, et al. An approach for interlaboratory comparison of conventional and real-time PCR assays for diagnosis of human leishmaniasis. Exp Parasitol. 2013;134:281–9.

Cunningham J, Hasker E, Das P, El Safi S, Goto H, Mondal D, for the WHO/TDR, et al. Visceral Leishmaniasis laboratory network. A global comparative evaluation of commercial immunochromatographic rapid diagnostic tests for visceral Leishmaniasis. Clin Infect Dis. 2012;55:1312–9.

Cupolillo E, Grimaldijr G, Momen H, Beverley SM. Intergenic region typing (IRT): a rapid molecular approach to the characterization and evolution of *Leishmania*. Mol Biochem Parasitol. 1995;73:145–55.

Curtis KA, Rudolph DL, Owen SM. Rapid detection of HIV-1 by reverse-transcription, loop-mediated isothermal amplification (RT-LAMP). J Virol Methods. 2008;151:264–70.

De Bruijn MH, Barker DC. Diagnosis of New World leishmaniasis: specific detection of species of the *Leishmania braziliensis* complex by amplification of kinetoplast DNA. Acta Trop. 1992;52:45–58.

de Paiva-Cavalcanti M, de Morais RC, Pessoa-E-Silva R, Trajano-Silva LA, Gonçalves-de-Albuquerque Sda C, Tavares Dde H, et al. Leishmaniases diagnosis: an update on the use of immunological and molecular tools. Cell Biosci. 2015;5:31.

de Vries PJ, van der Meide WF, Godfried MH, Schallig HD, Dinant HJ, Faber WR. Quantification of the response to miltefosine treatment for visceral leishmaniasis by QT-NASBA. Trans R Soc Trop Med Hyg. 2006;100:1183–6.

Deborggraeve S, Laurent T, Espinosa D, Van der Auwera G, Mbuchi M, Wasunna M, et al. A simplified and standardized polymerase chain reaction format for the diagnosis of leishmaniasis. J Infect Dis. 2008a;198:1565–72.

Deborggraeve S, Boelaert M, Rijal S, De Donker S, Dujardin JC, Herdewijn P, et al. Diagnostic accuracy of a new *Leishmania* PCR for clinical visceral leishmaniasis in Nepal and its role in diagnosis of disease. Trop Med Int Health. 2008b;13:1378–83.

Desjeux P, Ghosh RS, Dhalaria P, Strub-Wourgaft N, Zijlstra EE. Report of the Post Kala-azar Dermal Leishmaniasis (PKDL) Consortium Meeting, New Delhi, India, 27–29 June 2012. Parasit Vectors. 2013;6:196.

Di Muccio T, Scalone A, Bruno A, Marangi M, Grande R, Armignacco O, et al. Epidemiology of Imported Leishmaniasis in Italy: implications for a European endemic country. PLoS One. 2015;10:e0129418.

el Tai NO, Osman OF, el Fari M, Presber W, Schönian G. Genetic heterogeneity of ribosomal internal transcribed spacer in clinical samples of *Leishmania donovani* spotted on filter paper as revealed by single-strand conformation polymorphisms and sequencing. Trans R Soc Trop Med Hyg. 2000;94:575–9.

El-Beshbishy HA, Al-Ali KH, El-Badry AA. Molecular characterization of cutaneous leishmaniasis in Al-Madinah Al-Munawarah province, western Saudi Arabia. Int J Infect Dis. 2013;17:334–8.

Espir TT, Guerreiro TS, de Farias NM, de Paula FL, Veloso Soares F, da Silva SS, et al. Evaluation of different diagnostic methods of American Cutaneous Leishmaniasis in the Brazilian Amazon. Exp Parasitol. 2016;167:1–6.

Evans DA. *Leishmania*. In: Taylor AE, Baker JR, editors. In vitro methods for parasite cultivation. London: Academic Press; 1987. p. 52–75.

Gidwani K, Jones S, Kumar R, Boelaert M, Sundar S. Interferon-gamma release assay (modified QuantiFERON) as a potential marker of infection for *Leishmania donovani*, a proof of concept study. PLoS Negl Trop Dis. 2011;5:e1042.

Goto H, Lindoso JAL. Current diagnosis and treatment of cutaneous and mucocutaneous leishmaniasis. Expert Rev Anti-Infect Ther. 2010;8:419–33.

Gradoni L, Gramiccia M. *Leishmania infantum* tropism: strain genotype or host immune status? Parasitol Today. 1994;10:264–7.

Gradoni L, Gramiccia M. Leishmaniosis. In: OIE manual of diagnostic tests and vaccines for terrestrial animals (mammals, birds and bees). 7th ed. Paris: Office International des Epizooties; 2014. p. 1–12.

Gramiccia M, Gradoni L. Successful *in vitro* isolation and cultivation of Italian dermotropic strains of *Leishmania infantum sensu lato*. Trans R Soc Trop Med Hyg. 1989;83:76.

Guerbouj S, Victoir K, Guizani I, Seridi N, Nuwayri-Salti N, Belkaid M, et al. Gp63 gene polymorphism and population structure of *Leishmania donovani complex*: influence of the host selection pressure? Parasitology. 2001;122:25–35.

Herwaldt BL. Leishmaniasis. Lancet. 1999;354:1191–9.

Jara M, Adaui V, Valencia BM, Martinez D, Alba M, Castrillon C, et al. Real-time PCR assay for detection and quantification of *Leishmania* (*Viannia*) organisms in skin and mucosal lesions: exploratory study of parasite load and clinical parameters. J Clin Microbiol. 2013;51:1826–33.

Khan MG, Bhaskar KR, Salam MA, Akther T, Pluschke G, Mondal D. Diagnostic accuracy of loop-mediated isothermal amplification (LAMP) for detection of *Leishmania* DNA in buffy coat from visceral leishmaniasis patients. Parasit Vectors. 2012;5:280.

Khatri ML, Di Muccio T, Gramiccia M. Cutaneous leishmaniasis in North-Western Yemen: a clinicoepidemiologic study and *Leishmania* species identification by polymerase chain reaction-restriction fragment length polymorphism analysis. J Am Acad Dermatol. 2009;61:15–21.

Khatri ML, Di Muccio T, Fiorentino E, Gramiccia M. Ongoing outbreak of cutaneous leishmaniasis in northwestern Yemen: clinicoepidemiologic, geographic, and taxonomic study. Int J Dermatol. 2016;55(11):1210–8. https://doi.org/10.1111/ijd.13310.

Kohanteb J, Ardehali S. Cross-reaction of sera from patients with various infectious diseases with *Leishmania infantum*. Med Princ Pract. 2005;14:79–82.

Kumar R, Pai K, Pathak K, Sundar S. Enzyme-linked immunosorbent assay for recombinant K39 antigen in diagnosis and prognosis of Indian visceral leishmaniasis. Clin Diagn Lab Immunol. 2001;8:1220–4.

Lachaud L, Marchergui-Hammami S, Chabbert E, Dereure J, Dedet JP, Bastien P. Comparison of six PCR methods using peripheral blood for detection of canine visceral leishmaniasis. J Clin Microbiol. 2002;40:210–5.

Laurent T, Rijal S, Yardley V, Croft S, De Doncker S, Decuypere S, Khanal B, et al. Epidemiological dynamics of antimonial resistance in *Leishmania donovani*: genotyping reveals a polyclonal population structure among naturally-resistant clinical isolates from Nepal. Infect Genet Evol. 2007;7:206–12.

Loría-Cervera EN, Andrade-Narváez FJ. Animal model for the study of leishmaniasis. Rev Inst Med Trop Sao Paulo Rev. 2014;56:1–11.

Manson-Bahr PC. Diagnosis. In: Peters W, Killick-Kendrick R, editors. The Leishmaniases in biology and medicine. Vol. II. Clinical aspects and control. London: Academic Press; 1987. p. 703–29.

Marfurt J, Nasereddin A, Niederwieser I, Jaffe CL, Beck HP, Felger I. Identification and differentiation of *Leishmania* species in clinical samples by PCR amplification of the miniexon sequence and subsequent restriction fragment length polymorphism analysis. J Clin Microbiol. 2003;41:3147–53.

Mary C, Lamouroux D, Dunan S, Quilici M. Western blot analysis of antibodies to *Leishmania infantum* antigens: potential of the 14-kD & 16-kD antigens for diagnosis and epidemiologic purposes. Am J Trop Med Hyg. 1992;47:764–71.

Mary C, Faraut F, Lascombe L, Dumon H. Quantification of *Leishmania infantum* DNA by a real-time PCR assay with high sensitivity. J Clin Microbiol. 2004;42:5249–55.

Mauricio IL, Stothard JR, Miles MA. *Leishmania donovani* complex: genotyping with the ribosomal internal transcribed spacer and the mini-exon. Parasitology. 2004;128:263–7.

Meredith SE, Zijlstra EE, Schoone GJ, Kroon CC, van Eys GJ, Schaeffer KU, et al. Development and application of the polymerase chain reaction for the detection and identification of *Leishmania* parasites in clinical material. Arch Inst Pasteur Tunis. 1993;70:419–31.

Minodier P, Piarroux R, Gambarelli F, Joblet C, Dumon H. Rapid identification of causative species in patients with Old World leishmaniasis. J Clin Microbiol. 1997;35:2551–5.

Monge-Maillo B, Norman FF, Cruz I, Alvar J, López-Vélez R. Visceral leishmaniasis and HIV coinfection in the Mediterranean region. PLoS Negl Trop Dis. 2014;8:e3021.

Montalvo AM, Fraga J, El Safi S, Gramiccia M, Jaffe CL, Dujardin JC, et al. Direct *Leishmania* species typing in Old World clinical samples: evaluation of 3 sensitive methods based on the heat-shock protein 70 gene. Diagn Microbiol Infect Dis. 2014;80:35–9.

Moody A, Chiodini P. Methods for the detection of blood parasites. Clin Lab Haematol. 2000;22:189–01.

Morales MA, Cruz I, Rubio JM, Chicharro C, Cañavate C, Laguna F, et al. Relapses versus reinfections in patients coinfected with *Leishmania infantum* and human immunodeficiency virus type 1. J Infect Dis. 2002;185:1533–7.

Mugasa CM, Deborggraeve S, Schoone GJ, Laurent T, Leeflang MM, Ekangu RA, et al. Accordance and concordance of PCR and NASBA followed by oligochromatography for the molecular diagnosis of *Trypanosoma brucei* and *Leishmania*. Trop Med Int Health. 2010;15:800–5.

Murray HW, Berman JD, Davies CR, Saravia NG. Advances in Leishmaniases. Lancet. 2005;366:1561–77.

Nasereddin A, Bensoussan-Hermano E, Schonian G, Baneth G, Jaffe CL. Molecular diagnosis of old world cutaneous leishmaniasis and species identification by use of a reverse line blot hybridization assay. J Clin Microbiol. 2008;46:2848–55.

Neitzke-Abreu HC, Venazzi MS, Bernal MVZ, Reinhold-Castro KR, Vagetti F, Mota CA, et al. Detection of DNA from *Leishmania* (*Viannia*): accuracy of polymerase chain reaction for the diagnosis of cutaneous leishmaniasis. PLoS One. 2013;8:62473.

Notomi T, Okayama H, Masubuchi H, Yonekawa T, Watanabe K, Amino N, et al. Loop-mediated isothermal amplification of DNA. Nucleic Acids Res. 2000;28:63.

Oliveira JG, Novais FO, de Oliveira CI, da Cruz Junior AC, Campos LF, da Rocha AV, et al. Polymerase chain reaction (PCR) is highly sensitive for diagnosis of mucosal leishmaniasis. Acta Trop. 2005;94:55–9.

Ozensoy Toz S, Culha G, Yildiz ZF, Ertabaklar H, Alkan MZ, Vardarlı AT, et al. A real-time ITS1-PCR based method in the diagnosis and species identification of *Leishmania* parasite from human and dog clinical samples in Turkey. PLoS Negl Trop Dis. 2013;7:2205.

Pita-Pereira D, Lins R, Oliveira MP, Lima RB, Pereira BAS, Moreira OC, et al. SYBR green-based real-time PCR targeting kinetoplast DNA can be used to discriminate between the main etiologic agents of Brazilian cutaneous and visceral leishmaniasis. Parasit Vectors. 2012;12:5–15.

Quispe Tintaya KW, Ying X, Dedet JP, Rijal S, De Bolle X, Dujardin JC. Antigen genes for molecular epidemiology of leishmaniasis: polymorphism of cysteine proteinase B and surface metalloprotease glycoprotein 63 in the *Leishmania donovani* complex. J Infect Dis. 2004;189:1035–43.

Reithinger R, Dujardin JC. Molecular diagnosis of Leishmaniasis: current status and future applications. J Clin Microbiol. 2007;45:21–5.

Riera C, Fisa R, Lopez P, Ribera E, Carrió J, Falcó V, et al. Evaluation of a latex agglutination test (KAtex) for detection of *Leishmania* antigen in urine of patients with HIV-*Leishmania* coinfection: value in diagnosis and post-treatment follow-up. Eur J Clin Microbiol Infect Dis. 2004;23:899–04.

Rioux JA, Lanotte G, Serres E, Pratlong F, Bastien P, Perières J. Taxonomy of *Leishmania*. Use of isoenzymes. Suggestions for a new classification. Ann Parasitol Hum Comp. 1990;65:111–25.

Rodgers MR, Popper SJ, Wirth DF. Amplification of kinetoplast DNA as a tool in the detection and diagnosis of *Leishmania*. Exp Parasitol. 1990;71:267–75.

Rotureau B, Ravel C, Couppie P, Pratlong F, Nacher M, Dedet JP, et al. Use of PCR-restriction fragment length polymorphism analysis to identify the main new world *Leishmania* species and analyze their taxonomic properties and polymorphism by application of the assay to clinical samples. J Clin Microbiol. 2006;44:459–67.

Saad AA, Ahmed NG, Osman OS, Al-Basheer AA, Hamad A, Deborggraeve S, et al. Diagnostic accuracy of the *Leishmania* OligoC-TesT and NASBA-Oligochromatography for diagnosis of leishmaniasis in Sudan. PLoS Negl Trop Dis. 2010;4:776.

Sakkas H, Gartzonika C, Levidiotou S. Laboratory diagnosis of human visceral leishmaniasis. J Vector Borne Dis. 2016;53:8–16.

Salotra P, Sreenivas G, Pogue GP, Lee N, Nakhasi HL, Ramesh V, et al. Development of a species-specific PCR assay for detection of *Leishmania donovani* in clinical samples from patients with kala-azar and post-kala-azar dermal leishmaniasis. J Clin Microbiol. 2001;39:849–54.

Santos-Gomes G, Gomes-Pereira S, Campino L, Araujo MD, Abranches P. Performance of immunoblotting in diagnosis of visceral Leishmaniasis in human immunodeficiency virus-*Leishmania* sp.-coinfected patients. J Clin Microbiol. 2000;38:175–8.

Schaefer KU, Kurtzhals JAL, Kager PA, Gachihi GS, Gramiccia M, Kagaij M, et al. Studies on the prevalence of leishmanin skin test positivity in the Baringo district, Rift valley, Kenya. Am J Trop Med Hyg. 1994;50:78–84.

Schenkel K, Rijal S, Koirala SI, Koirala SH, Vanlerberghe V, Van der Stuyft P, et al. Visceral Leishmaniasis in South-Eastern Nepal: a cross sectional survey on *Leishmania donovani* infection and its risk factors. Trop Med Int Health. 2006;11:1792–9.

Schönian G, Nasereddin A, Dinse N, Schweynoch C, Schallig HD, Presber W, et al. PCR diagnosis and characterization of *Leishmania* in local and imported clinical samples. Diagn Microbiol Infect Dis. 2003;47:349–58.

Schönian G, Kuhls K, Mauricio IL. Molecular approaches for a better understanding of the epidemiology and population genetics of *Leishmania*. Parasitology. 2011;138:405–25.

Serin MS, Daglioglu K, Bagirova M, Allahverdiyev A, Uzun S, Vural Z, et al. Rapid diagnosis and genotyping of *Leishmania* isolates from cutaneous and visceral leishmaniasis by microcapillary cultivation and polymerase chain reaction–restriction fragment length polymorphism of miniexon region. Diagn Microbiol Infect Dis. 2005;53:209–14.

Shahbazi F, Shahabi S, Kazemi B, Mohebali M, Abadi AR, Zare Z. Evaluation of PCR assay in diagnosis and identification of cutaneous leishmaniasis: a comparison with the parasitological methods. Parasitol Res. 2008;103:1159–62.

Siddig M, Ghalib H, Shillington DC, Petersen EA. Visceral leishmaniasis in the Sudan: comparative parasitological methods of diagnosis. Trans R Soc Trop Med Hyg. 1988;82:66–8.

Singh OP, Sundar S. Developments in diagnosis of visceral Leishmaniasis in the elimination era. J Parasitol Res. 2015;2015:239469.

Sreenivas G, Ansari NA, Kataria J, Salotra P. Nested PCR assay for detection of *Leishmania donovani* in slit aspirates from post-kala-azar dermal Leishmaniasis lesions. J Clin Microbiol. 2004;42:1777–8.

Srivastava P, Dayama A, Mehrotra S, Sundar S. Diagnosis of visceral leishmaniasis. Trans R Soc Trop Med Hyg. 2011;105:1–6.

Srividya G, Kulshrestha A, Singh R, Salotra P. Diagnosis of visceral leishmaniasis: developments over the last decade. Parasitol Res. 2012;110:1065–78.

Strazzulla A, Cocuzza S, Pinzone MR, Postorino MC, Cosentino S, Serra A, et al. Mucosal leishmaniasis: an underestimated presentation of a neglected disease. Biomed Res Int. 2013;2013:805108.

Sundar S, Rai M. Laboratory diagnosis of visceral Leishmaniasis. Clin Diagn Lab Immunol. 2002;9:951–8.

Sundar S, Singh RK, Maurya R, Kumar B, Chhabra A, Singh V, et al. Serological diagnosis of Indian visceral leishmaniasis: direct agglutination test versus rK39 strip test. Trans R Soc Trop Med Hyg. 2006;100:533–7.

Takagi H, Itoh M, Islam MZ, Razzaque A, Ekram ARMS, Hashighuchi Y, et al. Sensitive, specific, and rapid detection of *Leishmania donovani* DNA by loop-mediated isothermal amplification. Am J Trop Med Hyg. 2009;81:578–82.

Turgay N, Balcioglu IC, Toz SO, Ozbel Y, Jones SL. Quantiferon-*Leishmania* as an epidemiological tool for evaluating the exposure to *Leishmania* infection. Am J Trop Med Hyg. 2010;83:822–4.

Vaish M, Singh OP, Chakravarty J, Sundar S. rK39 antigen for the diagnosis of visceral leishmaniasis by using human saliva. Am J Trop Med Hyg. 2012;86:598–00.

Van der Auwera G, Dujardin JC. Species typing in dermal leishmaniasis. Clin Microbiol Rev. 2015;28:265–94.

Van der Auwera G, Ravel C, Verweij JJ, Bart A, Schönian G, Felger I. Evaluation of four single-locus markers for *Leishmania* species discrimination by sequencing. J Clin Microbiol. 2014;52:1098–04.

Van der Auwera G, Bart A, Chicharro C, Cortes S, Davidsson L, Di Muccio T, et al. Comparison of *Leishmania* typing results in 16 European clinical laboratories. Euro Surveill. 2016;21:30418.

Van der Meide WF, Schoone GJ, Faber WR, Zeegelaar JE, de Vries HJ, Ozbel Y, et al. Quantitative nucleic acid sequence-based assay as a new molecular tool for the detection and quantification of *Leishmania* parasites in skin biopsies. J Clin Microbiol. 2005;43:5560–6.

Van der Meide W, Guerra J, Schoone G, Farenhorst M, Coelho L, Faber W, et al. Comparison between quantitative nucleic acid sequence-based amplification, real-time reverse transcriptase PCR, and real-time PCR for quantification of *Leishmania* parasites. J Clin Microbiol. 2008;46:73–8.

Van Eys GJ, Schoone GJ, Kroon NC, Ebeling SB. Sequence analysis of small subunit ribosomal RNA genes and its use for detection and identification of *Leishmania* parasites. Mol Biochem Parasitol. 1992;51:133–42.

Van Griensven JE, Carrillo ER, Lopez-Velez R, Lynen L, Moreno J. Leishmaniasis in immunosuppressed individuals. Clin Microbiol Infect. 2014;20:286–99.

Verma S, Kumar R, Katara GK, Singh LC, Negi NS, Ramesh V, et al. Quantification of parasite load in clinical samples of leishmaniasis patients: IL-10 level correlates with parasite load in visceral leishmaniasis. PLoS One. 2010;5:10107.

Victoir K, De Doncker S, Cabrera L, Alvarez E, Arevalo J, Llanos-Cuentas A, et al. Direct identification of *Leishmania* species in biopsies from patients with American tegumentary leishmaniasis. Trans R Soc Trop Med Hyg. 2003;97:80–7.

World Health Organization (WHO). Control of the leishmaniases, WHO Technical Report Series 949. Geneva: WHO; 2010. p. 1–186.

Zelazny AM, Fedorko DP, Li L, Neva FA, Fischer SH. Evaluation of 7SL RNA gene sequences for the identification of *Leishmania* spp. Am J Trop Med Hyg. 2005;72:415–20.

Zijlstra EE. PKDL and other dermal lesions in HIV co-infected patients with Leishmaniasis: review of clinical presentation in relation to immune responses. PLoS Negl Trop Dis. 2014;8:e3258.

Zijlstra EE, el-Hassan AM, Ismael A, Ghalib HW. Endemic kala-azar in eastern Sudan: a longitudinal study on the incidence of clinical and subclinical infection and post-kala-azar dermal leishmaniasis. Am J Trop Med Hyg. 1994;51:826–36.

Zijlstra EE, Musa AM, Khalil EA, el-Hassan IM, el-Hassan AM. Post-kala-azar dermal leishmaniasis. Lancet Infect Dis. 2003;3:87–98.

Treatment of Visceral Leishmaniasis

8

Begoña Monge-Maillo and Rogelio López-Vélez

8.1 General Treatment Considerations in Visceral Leishmaniasis

Untreated, advanced cases of visceral leishmaniasis (VL) can result in death mainly associated to progressive wasting, superinfection and/or haemorrhage. So, all persons with symptomatic VL should be treated with antileishmanial drugs. Other complementary measures are needed in many cases and can include nutritional support, treatment of other infectious diseases (e.g. tuberculosis, malaria or bacterial or parasitic dysentery) and blood transfusions. The therapeutic options for VL are diverse and depend on different factors, such as geographical area of the infection (Alvar et al. 2006); the *Leishmania* species involved, the development of failure to habitual treatments (Croft et al. 2006; Alvar et al. 2008), the evidence of HIV co-infection or other infections and the presence of malnourishment. The goal of the "best treatment option" is to cure the patient, to minimize the appearance of resistance and to decrease the duration of hospitalization, all at the lower cost (Alvar et al. 2006).

The treatment regimen recommended should also follow national and regional guidelines, if applicable. This drug policy in endemic countries and the therapeutic decisions should be based on the individual benefit–risk ratio of medicines, the health service setting, the availability of antileishmanial medicines and public health considerations, such as the prevention of drug resistance.

B. Monge-Maillo • R. López-Vélez (✉)
National Referral Unit for Tropical Diseases, Infectious Diseases Department, Ramón y Cajal University Hospital, IRICYS, Madrid, Spain
e-mail: rogelio.lopezvelez@salud.madrid.org

F. Bruschi, L. Gradoni (eds.), *The Leishmaniases: Old Neglected Tropical Diseases*, https://doi.org/10.1007/978-3-319-72386-0_8

8.2 Patient Evaluation Before Treatment

Treatment should be given always after confirmation of the infection. While there are several approaches to the diagnosis of VL, it is recommended to use different diagnostic approaches to maximize the likelihood of a positive *Leishmania* results. Methods employed are visualization of the characteristic amastigote in blood smears or aspirates from lymph nodes, bone marrow, liver or spleen (histopathology), parasite isolation by in vitro culture, molecular detection of parasite DNA and serologic testing (Mary et al. 2006; Sundar and Rai 2002a). Persons newly diagnosed with VL should also be assessed for concurrent HIV/AIDS or other causes of cell-mediated immunosuppression.

8.3 Treatment Options

Properties of antileishmanial drugs are shown in Appendix and end of Chap. 9 (Treatment of tegumentary forms of leishmaniasis).

The traditional treatment of VL used to be *pentavalent antimonials*, introduced in the 1940s. However, the development of resistance, especially in India, with failure rates of up to 60%, as well as their potential toxicity, made it necessary to research for new treatment options. Thus and since the 1980s, the use of *amphotericin B deoxycholate* has been introduced, especially in the more developed countries. Progressively, and due to their efficacy and lower toxicity, *lipid formulations of amphotericin B* have been gaining importance, becoming the first-choice treatment established by the US Food and Drug Administration. Nonetheless, their elevated cost reduces its use in less powerful nations. In countries of fewer resources, studies have been carried out demonstrating the efficacy of parenteral *paromomycin* as a cheap treatment with medium toxicity, commercialized in India and available in East Africa.

Within the range of oral treatments *miltefosine* had demonstrated very good cure rates in adults and children in India, Nepal and Bangladesh with VL by *L. donovani* (Sundar et al. 2002, 2006; Bhattacharya et al. 2007; Ritmeijer et al. 2006). However, currently, a high rate of clinical failures has been reported (Rijal et al. 2013; Sundar et al. 2012). Moderate efficacy has been observed in East Africa (Ritmeijer et al. 2006), while more data from Mediterranean countries and Latin America are needed.

Currently, *combination therapies* are considered the best regimens for treating VL in many parts of the world as dosing and duration of treatment are decreased, thereby decreasing toxicity, costs and drug resistance (Monge-Maillo and Lopez-Velez 2013).

8.4 Definition of Healing and Follow-Up

It has been shown that clinical parameters correlate well with parasitological response to VL treatment. Therefore, clinical parameters should be used to monitor the response to the VL treatment and to make the follow-up.

The confirmation of a parasitological response performed by repeating a bone marrow or spleen aspiration is not recommended in a patient with an adequate clinical response. The antibody levels are not useful to monitor the treatment response because they can persist positive for a long time (usually 6–8 months).

The clinical parameters that indicate a response to the VL treatment are the normalization of the temperature, an increase in appetite and weight and a decrease in the liver and spleen size. Blood test must show that the level of leukocytes, haemoglobin and platelets rises (Maru 1979; Kager et al. 1984).

Normally patients responding to treatment become afebrile in 5–7 days while visceromegaly usually resolves slower, within 3–6 months, although some decrease may be seen in approximately 10 days after initiation of treatment (Cascio et al. 2004). Leukopenia and thrombocytopenia generally normalize within a month, but resolution of anaemia may be slower taking from 6 to 12 months to recover (Kager et al. 1984; Berman et al. 1998).

There have been identified several factors associated with a higher risk of death which are immunosuppression, prolonged disease, malnutrition, concomitant infections, gastrointestinal symptoms, mucosal bleeding, jaundice, <1 year of age and laboratory signs such as severe anaemia, neutrophils <500 cells/μL and platelets <50,000 cells/μL (Werneck et al. 2003; Collin et al. 2004; Mueller et al. 2009).

Patients should be clinically evaluated at the end of treatment, at 1 month after and at 6 months after. Therapeutic failure is defined as a return of clinical signs and symptoms of VL in concert with parasitological confirmation. It can occur in patients with no immunodeficiency, and mostly 6–12 months after treatment. However, failure is more likely in those with HIV co-infection or compromised cell-mediated immunity for other reasons representing more of an immunologic failure rather than a drug failure.

8.5 Treatment According to the Country of Infection

The Oxford evidence grading system was applied when reviewing information:

(A) Randomized controlled trials in representative patient groups.
(B) Randomized controlled trials in less homogenous patient groups (small numbers, different species included) as well as cohort trials and case control studies in representative patient.
(C) Cohort trials or case control studies in less homogenous patient groups, as well as case series of representative patient groups.
(D) Case series of less homogenous patient groups and expert opinion were ranked.

8.5.1 The Mediterranean Region (Box 8.1)

VL is hypoendemic in the Mediterranean region, where it is caused by the protozoon *L. infantum*. This parasite is transmitted by the bite of infected phlebotomine female sandflies of the *Phlebotomus* genus and is maintained in a zoonotic cycle with dogs acting as the main reservoir (World Health Organization (WHO) 2010).

Cases in the Mediterranean region only contribute to 5–6% of the global burden of VL, with an estimated annual incidence of 1200–2000 cases (Alvar et al. 2012).

The incidence of VL has been declining in the last decades, mainly in areas where living standards have improved. VL associated with HIV infection is also declining in the past few years in Europe and the Mediterranean region.

The actual recommended therapeutic regimens for VL caused by *L. infantum* in the Mediterranean region by the World Health Organization are liposomal amphotericin B, up to a total dose of 18–21 mg/kg as first choice; pentavalent antimonials, 20 mg Sb^{v+}/kg per day IM or IV for 28 days as second choice and amphotericin B deoxycholate, 0.75–1.0 mg/kg for a total dose of 2–3 g, as third choice (World Health Organization (WHO) 2010).

In the Mediterranean region, evidence with pentavalent antimonials is not too strong, and therapeutic attitudes even vary from country to country. During the 1990s, antimonials were the first-line treatment in France, Greece, Italy, Malta, Spain, Portugal, Morocco, Algeria and Tunisia, with cure rates of 95% in immunocompetent patients (Gradoni et al. 2008). The information recollected from 11 countries of Southern Europe, Northern Africa and the Middle East in the twenty-first century reflects certain variations in the treatment recommendations (Gradoni et al. 1995): By this way in Morocco, Tunisia, Turkey and Palestine, the antimonials were the first-line treatment. Meanwhile in Portugal, Spain, Greece and Italy, antimonials and amphotericin B preparations were the two options for first-line treatment, even though antimonials were not administered in patients with severe immunosuppression and preparations of liposomal amphotericin B were recommended for the treatment of relapses after antimonials. In France, Italy and Cyprus, liposomal amphotericin B was the first-line treatment, and relapses were treated with different regimens of the same dug. Another study which recollects a total of 1210 cases of VL in children of between 0 and 14 years in Albania from 1995 to 2009 demonstrated that antimonials at a dose of 20 mg Sb^{v+}/kg/day for 21–28 days continue to be effective, with a cure rate of 99% (Petrela et al. 2010).

In the Mediterranean area there is scare experience with amphotericin B deoxycholate; in fact, the liposomal preparations of amphotericin B are preferred as the first-line drugs in those cases where antimonials had previously failed (Gradoni et al. 1995). A retrospective study of five cases of VL in Tunisia treated with amphotericin B deoxycholate obtained a 100% response (Toumi et al. 2007).

Although there are no randomized clinical trials performed in the Mediterranean region with liposomal amphotericin B, there are a high number of case series that give an important accumulation of evidence about its use. Response rates obtained were superior to 97% with total doses of between 18 and 24 mg/kg in different regimens of administration (Figueras Nadal et al. 2003; Cascio et al. 2004; di

Martino et al. 2004; Kafetzis et al. 2005; Minodier et al. 2005). Therefore, it has been proven that liposomal amphotericin B reduced the average duration of hospital care when compared to antimonials (Kafetzis et al. 2005), and it is effective in those cases where antimonials had previously failed (Minodier and Garnier 2000). For all these reasons, and despite the absence of randomized clinical trials, liposomal amphotericin B is considered a reference treatment in the case of VL in the Mediterranean countries in adults as well as in children (Rosenthal et al. 2009).

There is nearly no experience neither with pentamidine nor with paromomycin for VL in the Mediterranean area.

Reliable data on the efficacy of miltefosine in VL in the Mediterranean region has not been published. However, its oral administration makes it an attractive therapeutic option.

Box 8.1 Therapeutic Options for Visceral Leishmaniasis in the Mediterranean Region

- **Sodium stibogluconate or meglumine antimoniate:** (IM or IV) 20 mg Sb^{v+}/kg/day for 28–30 days [B]
- **Amphotericin B deoxycholate:** (IV) 0.7–1 mg/kg/day, on alternate days, for 15–20 doses [D]
- **Liposomal amphotericin B:** (IV) 3–5 mg/kg/day for 3–10 doses (total dose 18–30 mg/kg in adults and 15 mg/kg in children) [B]
- **Paromomycin:** (IM) 15–20 mg (11–15 mg base) kg/day for 21–28 days [D]
- **Miltefosine:** (oral dosing) for 28 days; 2.5 mg/kg/day in children aged 2–11 years; 50 mg/day in those aged ≥12 years with bodyweight <25 kg; 100 mg/day in those aged ≥12 years with bodyweight ≥25 kg; 150 mg/day in those aged ≥12 years with bodyweight ≥50 kg [D]

8.5.2 The Middle East and Central Asia Region (Box 8.2)

Incidence of visceral leishmaniasis in the Middle East to Central Asia from 2004 to 2008 has been estimated to be between 4500 and 9500 cases per year. The most affected countries are Iraq, China, Georgia and Iran. In the last years, countries with ample resources like Saudi Arabia have been able to take good measures to control the diseases and have manage to reduce the incidence of leishmaniasis. Meanwhile, other countries like Syria and Iraq due to the war they suffer from a lack of access to health care and from a compromised nutritional status putting the exposed population at greater risk of the disease. The *Leishmania* species mainly involved in the VL cases in the Middle East and Central Asia region is *L. infantum* except for Saudi Arabia where it is mostly caused by *L. donovani* (Alvar et al. 2012; Salam et al. 2014).

No randomized clinical trials have been performed in the Middle East and Central Asia evaluating the efficacy of the different therapeutic regimens for

VL. However pentavalent antimonials have been the drug of choice for more than 70 years now. Currently due to the emergence of drug resistance and toxicity of antimonials, liposomal amphotericin B, miltefosine and paromomycin are replacing antimonials. Since the species involved is mainly *L. infantum* as in the Mediterranean region, the treatment recommendations established for the Middle Eastern region could be compared to those for the Mediterranean region.

Box 8.2 Therapeutic Options for Visceral Leishmaniasis in the Middle East and Central Asia Region

- **Sodium stibogluconate or meglumine antimoniate:** (IM or IV) 20 mg Sb^{v+}/kg/day for 28–30 days [C]
- **Amphotericin B deoxycholate:** (IV) 0.7–1 mg/kg/day, on alternate days, for 15–20 doses [D]
- **Liposomal amphotericin B:** (IV) 3–5 mg/kg/day for 3–10 doses (total dose 18–30 mg/kg in adults and 15 mg/kg in children) [D]
- **Paromomycin:** (IM) 15–20 mg (11–15 mg base) kg/day for 21–28 days [D]
- **Miltefosine:** (orally) for 28 days; 2.5 mg/kg/day in children aged 2–11 years; 50 mg/day in those aged ≥12 years with bodyweight <25 kg; 100 mg/day in those aged ≥12 years with bodyweight ≥25 kg; 150 mg/day in those aged ≥12 years with bodyweight ≥50 kg [D]

8.5.3 The Indian Subcontinent and South Asia Region (Box 8.3)

The estimated incidence of visceral leishmaniasis in the Indian subcontinent and South-east between 2006 and 2010 was from 162,100 to 313,600 cases per year. India (where the state of Bihar accounts for the majority of the cases), Bangladesh and Nepal are the most affected countries. Visceral leishmaniasis in this area has an anthroponotic transmission and is caused by *L. donovani* (Alvar et al. 2012).

Many clinical trials have been performed in the Indian subcontinent which confers in most cases a high quality of evidence in the treatment recommendation. Pentavalent antimonials are considered a first-line drug due to their efficacy observed in several clinical trials in Bangladesh and Nepal. However, there are zones where resistance have developed, fundamentally in the state of Bihar (India) where the pentavalent antimonial rates of resistance reaches up to 60% of all cases (Sundar and Rai 2002b), and so they should not be used. In the beginning of the 1990s, the first clinical trials with amphotericin B were performed in India for VL obtaining response rates of 98–100% even in those cases where antimonials had previously failed (Jha et al. 1995; Thakur et al. 1993a).

Due to the high rate of resistance to antimonials in India and the high cure rates obtained with amphotericin B, this is established nowadays as one of the drugs of choice for VL in the Indian subcontinent. In India different regimens of liposomal amphotericin B have been tested with a diverse range of response rates (Sundar et al. 2004). Doses administered for several days at different total doses have been tested and more recently a regimen based on a single dose has demonstrated efficacy (Thakur 2001; Sundar et al. 2010). Therefore, due to is efficacy and because lipid formulation of amphotericin B has less renal toxicity than amphotericin B deoxycholate, liposomal amphotericin B is considered a first-line drug for VL in the Indian subcontinent.

Pentamidine was the second drug tested in VL-endemic areas in India when faced with the need to find alternative treatment to pentavalent antimonials. Already in the 1980s, pentamidine resulted as an efficient therapeutic regimen (Thakur 1984). However, the response rate started to decrease after the decade of the 1990s in certain areas of India (Mishra et al. 1992). Moreover, later studies compared it with other therapeutic options used in the area, as is the case with amphotericin B demonstrating that pentamidine was less effective (Das et al. 2009).

Paromomycin has been also tested in India mostly in Bihar where it presented high cure rates and good tolerance with the exception of an increase in liver function test parameters, which decrease towards baseline over time. So paromomycin had a reasonably safe profile and efficacy tested even in paediatric patients (Jha et al. 1998; Sundar et al. 2007; Sinha et al. 2011a).

Several clinical trials performed for VL in India and Bangladesh showed high cure rates with miltefosine regimens for 28 days (Sundar et al. 2002, 2006; Bhattacharya et al. 2007; Rahman et al. 2011). Also high cure rates have been obtained in paediatric clinical trials performed in India (Sundar et al. 2003; Bhattacharya et al. 2004). These initial results lead to propose miltefosine as a first-line drug for VL in India, Nepal and Bangladesh. However, more recent studies have revealed that after a decade of use of miltefosine in the Indian subcontinent, the relapse rate with miltefosine has increased significantly in these countries, and a development of tolerance and resistance to this drug is suspected. Therefore, the strength of the recommendation of miltefosine for the Indian subcontinent may decrease (Rijal et al. 2013; Dorlo et al. 2014).

In India, several studies published about combined therapy have obtained favourable results in patients. Combinations tested have been pentamidine and antimonials (Thakur et al. 1991), pentamidine and allopurinol (Das et al. 2001) or paromomycin and antimonials (Thakur et al. 1992, 2000). Several studies have demonstrated the efficacy of combinations of liposomal amphotericin B at a single dose followed by different regimens of oral miltefosine (Sundar et al. 2008). Another comparative study performed in India tested a single dose liposomal amphotericin B plus miltefosine or paromomycin or miltefosine plus paromomycin. These combined therapies were non-inferior to the standard treatment (amphotericin B for 30 days) and resulted in fewer adverse events than those assigned standard treatment (Sundar et al. 2011).

Box 8.3 Therapeutic Options for Visceral Leishmaniasis in the Indian Subcontinent and South-East Asian Region

- **Sodium stibogluconate or meglumine antimoniate**: (IM or IV) 20 mg Sb^{v+}/kg/day for 28–30 days
- Nepal and Bangladesh [A], India [not recommended]
- **Amphotericin B deoxycholate**: (IV) 0.7–1 mg/kg/day, daily on alternate days, for 15–20 doses [A]
- **Liposomal amphotericin B**: (IV) 5–10 mg/kg for 1–2 doses (up to total dose of 10 mg/kg) or 3–5 mg/kg/day for 3–5 doses (up to total dose of 15 mg/kg) [A]
- **Pentamidine isethionate**: (IM or IV) 4 mg/kg/day, on alternate days or three times a week, for 15–20 doses [not recommended]
- **Paromomycin**: (IM) 15 mg (11 mg base)/kg/day for 21 days [A]
- **Miltefosine**: (orally) for 28 days; 2.5 mg/kg/day in children aged 2–11 years; 50 mg/day in those aged ≥12 years with bodyweight <25 kg; 100 mg/day in those aged ≥12 years with bodyweight ≥25 kg; 150 mg/day in those aged ≥12 years with bodyweight ≥50 kg [A]
- **Combination therapy**:
  - **Liposomal amphotericin B** (IV), 5 mg/kg single dose + **Miltefosine** (oral), for 7–14 day; 2.5 mg/kg/day in children 2–11 years; 50 mg/day in ≥12 years old with weight < 25 kg; 100 mg/day in ≥12 years with body weight ≥ 25 kg; 150 mg/day in ≥12 years with body weight ≥ 50 kg [A]
  - **Liposomal amphotericin B** (IV), 5 mg/kg single dose + **Paromomycin** (IM), 15 mg (11 mg base)/kg/day for 10 days [A]
  - **Miltefosine** (oral), for 10 days, as above + **Paromomycin** (IM), 15 mg (11 mg base)/kg/day for 10 days [A]

8.5.4 East Africa Region (Box 8.4)

East Africa is one of the most affected regions by VL, only surpassed by the Indian subcontinent, with an estimated annual incidence rate of 29,400–56,700 cases (Alvar et al. 2012). The countries most affected are Sudan, South Sudan and Ethiopia. With much lower VL burden, endemic foci of VL are also found in Eritrea, Somalia, Kenya and Uganda (Alvar et al. 2012). Leishmaniasis affects mostly to poor communities that live in remote areas and that have poor health-care infrastructure. Visceral leishmaniasis in East Africa is caused by *L. donovani*.

Currently treatment in these countries is mostly provided by international organizations such as Médecins Sans Frontières (MSF), Drugs for Neglected Diseases initiative (DNDi) and the World Health Organization (WHO).

In Africa, the first trials with pentavalent antimonials were realized in Kenya in 1983 (Anabwani et al. 1983). Further on, few new studies have been done since then. In the 1990s in Sudan, sodium stibogluconate combined with paromomycin showed higher cure rates than pentavalent antimonials alone (Seaman et al. 1993). Other studies in Kenya and Sudan have analysed the efficacy of generic sodium stibogluconate versus patented versions, without observing any significant differences and with the advantage of a lower cost (Veeken et al. 2000; Moore et al. 2001). In Uganda, a comparative study was done between amphotericin B deoxycholate and reported a 95% cure rate with antimonials (Mueller et al. 2008). A new study, developed in Ethiopia, demonstrated differences in the cure rate after 6 months in patients from the North versus patients from the South (80% vs. 100%), which was justified by the different rates of confection by HIV (46.4% of the patients from the North were HIV positive, while no case was detected among the patients from the South). Thus, the efficacy of antimonials in Ethiopia in immunocompetent patients seems to be very high.

Experience with amphotericin B deoxycholate was obtained from a study where it was administrated at a dose of 1 mg/kg on alternate days for a period of 30 days, reaching similar cure rates than with antimonials and without any difference in the appearance of severe side effects (Mueller et al. 2008).

There is very little experience of liposomal amphotericin B in Eastern Africa, and the recommendations are based on results obtained in India. In Sudan, total doses of 20 mg/kg were tested with cure rates of 88%, but lower doses of 12 mg/kg only obtained a 50% response rate (Seaman et al. 1995). A reduced clinical trial in phase II in Kenya demonstrated that the efficacy of a total dose 14 mg/kg was higher that of 6 or 10 mg/kg (Berman et al. 1998). A randomized multicentre clinical trial conducted in Eastern Africa showed that a single dose on amphotericin B is not a suitable regimen for VL treatment across Eastern Africa (Khalil et al. 2014; Edwards et al. 2011). Thus, it is expected that, in Eastern Africa, higher doses are needed than in India.

In Eastern Africa, there is scare evidence on pentamidine, and it has proved to be effective in the treatment of patients in Sudan when pentavalent antimonials had previously failed (Khalil et al. 1998). However, in the Indian subcontinent, pentamidine for VL due to *L. donovani* had low cure rates.

In East Africa, the majority of studies executed are based on a comparison between paromomycin and antimonials, or on a combination of both. Good cure rates were obtained with regimens of paromomycin during 21 or 28 days in Kenya and Ethiopia with a lower response in Sudan (Seaman et al. 1993; Melaku et al. 2007; Musa et al. 2010).

A clinical trial performed in Ethiopia with miltefosine in immunocompetent patients registered a 75.6% cure rate (Ritmeijer et al. 2006).

About combined therapy, in Eastern Africa, more concretely in Sudan, two studies were carried out that demonstrated that paromomycin associated with antimonials increased the response rate in comparison with antimonials in monotherapy (Seaman et al. 1993; Melaku et al. 2007). In Kenya, a non-randomized trial drew a comparison between paromomycin and antimonials in monotherapy versus the combination of both, the latter being the most effective option (Chunge et al. 1990). Another clinical trial performed in Sudan, Ethiopia and Kenya demonstrated that combined therapy with paromomycin and antimonials was a safe regimen and just as efficient as antimonials on their own, thus being a good option for treatment in Eastern Africa (Musa et al. 2012). A further clinical trial has been performed in

East Africa to assess whether a short combination of antimonials plus a single dose of liposomal amphotericin B, miltefosine plus a single dose of liposomal amphotericin B and miltefosine alone, were effective in treating VL. None of the regimens tested showed cure rates sufficiently high to develop a phase III trial and to consider these regimens optimal for VL in East Africa (Wasunna et al. 2016).

Box 8.4 Therapeutic Options for Visceral Leishmaniasis in the East Africa Region

- **Sodium stibogluconate or meglumine antimoniate:** (IM or IV) 20 mg Sb^{v+}/kg/day for 28–30 days [A]
- **Amphotericin B deoxycholate:** (IV) 0.7–1 mg/kg/day, on alternate days, for 15–20 doses [C]
- **Liposomal amphotericin B:** (IV) 3–5 mg/kg/day for 6–10 doses (up to total dose of 30 mg/kg) [B]
- **Pentamidine isethionate:** (IM or IV) 4 mg/kg/day, on alternate days or three times a week, for 15–20 doses [not recommended]
- **Paromomycin:** (IM) 15–20 mg (11–15 mg base) kg/day for 21–28 days [B]
- **Miltefosine:** (orally) for 28 days; 2.5 mg/kg/day in children aged 2–11 years; 50 mg/day in those aged ≥12 years with bodyweight <25 kg; 100 mg/day in those aged ≥12 years with bodyweight ≥25 kg; 150 mg/day in those aged ≥12 years with bodyweight ≥50 kg [B]
- **Combination therapy:**
  - **Sodium stibogluconate or meglumine antimoniate** (IM or IV) 20 mg Sb^{v+}/kg/day for 17 day plus **Paromomycin** (IM) 15 mg (11 mg base)/kg/day for 17 days [A]
  - **Liposomal amphotericin B** (IV) 10 mg/kg single dose plus **Sodium stibogluconate** 20 mg/kg/day for 10 days [not recommended]
  - **Liposomal amphotericin B** (IV) 10 mg/kg single dose plus **miltefosine** (orally) 2.5 mg/kg/day for 10 days [not recommended]
  - **Miltefosine** (orally) 2.5 mg/kg/day for 28 days [not recommended]

8.5.5 Latin America Region (Box 8.5)

Visceral leishmaniasis in the Latin America affects mainly zones in the north-east of Brazil where 3000–5000 cases appear every year, usually in the early ages (Jeronimo et al. 2004; Wasunna et al. 2016; Hailu et al. 2010a; Musa et al. 2010). It is a zoonotic infection produced by *L. infantum/chagasi* that causes a high percentage of asymptomatic patients, as opposed to VL in India.

Classically, the treatment of VL in Brazil was based on the use of antimonials, with a cure rate of up to 95% (Santos et al. 2002). In fact, the Pan American Guide for the treatment of infectious diseases established as first-line treatment pentavalent antimonials at a dose of 20 mg Sb^{v+}/kg/day IM or IV for 20–28 days; if there is no response, they propose pentamidine, and if the patient is still not cured, amphotericin B should be used (Organización Panamericana de la Salud 2004). Evidence in Latin America with liposomal amphotericin B is very scarce. In Brazil, total dose of 20 mg/kg has proven to be effective (Berman et al. 1998). In Colombia two cases were published where the treatment with antimonials had failed but who responded to liposomal amphotericin B (Velez et al. 2009).

Although pentamidine is recommended by the Pan American Guide for the treatment of infectious diseases, there is little literature about its use in Latin America. Moreover a decrease of its efficacy has been observed which added to its serious and sometimes irreversible toxicity, and the development of other drugs has made that it is practically abandoned.

There is no data about paromomycin for visceral leishmaniasis in Latin America.

Box 8.5 Therapeutic Options for Visceral Leishmaniasis in the Latin America Region

- **Sodium stibogluconate or meglumine antimoniate:** (IM or IV) 20 mg Sb^{v+}/kg/day for 28–30 days [B]
- **Amphotericin B deoxycholate:** (IV) 0.7–1 mg/kg/day, on alternate days, for 15–20 doses [C]
- **Liposomal amphotericin B:** (IV) 3–5 mg/kg/day for 6–10 doses (up to total dose of 30 mg/kg) [C]
- **Pentamidine isethionate:** (IM or IV) 4 mg/kg/day, on alternate days or three times a week, for 15–20 doses [not recommended]
- **Miltefosine:** (orally) for 28 days; 2.5 mg/kg/day in children aged 2–11 years; 50 mg/day in those aged ≥12 years with bodyweight <25 kg; 100 mg/day in those aged ≥12 years with bodyweight ≥25 kg; 150 mg/day in those aged ≥12 years with bodyweight ≥50 kg [C]

8.6 Treatment of Failures and Relapses

There is scare experience to give a strong evidence recommendation of a therapeutic option for a VL infection that has initially failed to respond or that has relapse. They can be treated with another drug, or use the same drug in a different dose or for longer periods, or a combination therapy can be administrated. The selection of the drug must be based on the *Leishmania* species involved, on the immune situation of the patient and on the prevalence of therapeutic failure rates in the geographic area of acquisition.

8.7 Treatment of Visceral Leishmaniasis Under Special Conditions

8.7.1 Visceral Leishmaniasis and HIV Co-Infection

Leishmania and HIV co-infection have been reported in more than 35 countries. In the early 1990s, a rapid increase in the incidence of VL/HIV co-infection was noticed in the Mediterranean basin, coinciding with the peak of the HIV epidemic. The 85% of the countries where the WHO detected the first cases of co-infections were in the Mediterranean basin, with Spain in the lead (Alvar et al. 1997). The number of cases of co-infection reached its peak in 1997, and its incidence plateaued between 1998 and 2001. Since 2001, the incidence of VL/HIV co-infection has decreased significantly mainly due to the administration of antiretroviral treatments (ARTs) for HIV in the Mediterranean region (Alvar et al. 2008). On our days there are other geographical areas, mostly Ethiopia and Sudan, where the rate of VL/HIV co-infection is very high, probably due to the fact that ARTs are not so widespread. Interestingly, VL/HIV co-infection is increasing in other regions, such as in certain areas of India, where the incidence of HIV is low (<1%). The likely cause is population movements, and VL/HIV co-infection should be considered an emerging problem in these regions (Diro et al. 2014b; Singh 2014).

Patients with VL and HIV co-infection have usually a worse therapeutic response presenting frequent relapses especially among those patients with CD <200 cell/μL. Only a few clinical trials have been conducted on the efficacy of some drugs for VL/HIV co-infection, and the majorities have been carried out in Europe (infections caused by *L. infantum*) and East Africa. Many questions still remain unanswered, such as the optimal drug, dosage, duration of treatment and prophylaxis and the efficacy of combined therapies for VL/HIV co-infection (Cota et al. 2013).

8.7.1.1 The Mediterranean Region

The evidence currently available on the efficacy of pentavalent antimonials in HIV patients has been gathered mainly in European studies reporting varying cure rates ranging from 33 to 82%, with high relapse rates (Pintado and Lopez-Velez 2001). Specifically, two clinical trials have been performed comparing meglumine antimoniate with amphotericin B deoxycholate and amphotericin B lipid complex. The efficacy between pentavalent antimonials and amphotericin in the two evaluated presentation were similar. However, the toxicity of pentavalent antimonials was substantially higher (Laguna et al. 1999, 2003).

Experience with liposomal amphotericin B is based on studies performed in four European health centres, where VL was treated with liposomal amphotericin B in HIV patients, with a good initial clinical and parasitological response, although all patients who completed follow-up eventually relapsed (Russo et al. 1996).

In Germany, a study was performed with miltefosine in HIV patients in whom other previous treatment for VL had failed. Initially, the cure rates were high, but almost all patients finally relapsed when miltefosine was discontinued. However, miltefosine was well tolerated even in long-term treatment periods, suggesting that clinical relapse could be either treated by administering repeated courses of miltefosine or prevented with miltefosine in combination with other antileishmanial drugs (Sindermann et al. 2004). Another study performed in Spain described four cases of co-infected patients who were severely immunosuppressed and who had not responded to a previous treatment with amphotericin B or pentavalent antimonials and that where treated with miltefosine. Initially, all patients responded clinically but, when treatment was discontinued, all patients relapsed (Troya et al. 2008).

Combination therapy has been tested in several studies in the Mediterranean region for VL/HIV co-infected patients. In Spain, a study performed with 11 VL/

HIV co-infected patients due to *L. infantum*, meglumine antimoniate was combined with allopurinol, and good results were obtained (Laguna et al. 1994). Also in Spain, another case was reported of a co-infected patient who did not respond to previous monotherapies and who finally responded to a combined therapy of meglumine antimoniate plus paromomycin followed by maintenance therapy with itraconazole plus miltefosine given 1 month on and 2 months off until CD4 cell count was 350 cells/mm^3 for 3–6 months (Barragan et al. 2010). A case reported in Italy described a co-infected patient who received treatment with liposomal amphotericin B and the growth factor of rHuGM-CSF colonies (Mastroianni 2004). A German HIV-positive patient who had acquired VL after visiting several southern European countries did not response to liposomal amphotericin B and to miltefosine. He finally responded to a combination therapy with intravenous pentamidine and oral fluconazole for 3 weeks (Rybniker et al. 2010).

8.7.1.2 The Indian Subcontinent and South-East Asian Region

In India, the use of pentavalent antimonials is limited due to the high resistance rates reported, especially in the state of Bihar. VL infection in HIV patients—with lower cure rates and higher relapse rates as compared to immunocompetent patients—could be associated with higher resistance to antimonial drug (Chakravarty and Sundar 2010).

In a retrospective study performed in India, liposomal amphotericin B was given to recently diagnosed VL/HIV co-infected patients with a final cure rate obtained at 1- to 2-year follow-up of 85%, and the tolerance to the drug was excellent (Sinha et al. 2011b).

There are no specific studies performed in the Indian subcontinent for patients with VL and HIV co-infection treated with miltefosine. However recent studies performed with immunocompetent patients have revealed that after a decade of use of miltefosine in the Indian subcontinent, the relapse rate has increased and several risk factors for the development of tolerance and resistance to this drug have been identified. Therefore, HIV patients, who show higher relapse rates and more persistent asymptomatic parasitaemia than non-HIV patients (van Griensven et al. 2014), could be a group at a higher risk of developing resistance or tolerance to miltefosine in this area with anthroponotic transmission.

A retrospective study was carried out in India in a clinical cohort of 102 VL/HIV co-infected patients. The treatment administered was liposomal amphotericin B in combination with miltefosine. Cure rates at 6, 12 and 18 months did not reach 30% (Mahajan et al. 2015).

8.7.1.3 East African Region

Studies in East Africa with pentavalent antimonials for VL and HIV co-infected were performed mainly in Ethiopia, reporting heterogeneous cure rates, but most of them did not reach the 50% of patients cured (Ritmeijer et al. 2001, 2006; Hailu et al. 2010b). Moreover toxicity reached 21.1% of patients in some studies which made patients having to discontinue the treatment temporarily or permanently (Diro et al. 2014a).

In another study performed in Ethiopia liposomal amphotericin B was administered to a cohort of HIV-positive and HIV-negative patients reaching a 60% cure rate (Ritmeijer et al. 2011). In a retrospective study carried out in eastern Sudan, liposomal amphotericin B was administered to a cohort of VL patients. Although the cure rate for non-HIV patients was high, mortality in VL/HIV co-infected patients was substantial. The specific cure rate for HIV patients is not specifically reported in the study (Salih et al. 2014).

In Ethiopia, a randomized, open-label clinical trial was performed with oral miltefosine versus pentavalent antimonials in a population where HIV is highly prevalent. In this case, miltefosine was observed to be safer for HIV-infected patients, but less effective than pentavalent antimonials (Ritmeijer et al. 2006).

Box 8.6 Therapeutic Regimens for Visceral Leishmaniasis and HIV Co-Infected Patients

- **Mediterranean region**
  - Sodium stibogluconate or meglumine antimoniate: (IM or IV) 20 mg/Sb^{v+}/kg/day for 28 days [B]
  - Amphotericin B: (IV) 0.7 mg/kg/day for 28 days [A]
  - Amphotericin B lipid complex: (IV) total dose 30 mg/kg [B]
- **Indian Subcontinent and Central Asia region**
  - Liposomal Amphotericin B: (IV) total dose 20–30 mg/kg [C]
- **East Africa region**
  - Sodium stibogluconate or meglumine antimoniate: (IM or IV) 20 mg/Sb^{v+}/kg/day for 28 days [B]
  - Miltefosine (orally) 100 mg/day for 28 days [B]

8.7.1.4 Secondary Prophylaxis for VL in HIV Co-Infection (Box 8.7)

In VL/HIV co-infected patients after the patient has finished and response to the initial treatment for VL, there is some times the need to establish a secondary prophylaxis. There is a meta-analysis that included 1017 co-infected patients that reported that secondary prophylaxis reduces significantly the relapse rate of VL (OR 0.228). However, there is scarce information that can determine which is the best drug, the dose to be given and which is the most effective regimen (Cota et al. 2011).

The only randomized clinical trial performed took place in Spain, and maintenance therapy with amphotericin B lipid complex was compared with no maintenance therapy. Results demonstrated how maintenance therapy reduced the relapse rates from 22 to 50% (Lopez-Velez et al. 2004). Another prospective study evaluated the effectiveness of maintenance therapy with liposomal amphotericin B and reported up to 80% of patients free of diseases after 12 months follow-up (Molina et al. 2007). In another study maintenance therapy with pentavalent antimonials were evaluated, and the relapse rate reduced significantly more than in those patients who either did not receive any treatment or who received allopurinol as secondary prophylaxis (Ribera et al. 1996). Pentamidine was also evaluated, and there were no relapses during the follow-up period (Perez-Molina et al. 1996). Miltefosine was evaluated in Portugal as a maintenance therapy in three patients remaining free of relapse for a median period of 20 months (Marques et al. 2008).

Another oral drug such as azoles has been found effective but based only on a series of cases where itraconazole or a combination of itraconazole or fluconazole with allopurinol were evaluated (Lafeuillade et al. 1992; Raffi et al. 1995). The advantage of these drugs is their good tolerance and low toxicity, although there is a risk of developing resistant fungal infections (Angarano et al. 1998; Torrus et al. 1996).

There is not clear data about until when maintenance therapy should be kept. According to different authors, once the patients have recovered their immune function with ART and the VL is quiescent, suspension of the prophylaxis could then be considered when the CD4+ count is maintained >200 cells/μL for more than 6 months (Berenguer et al. 2000; Soriano et al. 2000).

Box 8.7 Therapeutic Regimens of Secondary Prophylaxis for Leishmaniasis and HIV Co-Infected Patients

- **Mediterranean region**
  - Amphotericin B lipid complex (IV) 3–5 mg/kg/day every 3 weeks [A]
  - Meglumine antimoniate (IM or IV) 850 mg Sb^{v+} every 4 weeks [B]
  - Pentamidine isethionate (IV) 4 mg/kg/day every 2–4 weeks [C]

8.7.1.5 Follow-Up and Detection of Relapse of VL and HIV Co-Infection

There are several factors that have been identified as possible risk factors for VL relapse among HIV patients: (a) CD 4 cell count <100 cells/mm^3 when VL is diagnosed, (b) a low scarce increase in the CD 4 cell count in response to ART and (c) absence of secondary prophylaxis and history of previous episodes of relapse (Cota et al. 2011). Relapse may occur even among those patients who have been treated correctly and are receiving ART and even with secondary prophylaxis, so probably these measures only partially protect the patients (Cota et al. 2011). Hence these patients have to be monitored, indefinitely identifying clinical data that can suggest a relapse which should be parasitologically confirmed. It has been reported that the evidence of only a positive nonquantitative polymerase chain reaction (PCR) for *Leishmania* is not enough to determinate a VL relapse. However, the use of an ultrasensitive quantitative *Leishmania* PCR to monitor the parasite load seems useful to predict the risk of relapse in VL/HIV co-infected patients (Molina et al. 2013).

8.7.2 Visceral Leishmaniasis and Pregnancy

There is little experience on the treatment of VL in pregnancy, and most of the published information is based on clinical cases, most of them from East Africa (Mueller et al. 2006; Adam et al. 2009).

Undoubtedly not treating pregnant women with VL can pose a risk for the health of the mother and the foetus much greater than the possible toxicity of the treatment. Fatal outcomes of VL during pregnancy have been described such as spontaneous abortion, small-for-birth date and congenital leishmaniasis (Nyakundi et al. 1988; Eltoum et al. 1992).

Among the different therapeutic options, amphotericin B and its lipid formulations seem to be the most indicated. No congenital transmission and no spontaneous abortion have been described during amphotericin B treatment regimens on pregnant women (Thakur et al. 1993b; Dereure et al. 2003; Mueller et al. 2006). Pentavalent antimonials do not seem to be safe during pregnancy due to its potential teratogenic effect (Paumgartten and Chahoud 2001). Moreover, although pentavalent antimonials have been described as efficient for VL in pregnant women and able to avoid vertical transmission, relapse and therapeutic failures of VL have also been described (Utili et al. 1995). Paromomycin is an aminoglycoside able to cross the placental barrier and can accrue in the foetus plasma and amniotic fluid. There are no data about its use for VL in pregnant women, but as other aminoglycosides, its use could cause ototoxicity to the foetus, so it should not be administrated during pregnancy (Davidson et al. 2009). Pentamidine is also contraindicated during pregnancy as well as miltefosine because they are both potentially embryotoxic and teratogenic. In fact, women in child-bearing age should be tested for pregnancy before administrating any of these teratogenic drugs, and in the case of miltefosine, contraception should be administrated during and for 3 months after treatment (Monge-Maillo and Lopez-Velez 2015).

References

Adam GK, Abdulla MA, Ahmed AA, Adam I. Maternal and perinatal outcomes of visceral leishmaniasis (kala-azar) treated with sodium stibogluconate in eastern Sudan. Int J Gynaecol Obstet. 2009;107(3):208–10.

Alvar J, Canavate C, Gutierrez-Solar B, Jimenez M, Laguna F, Lopez-Velez R, et al. Leishmania and human immunodeficiency virus coinfection: the first 10 years. Clin Microbiol Rev. 1997;10(2):298–319.

Alvar J, Croft S, Olliaro P. Chemotherapy in the treatment and control of leishmaniasis. Adv Parasitol. 2006;61:223–74.

Alvar J, Aparicio P, Aseffa A, Den Boer M, Canavate C, Dedet JP, et al. The relationship between leishmaniasis and AIDS: the second 10 years. Clin Microbiol Rev. 2008;21(2):334–59.

Alvar J, Velez ID, Bern C, Herrero M, Desjeux P, Cano J, et al. Leishmaniasis worldwide and global estimates of its incidence. PLoS One. 2012;7(5):e35671.

Anabwani GM, Ngira JA, Dimiti G, Bryceson AD. Comparison of two dosage schedules of sodium stibogluconate in the treatment of visceral leishmaniasis in Kenya. Lancet. 1983;1(8318):210–3.

Angarano G, Maggi P, Coppola SL, Cavaliere RL. Itraconazole as maintenance therapy for visceral leishmaniasis in HIV-infected patients. Eur J Clin Microbiol Infect Dis. 1998;17(5):365–7.

Barragan P, Lopez-Velez R, Olmo M, Podzamczer D. Visceral Leishmaniasis treated with antimonials/paromomycin followed by itraconazole/miltefosine after standard therapy failures in a human immunodeficiency virus-infected patient. Am J Trop Med Hyg. 2010;83(1):10–2.

Berenguer J, Cosin J, Miralles P, Lopez JC, Padilla B. Discontinuation of secondary anti-leishmania prophylaxis in HIV-infected patients who have responded to highly active antiretroviral therapy. AIDS. 2000;14(18):2946–8.

Berman JD, Badaro R, Thakur CP, Wasunna KM, Behbehani K, Davidson R, et al. Efficacy and safety of liposomal amphotericin B (AmBisome) for visceral leishmaniasis in endemic developing countries. Bull World Health Organ. 1998;76(1):25–32.

Bhattacharya SK, Jha TK, Sundar S, Thakur CP, Engel J, Sindermann H, et al. Efficacy and tolerability of miltefosine for childhood visceral leishmaniasis in India. Clin Infect Dis. 2004;38(2):217–21.

Bhattacharya SK, Sinha PK, Sundar S, Thakur CP, Jha TK, Pandey K, et al. Phase 4 trial of miltefosine for the treatment of Indian visceral leishmaniasis. J Infect Dis. 2007;196(4):591–8.

Cascio A, di Martino L, Occorsio P, Giacchino R, Catania S, Gigliotti AR, et al. A 6 day course of liposomal amphotericin B in the treatment of infantile visceral leishmaniasis: the Italian experience. J Antimicrob Chemother. 2004;54(1):217–20.

Chakravarty J, Sundar S. Drug resistance in leishmaniasis. J Glob Infect Dis. 2010;2(2):167–76.

Chunge CN, Owate J, Pamba HO, Donno L. Treatment of visceral leishmaniasis in Kenya by aminosidine alone or combined with sodium stibogluconate. Trans R Soc Trop Med Hyg. 1990;84(2):221–5.

Collin S, Davidson R, Ritmeijer K, Keus K, Melaku Y, Kipngetich S, et al. Conflict and kala-azar: determinants of adverse outcomes of kala-azar among patients in southern Sudan. Clin Infect Dis. 2004;38(5):612–9.

Cota GF, de Sousa MR, Rabello A. Predictors of visceral leishmaniasis relapse in HIV-infected patients: a systematic review. PLoS Negl Trop Dis. 2011;5(6):e1153.

Cota GF, de Sousa MR, Fereguetti TO, Rabello A. Efficacy of anti-leishmania therapy in visceral leishmaniasis among HIV infected patients: a systematic review with indirect comparison. PLoS Negl Trop Dis. 2013;7(5):e2195.

Croft SL, Sundar S, Fairlamb AH. Drug resistance in leishmaniasis. Clin Microbiol Rev. 2006;19(1):111–26.

Das VN, Ranjan A, Sinha AN, Verma N, Lal CS, Gupta AK, et al. A randomized clinical trial of low dosage combination of pentamidine and allopurinol in the treatment of antimony unresponsive cases of visceral leishmaniasis. J Assoc Physicians India. 2001;49:609–13.

Das VN, Siddiqui NA, Pandey K, Singh VP, Topno RK, Singh D, et al. A controlled, randomized nonblinded clinical trial to assess the efficacy of amphotericin B deoxycholate as compared to pentamidine for the treatment of antimony unresponsive visceral leishmaniasis cases in Bihar, India. Ther Clin Risk Manag. 2009;5(1):117–24.

Davidson RN, den Boer M, Ritmeijer K. Paromomycin. Trans R Soc Trop Med Hyg. 2009;103(7):653–60.

Dereure J, Duong Thanh H, Lavabre-Bertrand T, Cartron G, Bastides F, Richard-Lenoble D, et al. Visceral leishmaniasis. Persistence of parasites in lymph nodes after clinical cure. J Infect. 2003;47(1):77–81.

di Martino L, Gramiccia M, Occorsio P, Di Muccio T, Scalone A, Gradoni L. Infantile visceral leishmaniasis in the Campania region, Italy: experience from a Paediatric Referral Centre. Parassitologia. 2004;46(1–2):221–3.

Diro E, Lynen L, Mohammed R, Boelaert M, Hailu A, van Griensven J. High parasitological failure rate of visceral leishmaniasis to sodium stibogluconate among HIV co-infected adults in Ethiopia. PLoS Negl Trop Dis. 2014a;8(5):e2875.

Diro E, Lynen L, Ritmeijer K, Boelaert M, Hailu A, van Griensven J. Visceral Leishmaniasis and HIV coinfection in East Africa. PLoS Negl Trop Dis. 2014b;8(6):e2869.

Dorlo TP, Rijal S, Ostyn B, de Vries PJ, Singh R, Bhattarai N, et al. Failure of miltefosine in visceral leishmaniasis is associated with low drug exposure. J Infect Dis. 2014;210(1):146–53.

Edwards T, Omollo R, Khalil EA, Yifru S, Musa B, Musa A, et al. Single-dose liposomal amphotericin B (AmBisome(R)) for the treatment of visceral leishmaniasis in East Africa: study protocol for a randomized controlled trial. Trials. 2011;12:66.

Eltoum IA, Zijlstra EE, Ali MS, Ghalib HW, Satti MM, Eltoum B, et al. Congenital kala-azar and leishmaniasis in the placenta. Am J Trop Med Hyg. 1992;46(1):57–62.

Figueras Nadal MC, Garcia de Miguel MJ, Asensi Botet F, Velasco Bernardo R, Canals Baeza A, Ausin Aoiz I. Short course treatment for visceral leishmaniasis with liposomal amphotericin B in immunocompetent patients. An Pediatr (Barc). 2003;59(6):535–40.

Gradoni L, Bryceson A, Desjeux P. Treatment of Mediterranean visceral leishmaniasis. Bull World Health Organ. 1995;73(2):191–7.

Gradoni L, Soteriadou K, Louzir H, Dakkak A, Toz SO, Jaffe C, et al. Drug regimens for visceral leishmaniasis in Mediterranean countries. Trop Med Int Health. 2008;13(10):1272–6.

Hailu A, Musa A, Wasunna M, Balasegaram M, Yifru S, Mengistu G, et al. Geographical variation in the response of visceral leishmaniasis to paromomycin in East Africa: a multicentre, open-label, randomized trial. PLoS Negl Trop Dis. 2010a;4(10):e709.

Hailu W, Weldegebreal T, Hurissa Z, Tafes H, Omollo R, Yifru S, et al. Safety and effectiveness of meglumine antimoniate in the treatment of Ethiopian visceral leishmaniasis patients with and without HIV co-infection. Trans R Soc Trop Med Hyg. 2010b;104(11):706–12.

Jeronimo SM, Duggal P, Braz RF, Cheng C, Monteiro GR, Nascimento ET, et al. An emerging peri-urban pattern of infection with Leishmania chagasi, the protozoan causing visceral leishmaniasis in northeast Brazil. Scand J Infect Dis. 2004;36(6–7):443–9.

Jha TK, Giri YN, Singh TK, Jha S. Use of amphotericin B in drug-resistant cases of visceral leishmaniasis in north Bihar, India. Am J Trop Med Hyg. 1995;52(6):536–8.

Jha TK, Olliaro P, Thakur CP, Kanyok TP, Singhania BL, Singh IJ, et al. Randomised controlled trial of aminosidine (paromomycin) v sodium stibogluconate for treating visceral leishmaniasis in North Bihar, India. BMJ. 1998;316(7139):1200–5.

Kafetzis DA, Velissariou IM, Stabouli S, Mavrikou M, Delis D, Liapi G. Treatment of paediatric visceral leishmaniasis: amphotericin B or pentavalent antimony compounds? Int J Antimicrob Agents. 2005;25(1):26–30.

Kager PA, Rees PH, Manguyu FM, Bhatt KM, Wellde BT, Hockmeyer WT, et al. Clinical, haematological and parasitological response to treatment of visceral leishmaniasis in Kenya. A study of 64 patients. Trop Geogr Med. 1984;36(1):21–35.

Khalil EA, el Hassan AM, Zijlstra EE, Hashim FA, Ibrahim ME, Ghalib HW, et al. Treatment of visceral leishmaniasis with sodium stibogluconate in Sudan: management of those who do not respond. Ann Trop Med Parasitol. 1998;92(2):151–8.

Khalil EA, Weldegebreal T, Younis BM, Omollo R, Musa AM, Hailu W, et al. Safety and efficacy of single dose versus multiple doses of AmBisome for treatment of visceral leishmaniasis in eastern Africa: a randomised trial. PLoS Negl Trop Dis. 2014;8(1):e2613.

Lafeuillade A, Chaffanjon P, Delbeke E, Quilichini R. Maintenance itraconazole for visceral leishmaniasis in HIV infection. Am J Med. 1992;92(4):449.

Laguna F, Lopez-Velez R, Soriano V, Montilla P, Alvar J, Gonzalez-Lahoz JM. Assessment of allopurinol plus meglumine antimoniate in the treatment of visceral leishmaniasis in patients infected with HIV. J Infect. 1994;28(3):255–9.

Laguna F, Lopez-Velez R, Pulido F, Salas A, Torre-Cisneros J, Torres E, et al. Treatment of visceral leishmaniasis in HIV-infected patients: a randomized trial comparing meglumine antimoniate with amphotericin B. Spanish HIV-Leishmania Study Group. AIDS. 1999;13(9):1063–9.

Laguna F, Videla S, Jimenez-Mejias ME, Sirera G, Torre-Cisneros J, Ribera E, et al. Amphotericin B lipid complex versus meglumine antimoniate in the treatment of visceral leishmaniasis in patients infected with HIV: a randomized pilot study. J Antimicrob Chemother. 2003;52(3):464–8.

Lopez-Velez R, Videla S, Marquez M, Boix V, Jimenez-Mejias ME, Gorgolas M, et al. Amphotericin B lipid complex versus no treatment in the secondary prophylaxis of visceral leishmaniasis in HIV-infected patients. J Antimicrob Chemother. 2004;53(3):540–3.

Mahajan R, Das P, Isaakidis P, Sunyoto T, Sagili KD, Lima MA, et al. Combination treatment for visceral Leishmaniasis patients co-infected with human immunodeficiency virus in India. Clin Infect Dis. 2015;61(8):1255–62.

Marques N, Sa R, Coelho F, Oliveira J, Saraiva Da Cunha J, Melico-Silvestre A. Miltefosine for visceral leishmaniasis relapse treatment and secondary prophylaxis in HIV-infected patients. Scand J Infect Dis. 2008;40(6–7):523–6.

Maru M. Clinical and laboratory features and treatment of visceral leishmaniasis in hospitalized patients in Northwestern Ethiopia. Am J Trop Med Hyg. 1979;28(1):15–8.

Mary C, Faraut F, Drogoul MP, Xeridat B, Schleinitz N, Cuisenier B, et al. Reference values for Leishmania infantum parasitemia in different clinical presentations: quantitative polymerase chain reaction for therapeutic monitoring and patient follow-up. Am J Trop Med Hyg. 2006;75(5):858–63.

Mastroianni A. Liposomal amphotericin B and rHuGM-CSF for treatment of visceral leishmaniasis in AIDS. Infez Med. 2004;12(3):197–204.

Melaku Y, Collin SM, Keus K, Gatluak F, Ritmeijer K, Davidson RN. Treatment of kala-azar in southern Sudan using a 17-day regimen of sodium stibogluconate combined with paromomycin: a retrospective comparison with 30-day sodium stibogluconate monotherapy. Am J Trop Med Hyg. 2007;77(1):89–94.

Minodier P, Garnier JM. Childhood visceral leishmaniasis in Provence. Arch Pediatr. 2000;7(Suppl 3):572s–7s.

Minodier P, Robert S, Noel G, Blanc P, Retornaz K, Garnier JM. First-line liposomal amphotericin B for pediatric visceral leishmaniasis in southern France. Arch Pediatr. 2005;12(7):1102–8.

Mishra M, Biswas UK, Jha DN, Khan AB. Amphotericin versus pentamidine in antimony-unresponsive kala-azar. Lancet. 1992;340(8830):1256–7.

Molina I, Falco V, Crespo M, Riera C, Ribera E, Curran A, et al. Efficacy of liposomal amphotericin B for secondary prophylaxis of visceral leishmaniasis in HIV-infected patients. J Antimicrob Chemother. 2007;60(4):837–42.

Molina I, Fisa R, Riera C, Falco V, Elizalde A, Salvador F, et al. Ultrasensitive real-time PCR for the clinical management of visceral leishmaniasis in HIV-infected patients. Am J Trop Med Hyg. 2013;89(1):105–10.

Monge-Maillo B, Lopez-Velez R. Therapeutic options for visceral Leishmaniasis. Drugs. 2013;73(17):1863–88.

Monge-Maillo B, Lopez-Velez R. Miltefosine for visceral and cutaneous leishmaniasis: drug characteristics and evidence-based treatment recommendations. Clin Infect Dis. 2015;60(9):1398–404.

Moore E, O'Flaherty D, Heuvelmans H, Seaman J, Veeken H, de Wit S, et al. Comparison of generic and proprietary sodium stibogluconate for the treatment of visceral leishmaniasis in Kenya. Bull World Health Organ. 2001;79(5):388–93.

Mueller M, Balasegaram M, Koummuki Y, Ritmeijer K, Santana MR, Davidson R. A comparison of liposomal amphotericin B with sodium stibogluconate for the treatment of visceral leishmaniasis in pregnancy in Sudan. J Antimicrob Chemother. 2006;58(4):811–5.

Mueller Y, Nguimfack A, Cavailler P, Couffignal S, Rwakimari JB, Loutan L, et al. Safety and effectiveness of amphotericin B deoxycholate for the treatment of visceral leishmaniasis in Uganda. Ann Trop Med Parasitol. 2008;102(1):11–9.

Mueller Y, Mbulamberi DB, Odermatt P, Hoffmann A, Loutan L, Chappuis F. Risk factors for in-hospital mortality of visceral leishmaniasis patients in eastern Uganda. Trop Med Int Health. 2009;14(8):910–7.

Musa AM, Younis B, Fadlalla A, Royce C, Balasegaram M, Wasunna M, et al. Paromomycin for the treatment of visceral leishmaniasis in Sudan: a randomized, open-label, dose-finding study. PLoS Negl Trop Dis. 2010;4(10):e855.

Musa A, Khalil E, Hailu A, Olobo J, Balasegaram M, Omollo R, et al. Sodium stibogluconate (SSG) & paromomycin combination compared to SSG for visceral leishmaniasis in East Africa: a randomised controlled trial. PLoS Negl Trop Dis. 2012;6(6):e1674.

Nyakundi PM, Muigai R, Were JB, Oster CN, Gachihi GS, Kirigi G. Congenital visceral leishmaniasis: case report. Trans R Soc Trop Med Hyg. 1988;82(4):564.

Organización Panamericana de la Salud. Guía para el tratamiento de las enfermedades infecciosas. Washington: OPS; 2004. http://www.ops-oms.org/common/Display.asp?Lang=S&RecID=9629.

Paumgartten FJ, Chahoud I. Embryotoxicity of meglumine antimoniate in the rat. Reprod Toxicol. 2001;15(3):327–31.

Perez-Molina JA, Lopez-Velez R, Montilla P, Guerrero A. Pentamidine isethionate as secondary prophylaxis against visceral leishmaniasis in HIV-positive patients. AIDS. 1996;10(2):237–8.

Petrela R, Kuneshka L, Foto E, Zavalani F, Gradoni L. Pediatric visceral leishmaniasis in Albania: a retrospective analysis of 1,210 consecutive hospitalized patients (1995–2009). PLoS Negl Trop Dis. 2010;4(9):e814.

Pintado V, Lopez-Velez R. HIV-associated visceral leishmaniasis. Clin Microbiol Infect. 2001;7(6):291–300.

Raffi F, Merrien D, Le Pape P, Reliquet V. Use of an Itraconazole/allopurinol combination for the treatment of visceral leishmaniasis in a patient with AIDS. Clin Infect Dis. 1995;21(5):1338–9.

Rahman M, Ahmed BN, Faiz MA, Chowdhury MZ, Islam QT, Sayeedur R, et al. Phase IV trial of miltefosine in adults and children for treatment of visceral leishmaniasis (kala-azar) in Bangladesh. Am J Trop Med Hyg. 2011;85(1):66–9.

Ribera E, Ocana I, de Otero J, Cortes E, Gasser I, Pahissa A. Prophylaxis of visceral leishmaniasis in human immunodeficiency virus-infected patients. Am J Med. 1996;100(5):496–501.

Rijal S, Ostyn B, Uranw S, Rai K, Bhattarai NR, Dorlo TP, et al. Increasing failure of miltefosine in the treatment of kala-azar in Nepal and the potential role of parasite drug resistance, reinfection, or noncompliance. Clin Infect Dis. 2013;56(11):1530–8.

Ritmeijer K, Veeken H, Melaku Y, Leal G, Amsalu R, Seaman J, et al. Ethiopian visceral leishmaniasis: generic and proprietary sodium stibogluconate are equivalent; HIV co-infected patients have a poor outcome. Trans R Soc Trop Med Hyg. 2001;95(6):668–72.

Ritmeijer K, Dejenie A, Assefa Y, Hundie TB, Mesure J, Boots G, et al. A comparison of miltefosine and sodium stibogluconate for treatment of visceral leishmaniasis in an Ethiopian population with high prevalence of HIV infection. Clin Infect Dis. 2006;43(3):357–64.

Ritmeijer K, ter Horst R, Chane S, Aderie EM, Piening T, Collin SM, et al. Limited effectiveness of high-dose liposomal amphotericin B (AmBisome) for treatment of visceral leishmaniasis in an Ethiopian population with high HIV prevalence. Clin Infect Dis. 2011;53(12):e152–8.

Rosenthal E, Delaunay P, Jeandel PY, Haas H, Pomares-Estran C, Marty P. Liposomal amphotericin B as treatment for visceral leishmaniasis in Europe, 2009. Med Mal Infect. 2009;39(10):741–4.

Russo R, Nigro LC, Minniti S, Montineri A, Gradoni L, Caldeira L, et al. Visceral leishmaniasis in HIV infected patients: treatment with high dose liposomal amphotericin B (AmBisome). J Infect. 1996;32(2):133–7.

Rybniker J, Goede V, Mertens J, Ortmann M, Kulas W, Kochanek M, et al. Treatment of visceral leishmaniasis with intravenous pentamidine and oral fluconazole in an HIV-positive patient with chronic renal failure—a case report and brief review of the literature. Int J Infect Dis. 2010;14(6):e522–5.

Salam N, Al-Shaqha WM, Azzi A. Leishmaniasis in the middle East: incidence and epidemiology. PLoS Negl Trop Dis. 2014;8(10):e3208.

Salih NA, van Griensven J, Chappuis F, Antierens A, Mumina A, Hammam O, et al. Liposomal amphotericin B for complicated visceral leishmaniasis (kala-azar) in eastern Sudan: how effective is treatment for this neglected disease? Trop Med Int Health. 2014;19(2):146–52.

Santos MA, Marques RC, Farias CA, Vasconcelos DM, Stewart JM, Costa DL, et al. Predictors of an unsatisfactory response to pentavalent antimony in the treatment of American visceral leishmaniasis. Rev Soc Bras Med Trop. 2002;35(6):629–33.

Seaman J, Pryce D, Sondorp HE, Moody A, Bryceson AD, Davidson RN. Epidemic visceral leishmaniasis in Sudan: a randomized trial of aminosidine plus sodium stibogluconate versus sodium stibogluconate alone. J Infect Dis. 1993;168(3):715–20.

Seaman J, Boer C, Wilkinson R, de Jong J, de Wilde E, Sondorp E, et al. Liposomal amphotericin B (AmBisome) in the treatment of complicated kala-azar under field conditions. Clin Infect Dis. 1995;21(1):188–93.

Sindermann H, Engel KR, Fischer C, Bommer W, Miltefosine Compassionate Use Program. Oral miltefosine for leishmaniasis in immunocompromised patients: compassionate use in 39 patients with HIV infection. Clin Infect Dis. 2004;39(10):1520–3.

Singh S. Changing trends in the epidemiology, clinical presentation, and diagnosis of Leishmania-HIV co-infection in India. Int J Infect Dis. 2014;29C:103–12.

Sinha PK, Jha TK, Thakur CP, Nath D, Mukherjee S, Aditya AK, et al. Phase 4 pharmacovigilance trial of paromomycin injection for the treatment of visceral leishmaniasis in India. J Trop Med. 2011a;2011:645203.

Sinha PK, van Griensven J, Pandey K, Kumar N, Verma N, Mahajan R, et al. Liposomal amphotericin B for visceral leishmaniasis in human immunodeficiency virus-coinfected patients: 2-year treatment outcomes in Bihar, India. Clin Infect Dis. 2011b;53(7):e91–8.

Soriano VF, Dona CF, Rodriguez-Rosado RF, Barreiro PF, Gonzalez-Lahoz J. Discontinuation of secondary prophylaxis for opportunistic infections in HIV-infected patients receiving highly active antiretroviral therapy. AIDS. 2000;14(4):383–6.

Sundar S, Rai M. Laboratory diagnosis of visceral leishmaniasis. Clin Diagn Lab Immunol. 2002a;9(5):951–8.

Sundar S, Rai M. Advances in the treatment of leishmaniasis. Curr Opin Infect Dis. 2002b;15(6):593–8.

Sundar S, Jha TK, Thakur CP, Engel J, Sindermann H, Fischer C, et al. Oral miltefosine for Indian visceral leishmaniasis. N Engl J Med. 2002;347(22):1739–46.

Sundar S, Jha TK, Sindermann H, Junge K, Bachmann P, Berman J. Oral miltefosine treatment in children with mild to moderate Indian visceral leishmaniasis. Pediatr Infect Dis J. 2003;22(5):434–8.

Sundar S, Mehta H, Suresh AV, Singh SP, Rai M, Murray HW. Amphotericin B treatment for Indian visceral leishmaniasis: conventional versus lipid formulations. Clin Infect Dis. 2004;38(3):377–83.

Sundar S, Jha TK, Thakur CP, Bhattacharya SK, Rai M. Oral miltefosine for the treatment of Indian visceral leishmaniasis. Trans R Soc Trop Med Hyg. 2006;100(Suppl 1):S26–33.

Sundar S, Jha TK, Thakur CP, Sinha PK, Bhattacharya SK. Injectable paromomycin for visceral leishmaniasis in India. N Engl J Med. 2007;356(25):2571–81.

Sundar S, Rai M, Chakravarty J, Agarwal D, Agrawal N, Vaillant M, et al. New treatment approach in Indian visceral leishmaniasis: single-dose liposomal amphotericin B followed by short-course oral miltefosine. Clin Infect Dis. 2008;47(8):1000–6.

Sundar S, Chakravarty J, Agarwal D, Rai M, Murray HW. Single-dose liposomal amphotericin B for visceral leishmaniasis in India. N Engl J Med. 2010;362(6):504–12.

Sundar S, Sinha PK, Rai M, Verma DK, Nawin K, Alam S, et al. Comparison of short-course multidrug treatment with standard therapy for visceral leishmaniasis in India: an open-label, non-inferiority, randomised controlled trial. Lancet. 2011;377(9764):477–86.

Sundar S, Singh A, Rai M, Prajapati VK, Singh AK, Ostyn B, et al. Efficacy of miltefosine in the treatment of visceral leishmaniasis in India after a decade of use. Clin Infect Dis. 2012;55(4):543–50.

Thakur CP. Epidemiological, clinical and therapeutic features of Bihar kala-azar (including post kala-azar dermal leishmaniasis). Trans R Soc Trop Med Hyg. 1984;78(3):391–8.

Thakur CP. A single high dose treatment of kala-azar with Ambisome (amphotericin B lipid complex): a pilot study. Int J Antimicrob Agents. 2001;17(1):67–70.

Thakur CP, Kumar M, Pandey AK. Comparison of regimes of treatment of antimony-resistant kala-azar patients: a randomized study. Am J Trop Med Hyg. 1991;45(4):435–41.

Thakur CP, Olliaro P, Gothoskar S, Bhowmick S, Choudhury BK, Prasad S, et al. Treatment of visceral leishmaniasis (kala-azar) with aminosidine (= paromomycin)-antimonial combinations, a pilot study in Bihar, India. Trans R Soc Trop Med Hyg. 1992;86(6):615–6.

Thakur CP, Sinha GP, Pandey AK, Barat D, Sinha PK. Amphotericin B in resistant kala-azar in Bihar. Natl Med J India. 1993a;6(2):57–60.

Thakur CP, Sinha GP, Sharma V, Barat D. The treatment of kala-azar during pregnancy. Natl Med J India. 1993b;6(6):263–5.

Thakur CP, Kanyok TP, Pandey AK, Sinha GP, Zaniewski AE, Houlihan HH, et al. A prospective randomized, comparative, open-label trial of the safety and efficacy of paromomycin (aminosidine) plus sodium stibogluconate versus sodium stibogluconate alone for the treatment of visceral leishmaniasis. Trans R Soc Trop Med Hyg. 2000;94(4):429–31.

Torrus D, Boix V, Massa B, Portilla J, Perez-Mateo M. Fluconazole plus allopurinol in treatment of visceral leishmaniasis. J Antimicrob Chemother. 1996;37(5):1042–3.

Toumi A, Kilani B, Ammari L, Tiouiri H, Kanoun F, Belhadj S, et al. Demographic, clinical and therapeutic features of adult visceral leishmaniasis at the Rabta hospital in Tunis (Tunisia) from 1983 to 2002. Bull Soc Pathol Exot. 2007;100(4):282–6.

Troya J, Casquero A, Refoyo E, Fernandez-Guerrero ML, Gorgolas M. Long term failure of miltefosine in the treatment of refractory visceral leishmaniasis in AIDS patients. Scand J Infect Dis. 2008;40(1):78–80.
Utili R, Rambaldi A, Tripodi MF, Andreana A. Visceral leishmaniasis during pregnancy treated with meglumine antimoniate. Infection. 1995;23(3):182–3.
van Griensven J, Carrillo E, Lopez-Velez R, Lynen L, Moreno J. Leishmaniasis in immunosuppressed individuals. Clin Microbiol Infect. 2014;20(4):286–99.
Veeken H, Ritmeijer K, Seaman J, Davidson R. A randomized comparison of branded sodium stibogluconate and generic sodium stibogluconate for the treatment of visceral leishmaniasis under field conditions in Sudan. Trop Med Int Health. 2000;5(5):312–7.
Velez ID, Colmenares LM, Munoz CA. Two cases of visceral leishmaniasis in Colombia resistant to meglumine antimonial treatment. Rev Inst Med Trop Sao Paulo. 2009;51(4):231–6.
Wasunna M, Njenga S, Balasegaram M, Alexander N, Omollo R, Edwards T, Dorlo TP, Musa B, Ali MH, Elamin MY, Kirigi G, Juma R, Kip AE, Schoone GJ, Hailu A, Olobo J, Ellis S, Kimutai R, Wells S, Khalil EA, Strub Wourgaft N, Alves F, Musa A. Efficacy and safety of AmBisome in combination with sodium stibogluconate or miltefosine and miltefosine monotherapy for African visceral leishmaniasis: phase II randomized trial. PLoS Negl Trop Dis. 2016;10(9):e0004880.
Werneck GL, Batista MS, Gomes JR, Costa DL, Costa CH. Prognostic factors for death from visceral leishmaniasis in Teresina, Brazil. Infection. 2003;31(3):174–7.
World Health Organization (WHO). Control of the leishmaniases: report of a meeting of the WHO Expert Committee on the Control of Leishmaniases, Geneva, 22–26 March 2010. Geneva: WHO; 2010. WHO technical report series; no. 949. Available from: http://whqlibdoc.who.int/trs/WHO_TRS_949_eng.pdf.

Treatment of Tegumentary Forms of Leishmaniasis

9

Johannes Blum, Andreas Neumayr, and Diana Lockwood

9.1 General Treatment Considerations in Cutaneous Leishmaniasis (CL) and Mucosal Leishmaniasis (ML)

Species-Oriented Treatment

CL lesions range from a single limited skin lesion that may heal spontaneously to large and multiple locally destructive skin lesions, which may spread to or involve mucosa. Choosing a treatment regimen is influenced by the size, number and location of the lesion(s) but mainly determined by the infecting *Leishmania* species. Since *Leishmania* parasites can now be genotyped by polymerase chain reaction (PCR) techniques followed by sequencing, the species-specific treatment response of the different *Leishmania* species became assessable and evidence-based species-specific treatment guidelines are increasingly replacing the previously used treatment recommendations solely based on the geographical background of the infection (Blum et al. 2014; Blum et al. 2004).

Patient Evaluation Before Treatment

All lesions have to be well documented (measuring the diameter of the ulcer and of the complete lesion plus photo documentation). Since mucosal spread may affect the nasal as well as the oral mucosa, an ear-nose-throat (ENT) examination is indicated, and referral to an ENT specialist may be warranted. The possibility of CL being part of visceral leishmaniasis (VL) occurs rarely but should be considered if

J. Blum (✉) • A. Neumayr
Swiss Tropical and Public Health Institute, Basel, Switzerland

University of Basel, Basel, Switzerland
e-mail: johannes.blum@unibas.ch

D. Lockwood
London School of Hygiene & Tropical Medicine, London, UK

F. Bruschi, L. Gradoni (eds.), *The Leishmaniases: Old Neglected Tropical Diseases*, https://doi.org/10.1007/978-3-319-72386-0_9

the patient has fever and hepatosplenomegaly and laboratory markers of VL infection (pancytopenia, positive *Leishmania* antibody titres). This clinical presentation is more likely if a patient has underlying immune suppression.

Treatment Options
Treatment options include systemic treatment with antileishmanial drugs, local topical treatment with antileishmanial ointments/creams, local intralesional injection of antileishmanial drugs and local physical treatment (cryotherapy, thermotherapy).

9.1.1 Systemic Treatment

The available antileishmanial drugs, including their dosing, mode of action, mechanism of elimination, adverse effects/toxicity profile and recommendations for treatment monitoring are described in Tables 9.1. Since most systemic drugs are associated with relevant adverse effects, the option and feasibility of local treatment should always be evaluated (see below).

9.1.2 Local Treatment with Ointment Containing 15% Paromomycin

An ointment with 15% paromomycin plus 12% methylbenzethonium chloride was tested in Old and New World cutaneous leishmaniasis with success, but was associated with local severe irritancy and intolerance in up to 75% of patients (Ben Salah et al. 2013). This combination appears to be more effective than the combination of 15% paromomycin plus 10% urea, but it causes more local inflammatory reactions (Bryceson et al. 1994). A cream containing 15% paromomycin sulphate plus 0.5% gentamicin sulphate in a complex base (called WR 279,396) was not superior to a cream containing paromomycin 15% without a second drug but to placebo (Ben Salah et al. 2013). The limited availability of topical paromomycin preparations in most countries and the frequent local irritations limit their use.

9.1.3 Thermotherapy and Cryotherapy

Cutaneous leishmaniasis has been treated at all ages using a wide range of physical methods including cauterization, surgical excision, cryotherapy and the application of local heat.

Thermotherapy Since decades it is known that it is possible to induce healing of lesions by applying local heat. Ethnomedical studies have also shown that, in rural communities of South America and Africa, the empiric application of caustic materials (powder, hot brown sugar, silver nitrate, oil, battery) or the cauterization of the

Table 9.1 Drugs and follow-up for treatment of tegumentary forms of leishmaniasis

Drug	Special precautions/ pregnancy categories	Renal function	Adverse effects	Monitoring of drug toxicity during treatment
Systemic pentavalent antimonials	– Age >60 years – Cardiomyopathy – Liver disease – Renal impairment – Pancreatitis	Renal excretion >80%	Cardiac toxicity with reversible ECG alterations is seen in 30–60% – repolarization alterations affecting T wave and ST segment – prolongation of the corrected QT interval Fatal arrhythmias have not been documented in CL patients treated with the usual dose ≤20 mg Sb/kg (Herwaldt and Berman 1992; Antezana et al. 1992; Wise et al. 2012; Ribeiro et al. 1999; Lawn et al. 2006) – hypokalaemia associated with risk of arrhythmias	ECG checks one to two times every week: interruption of treatment if – significant arrhythmias – QTc longer 0.5 s (age-adapted limits in children) – QTc longer than 0,45 second: monitoring/ dose reduction – concave ST segment – monitor potassium weekly
	Pregnancy category: unknown	Dose adjustment (Buffet et al. 1995)	Hepatotoxicity is seen in 50%, reversible	Transaminases weekly
				Treatment interruption if transaminases higher than five times the upper limit of normal value (ULN) (Hepburn et al. 1993)
			Haematotoxicity (anaemia, leucopaenia, thrombocytopaenia) (Hepburn 1993)	Haemoglobin, leucocytes and platelets weekly
			Pancreatitis can occur either very early in therapy (and is then often symptomatic) or more progressively during the course of therapy. Serum levels of amylase and lipase may decline despite continued treatment with antimonials	Amylase and lipase after 48 h of treatment then weekly
				Treatment interruption if serum amylase levels became >4 times the ULN or lipase levels of >15 times the ULN, regardless of symptoms. Therapy can be resumed, once these values tend significantly towards normal (Aronson et al. 1998; Gasser et al. 1994)

(continued)

Table 9.1 (continued)

Drug	Special precautions/ pregnancy categories	Renal function	Adverse effects	Monitoring of drug toxicity during treatment
			Subjective complaints: musculoskeletal symptoms, headache, gastrointestinal complaints, pain at the injection site	
			Rare complications: glomerulonephritis, acute renal failure, (Rodrigues et al. 1999) peripheral nephritis (Brummitt et al. 1996), exfoliate dermatitis, herpes zoster (Wortmann et al. 1998), hypersensitivity syndrome	Creatinine and examination of urine weekly
Pentamidine	– Renal impairment – Liver disease – Pancreatitis – Diabetes	Renal excretion <10%	Aseptic abscess (accidental contact of pentamidine with the subcutaneous tissue)	Pentamidine has to be given by infusion or injected slowly and strictly intramuscular with a long needle (50 mm)
	Pregnancy category C: Foetal risk is unknown	Dose adjustment not necessary	Hypoglycaemia, diabetes, proteinuria	Fasting glycaemia and urine for proteinuria and glycosuria have to be checked before every injection and at 3 weeks and at 2 months after the last injection (Hellier et al. 2000)
			Rhabdomyolysis (Lightburn et al. 2003; Delobel and Pradinaud 2003)	Monitor CK in the case of clinical signs of rhabdomyolysis such as myalgia or kidney failure
			Hypotension (Soto-Mancipe et al. 1993; Nacher et al. 2001)	The blood pressure and heart rate have to be measured before and after the injection (every 15 min for 1 h) (Hellier et al. 2000); hypotension is less frequent when administered by slow infusion
			Subjective complaints: myalgia, nausea and gustative abnormalities, headache, pain at the injection site, abdominal pain (Nacher et al. 2001)	

Miltefosine	Teratogenic: Pregnancy is contraindication subtherapeutic concentrations in the blood >4 months after treatment: contraception required until 4 months post-treatment	Renal excretion <1%	Subjective complaints: nausea (36%), vomiting up to 40% often during the first week, motion sickness (29%), headache (27%), diarrhoea (6–16%), vomiting (32–38%) (Soto and Berman 2006; Sundar et al. 2006)	Intake of (fatty) food just before intake of miltefosine reduces gastrointestinal side effects without effect on bioavailability
		Dose adjustment not necessary	Impaired renal function: Creatinine increased above the normal range in 32%, in 31% < 1.5 times the upper limit of normal und in 1% between 1.5 and 3 times the upper limit of normal (Soto and Berman 2006)	Creatinine weekly
			Hepatotoxicity: The AST was elevated in 8% and the ALT in 10% but always less than 2.5 times the upper limit of normal (Soto and Berman 2006)	Transaminases weekly
			Discoloration of sperm	
Ketoconazole fluconazole	Liver disease	Ketoconazole	Hepatotoxicity reversible, usually mild (Saenz et al. 1990), sometimes severe	Transaminases weekly
				Treatment interruption if transaminases higher than 5 times ULN
	Pregnancy category C: Foetal risk is unknown	Dose adjustment not necessary	Diminution of testosterone values (70%), but without diminution of libido or beard growth: ketoconazole (Saenz et al. 1990)	Reversible, no controls needed
		Fluconazole	Subjective complaints: abdominal pain, headache, nausea, fever and malaise (Saenz et al. 1990)	
		CrCl <50 mL/min: 50% of daily dose	Allergic reactions (skin, angioedema)	
			Haematotoxicity (anaemia, leucopaenia, thrombocytopaenia)	Haemoglobin, leucocytes and platelets weekly
Liposomal amphotericin B	Pregnancy category B: relatively safe to use during pregnancy	Close monitoring of renal function, if progressive: dose reduction (e.g. 50%)	Nephrotoxicity, hypokalaemia	Creatinine and potassium before each infusion Avoid concomitant administration of potentially nephrotoxic drugs
			Infusion related reactions including chest pain, flank pain, dyspnoea, flushing urticaria	May be partially prevented by hydrocortisone
			Nausea, anorexia, vomiting	

(continued)

Table 9.1 (continued)

Drug	Special precautions/ pregnancy categories	Renal function	Adverse effects	Monitoring of drug toxicity during treatment
Paromomycin parenteral (aminoglycoside)	Liver disease		Nephrotoxicity (1%)(Sundar et al. 2007; Jamil et al. 2015)	Creatinine and potassium weekly
	Renal impairment		Ototoxicity (1%) (Sundar et al. 2007; Jamil et al. 2015)	Hearing test weekly
	Pregnancy category C: Foetal risk is unknown		Hepatotoxicity: ASAT >3 ULN (6%), ASAT >5 ULN (2%) (Sundar et al. 2007; Jamil et al. 2015)	Transaminases weekly
				Treatment interruption if transaminases higher than 5 times ULN
			Gastrointestinal problems including pain, diarrhoea, vomiting (5%) (Jamil et al. 2015)	
			Local pain at injection site (16%)	
15% paromomycin plus 12% methylbenzethonium chloride			Local reaction, inflammation, burning sensation, pain	Weekly control
Intralesional antimonials in combination with cryotherapy (Figs 9.1 and 9.2)			Local reaction, inflammation, burning sensation, pain Allergic reactions such as urticaria or even anaphylactic shock are described (Esfandiarpour et al. 2012)	Weekly control

Pregnancy categories: Category A, no foetal risk; Category B, relatively safe to use during pregnancy; Category C, foetal risk is unknown; Category D, some evidence of foetal risk; Category X, causes abnormalities lesions with hot metal objects (spoons, knifes) is very common. The skin lesions are exposed to temperatures of 50–55 °C for 30 s or longer. Thermotherapy has local adverse events such as cellulitis due to bacterial superinfection, erythema and pain but had in contrast to systemic treatment practically no systemic adverse events (Cardona-Arias et al. 2015). In the last years several methods of thermotherapy were described; some of them are still experimental:

- Two sessions with local heat (55 °C during 5 min) provided by an infrared lamp (Junaid 1986).
- A device called the Hand-Held Exothermic Crystallization Therapy for CL (HECT-CL), a customised heating pack that delivers controlled conduction heat starting at 52 °C and decreasing to 48 °C over a 3-min period when applied over a CL lesion. The HECT-CL device is a vinyl pouch containing a super-saturated sodium acetate salt solution and a flexible 2-cm stainless-steel disc at room temperature. Flexing the disc triggers an exothermic liquid-to-solid phase crystallization reaction that in 10 s achieves an initial predictable 52 °C. The device is applied 3 min daily for 7 days (Shah et al. 2014; Valencia et al. 2013). This method is very easy to apply especially in children, does not need any sophisticated technology, but is not yet evaluated in larger populations, and the heating pack is not yet commercially available.
- Photodynamic therapy is widely used in dermatology for the treatment of premalignant and malignant skin lesions. It involves topical administration of a photosensitizer (usually aminolevulinic acid or methyl aminolaevulinate) and subsequent irradiation of the target by light of the appropriate wavelength in the presence of oxygen (Asilian and Davami 2006). The use of daylight as a light source is a novel and evolving modality successfully used for the treatment of actinic keratosis and of actinic cheilitis. It was successfully used in 31 patients with Old World CL (cure rate 89%) and could become a promising treatment option (Enk et al. 2015).
- One radiofrequency-induced heat therapy (RFHT) application during 30–60 s had comparable cure rates (94%) as seven intralesional injections of sodium stibogluconate (Safi et al. 2012; Bumb et al. 2013; Reithinger et al. 2005; Sadeghian et al. 2007; Jebran et al. 2014). Thermomed® is a best studied device for thermotherapy, is approved by FDA and was evaluated in controlled trials including 667 patients (Cardona-Arias et al. 2015). The Thermomed® is an operator, with special devices which achieve and maintain a temperature of 50 °C. The electrodes are placed locally in the lesion for 30 s; the device produces heat waves through radio-frequency technology, which extends them to deeper layers of the skin, causing destruction of the amastigotes. Cure rates with Thermomed® are comparable to the ones of systemic antimonials or miltefosine for Old World and New World CL (Cardona-Arias et al. 2015; Aronson et al. 2010; Lopez et al. 2012; Lopez et al. 2013). To avoid local bacterial infections, an antibiotic ointment (fusidic acid) has to be applied over the lesions over the next 10 days. The high price of the Thermomed® and its reduced availability limit its use.

Cryotherapy Cryotherapy is often used in combination with intralesional injection of antimonials. The procedure is described in Fig. 9.2. Since combining cryotherapy with intralesional injection with antimonials (Figs. 9.1 and 9.2) had better cure rates (89–91%) than either cryotherapy (57–68%) or intralesional antimonials alone (44–75%) (Asilian et al. 2004; Eldarouti and Alrubaie 1990; Salmanpour et al. 2006), we suggest using cryotherapy mainly in combination with intralesional treatment with antimonials.

9.1.4 Intralesional Treatment with Antimonials/Pentamidine

Local infiltration of lesions with pentavalent antimonials produces the maximum concentration in the lesions, but does not reach metastatic leishmaniasis. There are only few systemic side effects, but intralesional infiltrations are painful, may cause local irritation and inflammation and require some experience. However, allergic reactions such as urticaria or even anaphylactic shock are described (Esfandiarpour et al. 2012). Antimony as formulated for parenteral administration by the

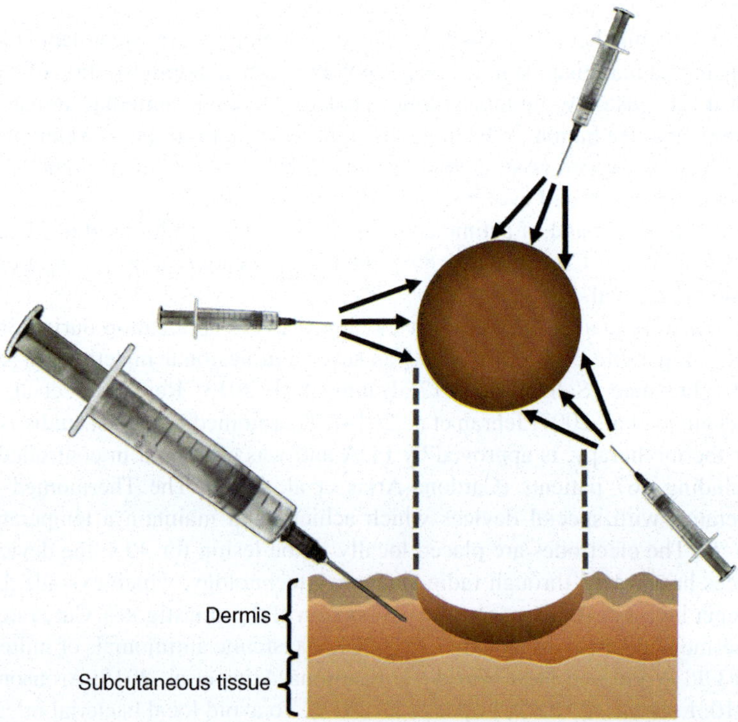

Fig. 9.1 Procedure for intralesional treatment with pentavalent antimony (Blum et al. 2004). Advance the needle whilst injecting under pressure in the dermis, covering the whole lesion including the centre

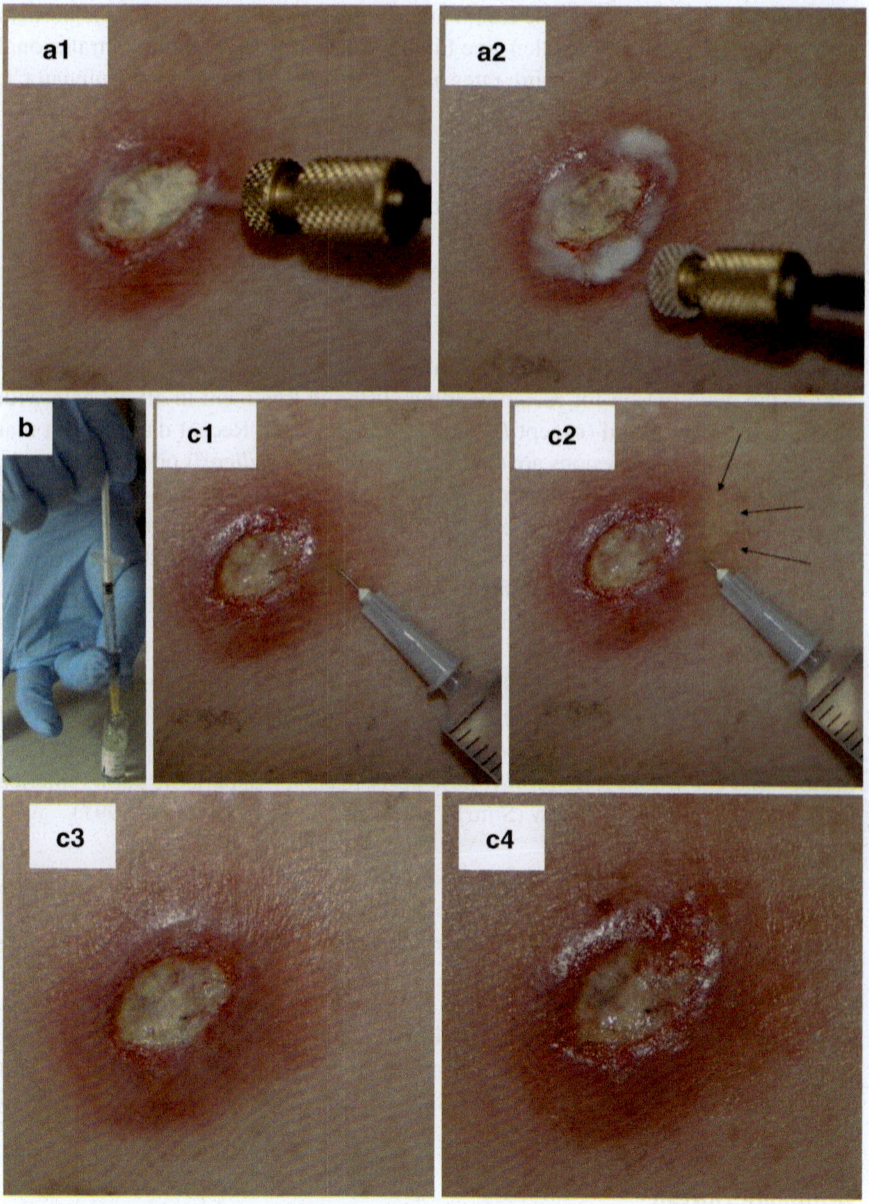

Fig. 9.2 Procedures for superficial cryotherapy and/or intralesional injection of antimony. The lesion is first swabbed with antiseptics several minutes before starting the procedure. (**A**) Cryotherapy: Cryotherapy with liquid nitrogen is then applied on the lesion (**A1**) and immediate borders (**A2**)—ideally with a sprayer—3 to 5 s blanching is obtained. (**B** and **C**) Intralesional injection: Antimony as formulated for parenteral administration by the manufacturer (**B**) is injected into the lesion (**C1**) and should induce blanching of the borders (**C2**, arrows), until the lesion is entirely swollen (before procedure **C3**, end of procedure **C4**). The procedure is usually repeated two to ten times at 2–8 days intervals. (This figure was contributed by Pierre Buffet, France)

manufacturer is injected into the lesion and should induce blanching of the borders until the lesion is entirely swollen (see Figs. 9.1 and 9.2). In one study intralesional pentamidine had comparable cure rates as intralesional antimonials in Bolivian CL (70%) (Soto et al. 2016).

9.1.5 General Considerations: Local Versus Systemic Treatment

The choice for topical or systemic treatment is determined by the following factors:

Risk of Developing Mucosal Leishmaniasis

This is the main reason for recommending systemic treatment in all patients with CL from the New World (except *L. mexicana* infections). Recent data suggest that the risk is higher when lesions are (1) infected with *L. braziliensis* or *L. panamensis*, (2) acquired in Bolivia, (3) multiple (>4), large (>4–6 cm^2), (4) present for >4 months, (5) localized above the belt, (6) associated with acquired or induced immunosuppression and (7) treated inappropriately (Blum et al. 2012). Whether local treatment predisposes patients to ML (compared to systemic treatment) has never systematically been studied, but there are no reports on ML developing in New World CL (NWCL) patients treated with paromomycin/methylbenzethonium chloride ointment or with local infiltration with antimonials (Blum et al. 2012).

If none of the above risk factors are present in patients with NWCL, the risk of developing ML is probably low. Local treatment is thus an option for those who can comply and for whom long-term follow-up is feasible. Experts in Latin America have recently adopted this view (Soto et al. 2013; Oliveira-Neto et al. 1997).

Failure of Prior Local Treatment

Local treatment includes topical treatment with ointment, cryotherapy and intralesional injection with antimonials. Failure to respond may indicate the need for systemic treatment.

Size, Number and Localization of Lesions

Lesions that are multiple and large that affect the nose, lips, eyelids or ears, or that are located close to small joints, are, for practical reasons, less suited for local therapy.

Lymphatic Spread

It is not clear whether local lymphadenopathy or lymphangitis is an absolute indication for systemic treatment. It may indicate extra-dermal parasite spread and thus a risk of subsequent ML. In studies with local treatment, concomitant lymphadenopathy was either an exclusion criterion (Armijos et al. 2004; Soto et al. 2002) or was not reported (Oliveira-Neto et al. 1997; Krause and Kroeger 1994; Arana et al. 2001). Therefore, it is currently unknown whether lymphatic spread of leishmaniasis responds to local treatment.

Toxicity of Local and Systemic Treatment

Local treatment may be painful and cause local inflammation (see above). Table 9.1 summarizes the adverse events associated with current systemic and local treatment options, based on data from CL and ML studies conducted mainly in young and otherwise healthy patients. Adverse events may be more severe and frequent in patients with comorbidities such as cardiac, renal or hepatic disease, diabetes mellitus or immune suppression. Miltefosine has a very long half-life and is still detectable in blood samples 6 months after a standard 28-day treatment (Dorlo et al. 2008). Women of childbearing age must adopt contraceptive measures during treatment and for at least 4 months after treatment completion (Dorlo et al. 2012a).

Availability and Price of Drugs

The reduced availability and the high prices often influence the choice of treatment. While some drugs are not readily available (e.g. paromomycin/methylbenzethonium ointment) or prohibitively expensive (e.g. miltefosine, liposomal amphotericin B), even available drugs are often neither registered nor approved for the treatment of leishmaniasis in many countries. Especially in nonendemic countries, most treatment options for tegumentary leishmaniasis are off label, and this is unlikely to change.

9.1.6 Definition of Healing and Follow-Up

Cutaneous lesions usually heal within 1 month after starting treatment with pentavalent antimonials, irrespective of local or systemic treatment, but large ulcers may take longer. Treatment failure is present when reepithelialization is incomplete 3 months after starting therapy. A relapse is defined as the reappearance of the ulcer after complete healing or a renewed increase in the indurated area of a nodular lesion. Parasitological confirmation is not required, except in clinically complex cases. In such cases, parasite identification (by microscopy and or culture) is preferred, as *Leishmania* DNA can be detected by PCR in lesions several years after successful treatment (Schubach et al. 1998; Mendonca et al. 2004). A follow-up visit at 3 and at 12 months is required to ascertain complete cure.

9.2 Species-Oriented Treatment for CL

The Oxford evidence grading system was applied when reviewing information. The highest ranking **(A)** was assigned to randomized controlled trials in representative patient groups. Randomized controlled trials in less homogenous patient groups (small numbers, different species included) as well as cohort trials and case-control studies in representative patient groups were given the ranking **(B).** Cohort trials or case-control studies in less homogenous patient groups, as well as case series of representative patient groups were given the ranking **(C)**. Case series of less homogenous patient groups and expert opinion were ranked **(D).**

9.2.1 Treatment of *L. major*

a. *Up to three lesions, not cosmetically disfiguring, patients not immunosuppressed, option acceptable to patient:*
   No antileishmanial treatment, simple wound care
b. *Up to three lesions with diameters ≤30 mm—local treatment:*
   1. Cryotherapy plus local infiltration with antimonials: two to ten times at 2–8 day intervals **(A)** (Asilian et al. 2004; Faris et al. 1993; Tallab et al. 1996; Eldarouti and Alrubaie 1990; Salmanpour et al. 2006)
   2. 15% paromomycin/12% methylbenzethonium chloride ointment bid for 10–20 days **(A)** (Ben Salah et al. 2013; el On et al. 1992; Ben Salah et al. 2009; el On et al. 1985; Asilian et al. 2003)
   3. Local heat therapy (50 °C for 30 s) **(A)** (Safi et al. 2012; Bumb et al. 2013; Sadeghian et al. 2007; Aronson et al. 2010)
c. *More than three lesions, diameter >30 mm, delicate location, and/or refractory to local treatment:*
   1. Miltefosine (50 mg tid for 28 days) **(B)** (van Thiel et al. 2010a; Mohebali et al. 2007; Rahman et al. 2007)
   2. Fluconazole (200 mg bid for 6 weeks) **(C)** (Alrajhi et al. 2002; Morizot et al. 2007; Emad et al. 2011)
   3. Liposomal amphotericin B (18 mg/kg total dose): 3 mg/kg/d on days 1–5 and 10) **(D)** (Wortmann et al. 2010)
   4. Systemic pentavalent antimonial (Sb 20 mg/kg/d) and pentoxifylline (3 × 400 mg/d for 20 days) **(A)** or systemic pentavalent antimonial (Sb 20 mg/kg/d for 10–20 days) **(C)** (Aronson et al. 2010; Mohebali et al. 2007; Sadeghian and Nilforoushzadeh 2006; Momeni et al. 2002; Firdous et al. 2009; Mohammadzadeh et al. 2013)

Watchful waiting is a feasible option. Studies reported spontaneous cure rates of 53% at 8 weeks (Nassiri-Kashani et al. 2005), from 40 to 90% at 3 months (Morizot et al. 2013) and close to 100% at 12 months (Bailey and Lockwood 2007). However, CL acquired in Afghanistan often does not heal spontaneously and may require systemic treatment (van Thiel et al. 2010a).

In a large study of 634 patients with CL (*L. major* or *L. tropica*), combining cryotherapy with intralesional injection with antimonials (Figs. 9.1 and 9.2) had better cure rates (89–91%) than either cryotherapy (57–68%) or intralesional antimonials alone (44–75%) (Asilian et al. 2004; Eldarouti and Alrubaie 1990; Salmanpour et al. 2006). Local heat therapy (50 °C for 30 s) had a cure rate comparable to that of systemic pentavalent antimonials (Sb 20 mg/kg for 10 days) (48 vs. 54%, $n = 54$) (Aronson et al. 2010). Compared to intralesional antimonials, the cure rates of local heat were superior (81 vs. 55%, $n = 116$) (Sadeghian et al. 2007) (83 vs. 74%, $n = 382$) (Safi et al. 2012) or similar (98 vs. 94%, $n = 100$) (Bumb et al. 2013). Local heat therapy is a promising method for local treatment and is a valuable option for centres with the necessary equipment (e.g. Thermomed® device). Topical application of an ointment containing 15% paromomycin and 12% methylbenzethonium chloride appears to be more effective than an ointment with 15% paromomycin plus 10% urea but also causes more local inflammation (Bryceson et al. 1994). A newly developed topical aminoglycoside formulation is more effective than placebo among Tunisian patients and French travellers (*L. major*), with cure rates consistently above 80% at 3 months (Ben Salah et al. 2013; Ben Salah et al. 2009).

Miltefosine at 150 mg daily for 28 days is a treatment option for patients with contraindications for local treatment (Mosimann et al. 2016). In treatment studies of *L. major* CL (three studies, $n = 81$), cure rates of miltefosine had a mean of 93% (range 87–100%) (van Thiel et al. 2010a; Mohebali et al. 2007; Rahman et al. 2007). This was somewhat superior to the 85% cure rates of systemic meglumine antimoniate (20 mg/kg/d for 14 days) (Mohebali et al. 2007). In Old World CL (OWCL) the efficacy of systemic pentavalent antimony is poorly documented (Khatami et al. 2007). In an open, uncontrolled study (pentavalent Sb 20 mg/kg/d for 10 days), the cure rate ranged from 52 to 87% at 3 weeks (Aronson et al. 2010; Mohebali et al. 2007; Sadeghian and Nilforoushzadeh 2006; Momeni et al. 2002; Firdous et al. 2009; Mohammadzadeh et al. 2013) and 90% at 12 months (Aronson et al. 2010). Systemic pentavalent antimonial (Sb) treatments had the same cure rate as placebo (Belazzoug and Neal 1986). Adding allopurinol (15–20 mg/kg/d for 20 days) produced only marginally better cure rates than Sb alone (80 vs. 74%) (Momeni et al. 2002), but when used in combination with pentoxifylline 3 × 400 mg daily for 20 days, the cure rate improved significantly (26/32 = 81 vs. 16/31 = 52%) (Sadeghian and Nilforoushzadeh 2006). Experts are often reluctant to use antimonials for *L. major* CL considering its mostly benign character and the toxicity of antimonials.

Fluconazole (200 mg/d for 6 weeks) was a well-tolerated treatment for *L. major* leishmaniasis in Saudi Arabia, with a cure rate of 79% (63/80) vs. 34% (22/65) for the placebo group at 3 months (Alrajhi et al. 2002). Unfortunately, this favourable result could not be reproduced elsewhere (Morizot et al. 2007). Increasing the dosage of fluconazole to 400 mg daily produced a higher cure rate (81%) than fluconazole 200 mg/d (48%) at 2 months but with increased adverse events rates. Adverse events leading to treatment interruption included raised serum creatinine or liver enzymes (4%), cheilitis (45%) and nausea (10%) (Emad et al. 2011). Ketoconazole, another imidazole compound, showed an acceptable cure rate of 70% (5/8) in a small case series (Weinrauch et al. 1983). It was superior to intralesional antimonials in a study with *L. major* and *L. tropica* CL, with cure rates of 89% (57/64) and 72% (23/32), respectively (Salmanpour et al. 2001). So far no placebo-controlled studies with ketoconazole have been conducted.

9.2.2 Treatment of *L. tropica*

a. *Up to three lesions, not cosmetically disfiguring and patients not immunosuppressed, option acceptable to patient:*
   Simple wound care
b. *Up to three lesions with diameter <30 mm—local treatment:*
   1. Local infiltration with antimonials with or without cryotherapy two to ten times at 2–8 day intervals **(A)** (Asilian et al. 2004; Eldarouti and Alrubaie 1990; Salmanpour et al. 2006)
   2. 15% paromomycin/12% methylbenzethonium chloride ointment bid for 10–20 days **(D)**
   3. Local heat therapy (50 °C for 30 s) **(A)** (Reithinger et al. 2005; Sadeghian et al. 2007)
c. *More than three lesions, diameter >30 mm, delicate location, and/or refractory to local treatment:*
   1. Liposomal amphotericin B (18 mg/kg total dose: 3 mg/kg/d on days 1–5 and 10) **(C)** (Wortmann et al. 2010; Solomon et al. 2011)
   2. Miltefosine (50 mg tid for 28 days) **(D)** (Mosimann et al. 2016; Neub et al. 2008; Tappe et al. 2010; Killingley et al. 2009)
   3. Pentavalent antimonials (Sb 20 mg/kg/d for 10–20 days) (+/− allopurinol 15–20 mg/kg/day for 20 days) **(C)** (Reithinger et al. 2005; Firooz et al. 2006; Munir et al. 2008; Zerehsaz et al. 1999)

Spontaneous cure for *L. tropica* CL is estimated at 1–10% at 3 months, 68% at 12 months and close to 100% in 6 months to 3 years (Bailey and Lockwood 2007; Munir et al. 2008). As mentioned above, cryotherapy combined with intralesional antimonials (see Figs. 9.1 and 9.2) produced excellent cure rates in *L. major* or *L. tropica* CL (Asilian et al. 2004; Eldarouti and Alrubaie 1990; Salmanpour et al. 2006; Negera et al. 2012). Thermotherapy and photodynamic therapy was effective in *L. tropica* and *major* CL and is a valuable option for centres with the necessary equipment (e.g. Thermomed® device) (Junaid 1986; Sadeghian et al. 2007; Aronson et al. 2010).

Liposomal amphotericin (AmBisome®, 3 mg/kg/d for 5 consecutive days and at day 10, with a total dose of 18 mg/kg) had a cure rate of 84% in 13 travellers and immigrants with *L. tropica* CL (Solomon et al. 2011).

For *L. tropica, L. major* and *L. infantum/donovani* CL, experience with miltefosine is limited to case reports (Neub et al. 2008; Tappe et al. 2010; Killingley et al. 2009; Faber et al. 2009; Poeppl et al. 2011) and small case series, with all reported patients cured (Rahman et al. 2007). In a case series of 24 patients with complicated Old World CL or ML treated with miltefosine, all patients were cured (Mosimann et al. 2016; Neumayr et al. 2012; Stoeckle et al. 2013). Cure rates of systemic antimonials in *L. tropica* CL ranged from 24 to 55% (Reithinger et al. 2005; Firooz et al. 2006; Munir et al. 2008; Zerehsaz et al. 1999). For *L. tropica* CL, adding allopurinol (15–20 mg/kg/day for 20 days) increased cure rates to 46%, compared with 24% in the antimony-only group (Esfandiarpour and Alavi 2002). According to

European experts, systemic pentavalent antimonials are effective for treating complex CL lesions and are recommended in some national guidelines (Bailey and Lockwood 2007; Boecken et al. 2009; Blum and Hatz 2009).

9.2.3 Treatment of *L. infantum/donovani*

a. *Up to three lesions, not cosmetically disfiguring and patients not immunosuppressed, option acceptable to patient*:
   Simple wound care
b. *Up to three lesions with diameter ≤30 mm—local treatment*:
   1. Local infiltration with antimonials with or without cryotherapy two to ten times at 2–8 day intervals **(A)** (Asilian et al. 2004; Eldarouti and Alrubaie 1990; Salmanpour et al. 2006)
   2. 15% paromomycin/12% methylbenzethonium chloride ointment bid for 10–20 days **(D)**
   3. Local heat therapy (50 °C for 30 s) **(A)** (Reithinger et al. 2005; Sadeghian et al. 2007)
c. *More than three lesions, diameter >30 mm, delicate location, and/or refractory to local treatment*:
   1. Liposomal amphotericin B (18 mg/kg total dose: 3 mg/kg/d on days 1–5 and10) **(C)** (Wortmann et al. 2010; Solomon et al. 2011)
   2. Miltefosine (50 mg tid for 28 days) **(D)** (Mosimann et al. 2016; Neub et al. 2008; Tappe et al. 2010; Killingley et al. 2009)
   3. Pentavalent antimonials (Sb 20 mg/kg for 10–20 days) (+/− allopurinol 15–20 mg/kg/day for 20 days) **(C)** (Reithinger et al. 2005; Firooz et al. 2006; Munir et al. 2008; Zerehsaz et al. 1999)

There are no studies on treatment of *L. infantum* and no studies on the spontaneous cure rate. However, the experience in many centres shows that the treatment recommendation for *L. tropica* can be successfully applied (see above) (Mosimann et al. 2016; Harms et al. 2003).

9.2.4 Treatment of *L. aethiopica*

a. *Up to three lesions with diameter ≤30 mm—local treatment*:
   1. Local infiltration with antimonials with cryotherapy two–ten times at 2–8 day intervals **(C)** (Negera et al. 2012; van Griensven et al. 2016)
   2. 15% paromomycin/12% methylbenzethonium chloride ointment bid for 10–20 days **(D)**
   3. Local heat therapy (50 °C for 30 s) **(D)**

b. *More than three lesions, diameter >30 mm, delicate location, and/or refractory to local treatment*:
   1. Miltefosine (50 mg tid for 28 days) **(D)** (Mosimann et al. 2016; van Griensven et al. 2016)
   2. Liposomal amphotericin B (18 mg/kg total dose: 3 mg/kg/d on days 1–5 and10) **(D)** (van Griensven et al. 2016; Zanger et al. 2011a)
   3. Pentavalent antimonials (Sb 20 mg/kg/d for 10–20 days) **(C)** (Negera et al. 2012; van Griensven et al. 2016)
   4. Pentamidine isethionate (4 mg/kg base/d: three to four times within 7 days) **(D)** (van Griensven et al. 2016)

A systemic review of the treatment of CL caused by *L. aethiopica* showed that most published studies are case reports or case series and that the only two small randomized trials have an overall poor study quality (van Griensven et al. 2016). With cryotherapy, cure rates were 60–93% and with antimonials 69–90% (Negera et al. 2012; van Griensven et al. 2016). Pentamidine appeared effective against complicated CL, also in cases nonresponsive to antimonials (van Griensven et al. 2016). The available in vitro data suggest a good susceptibility of *L. aethiopica* to miltefosine, paromomycin, pentamidine and amphotericin B. Of interest, miltefosine has been successfully used in more than 50 CL cases due to *L. aethiopica* in Addis Ababa (unpublished data, personal communication with Dr. Asrat Hailu (van Griensven et al. 2016), and one case of successful miltefosine treatment was reported from Switzerland (Mosimann et al. 2016). Liposomal amphotericin B was found effective in one immunosuppressed Eritrean patient treated in Germany (Zanger et al. 2011a).

9.2.5 Treatment of *L. panamensis*

Although many experts consider *L. guyanensis* and *L. panamensis* as a single species complex, we analysed them separately, since trials were performed on each species. As knowledge of the taxonomy of *Leishmania* increases, there may be justification for merging recommendations in the future.

a. *Single or few lesion(s), not cosmetically disfiguring, lesions with diameter <30 mm, no lymphatic spread, option acceptable to patient—local treatment considered*:
   1. 15% paromomycin/12% methylbenzethonium chloride ointment bid for 30 days **(B)** (Armijos et al. 2004; Krause and Kroeger 1994)
   2. Local heat therapy **(A)** (Lopez et al. 2012)

b. *Multiple lesions or large single lesion—systemic treatment*:
   1. Miltefosine (50 mg tid for 28 days) **(A)** (Soto et al. 2001; Rubiano et al. 2012; Soto et al. 2004; Velez et al. 2010)
   2. Pentamidine isethionate (4 mg/kg base/d: three to four times within 7 days) **(A)** (Soto-Mancipe et al. 1993; Soto et al. 1994)
   3. Ketoconazole (600 mg/d for 28 days) **(B)** (Saenz et al. 1990)
   4. Pentavalent antimonials (Sb 20 mg/kg/d for 20 days) **(A)** (Lopez et al. 2012; Soto-Mancipe et al. 1993; Ballou et al. 1987; Wortmann et al. 2002)

Data on local treatment are scarce. In a group of 52 patients infected mainly with *L. panamensis*, topical treatment with 15% paromomycin/12% methylbenzethonium chloride ointment once or twice daily for 10–20 days produced cure rates of 90% at 3 months and of 85% after 1 year (Krause and Kroeger 1994), and the combination of paromomycin 15% and gentamycin 0.5% achieved comparable cure rates (87%) (Sosa et al. 2013). This ointment (once daily for 30 days; $n = 29$; cure rate 79%) was inferior to systemic treatment with systemic pentavalent antimonials (20 mg/kg/d for 10 days; $n = 36$; cure rate 92%) (Armijos et al. 2004). However, because of toxicity and the lack of superiority to other drug regimens, systemic pentavalent antimonials are no longer the treatment of choice. Cure rates with miltefosine were variable (60–94%) (Soto et al. 2001; Rubiano et al. 2012; Soto et al. 2004; Velez et al. 2010), superior to placebo (91 vs. 38%) (Soto et al. 2004) and comparable to antimonials (63–72%) (Rubiano et al. 2012; Velez et al. 2010) and thermotherapy (58%) (Lopez et al. 2013). Pentamidine was tested in different dosages (two to six injections of 2–4 mg/kg) in patients with predominantly *L. panamensis* CL. Cure rates were comparable with pentavalent antimonials (96 vs. 91%) and were highest (96%) with dosages of 4 mg/kg/d given three (to five) times within 7 days (Soto-Mancipe et al. 1993; Soto et al. 1994).

9.2.6 Treatment of *L. guyanensis*

a. *Single lesion, not cosmetically disfiguring, no lymphatic spread and infection not acquired in Bolivia*:
   No data on local treatment: no recommendation possible
b. *All other cases—systemic treatment*:
   1. Pentamidine isethionate (4 mg/kg base/d: three to four times within 7 days) **(A)** (Soto-Mancipe et al. 1993; Soto et al. 1994; Lai et al. 2002; Neves et al. 2011; Roussel et al. 2006; Nacher et al. 2001; Lightburn et al. 2003)
   2. Miltefosine (50 mg tid for 28 days) **(B)** (Chrusciak-Talhari et al. 2011)

Pentamidine is the first-line treatment for *L. guyanensis* CL in French Guyana, Surinam and Brazil, with cure rates of around 90% (Soto-Mancipe et al. 1993; Soto et al. 1994; Lai et al. 2002; Roussel et al. 2006; Nacher et al. 2001; Lightburn et al. 2003). Although these studies included many patients (>2000 patients), most are retrospective observations, and different dosages were used. The cure rate was lower (77%) in a study from Surinam, possibly due to a very low follow-up rate (van der Meide et al. 2009). For *L. guyanensis* acquired in Northeast Brazil, the cure rate was higher with miltefosine (40/56; 71%) than with meglumine antimoniate (16/28; 57%) (Chrusciak-Talhari et al. 2011), and in another study meglumine antimoniate (41/66 = 60%) and pentamidine 44/69 = 62%) had comparable cure rates (Neves et al. 2011).

9.2.7 Treatment of *L. braziliensis* and *L. peruviana*

L. braziliensis and *L. peruviana* species are genetically very similar. Data on treatment come from studies of *L. braziliensis CL.*

a. *Single or few lesion(s), not cosmetically disfiguring, lesion with diameter <30 mm, no lymphatic spread, not from Bolivia—local treatment possible*:
   1. Local infiltration with antimonials + cryotherapy two to ten times at 2–8 day intervals **(C)** (Soto et al. 2016; Oliveira-Neto et al. 1997)
   2. 15% paromomycin/12% methylbenzethonium chloride ointment bid for 30 days **(B)** (Arana et al. 2001; Weinrauch et al. 1993)
   3. Thermotherapy **(A)** (Lopez et al. 2012)
b. *All other cases—systemic treatment*:
   1. Pentavalent antimonials (Sb 20 mg/kg/d for 20 days) **(A)** (Lopez et al. 2012; Soto et al. 2004; Velez et al. 2010; Andersen et al. 2005; Navin et al. 1992; Soto et al. 2008; Soto and Berman 2006)
   2. Liposomal amphotericin B (18 mg/kg total dose: 3 mg/kg/d on days 1–5 and 10) **(B)** (Wortmann et al. 2010; Solomon et al. 2013)
   3. Miltefosine (only Bolivia, Brazil) (50 mg tid for 28 days) **(C)** (Soto et al. 2008; Machado et al. 2010)

The spontaneous cure rate of *L. braziliensis* is low (18%) (Soto et al. 2013). Local treatment with intralesional antimonials was only reported in three studies involving a total of 164 patients with *L. braziliensis* CL. Cure rate without relapse or development of ML was 70–80% (Soto et al. 2013, 2016; Oliveira-Neto et al. 1997). For CL due to *L. braziliensis* (75%) and *L. mexicana* (25%), topical treatment with 15% paromomycin/12% methylbenzethonium chloride ointment ($n = 35$)

was more effective than placebo (*n* = 33; response rate at 12 weeks was 91 vs. 39%) (Arana et al. 2001) and had a cure rate of 76% after 8 weeks (*n* = 53) (Weinrauch et al. 1993) but was not compared to systemic treatment with pentavalent antimonials. In studies of topical treatment of NWCL, patients were followed either until healed (Weinrauch et al. 1993; Soto et al. 1998) or until 1 year after treatment (Armijos et al. 2004; Krause and Kroeger 1994; Arana et al. 2001). Although none developed ML, the observation periods were too short, and the sample size is too small to assess that risk accurately. Since cure rates of thermotherapy were mediocre (31/95 = 53%) and inferior to those of systemic meglumine antimoniate (34/52 = 65%), thermotherapy cannot yet be proposed as a first-line treatment (Lopez et al. 2012).

Pentavalent antimonials are the principal treatment for *L. braziliensis* CL (Marsden 1986; Jones et al. 1987; Aronson et al. 1998; Herwaldt and Berman 1992; Herwaldt 1999). Cure rates range from low (50%) (Romero et al. 2001; Almeida et al. 1999) to excellent (96–100%) (Wortmann et al. 2002; Navin et al. 1992; Seaton et al. 1999). The variation may be attributed to strain and infection site differences (Romero et al. 2001). In CL patients infected with *L. peruviana* (*n* = 46), 76% were cured with systemic antimonials (Arevalo et al. 2007; Llanos-Cuentas et al. 2008).

Liposomal amphotericin B has been used after treatment failure in immunocompromised patients and when pentavalent antimonials are contraindicated. Case series of travellers and immigrants showed that AmBisome® (3–5 mg/kg/d for 5 consecutive days and a sixth dose on day 10 = cumulative doses 18–30 mg/kg) cured 29/34 (85%) of patients with *L. braziliensis* CL in Israel (Solomon et al. 2007, 2013) and 12/14 (86%) in Germany (Harms et al. 2011). Using similar cumulative doses, AmBisome® had a cure rate of about 84% in patients with OWCL (*n* = 10) and NWCL lesions (*n* = 10) alike (Wortmann et al. 2010).

Treatment with miltefosine has been disappointing, with cure rates varying with the geographical origin of the infection. A small series from Guatemala had unacceptably low cure rates (33%) compared to placebo (8%) (Soto et al. 2004). However cure rates were comparable to that of pentavalent antimonial treatment in Colombia (60 vs. 65%, *n* = 93) (Velez et al. 2010) and Bolivia (88 vs. 94%, *n* = 57) (Soto et al. 2008) and slightly better in Brazil (75 vs. 53%, *n* = 90) (Machado et al. 2010). Differences in drug susceptibility in some subspecies of *L. braziliensis* may account for the wide cure rate variation observed.

Fluconazole has only been evaluated in a small series of *L. braziliensis* CL patients, and different dosage schemes have been used. Cure rates increased with dosage, from 75% at 5 mg/kg (*n* = 8) to 93% at 6.5 mg/kg (*n* = 14) and to 100% at 8 mg/kg (*n* = 8), respectively (Sousa et al. 2011). Surprisingly, no significant adverse events were reported (Sousa et al. 2011). Because of a lack of solid evidence and intolerance at higher doses (400 mg/day) reported in patients with *L. major*, experts are currently reluctant to recommend fluconazole for *L. braziliensis* CL.

9.2.8 Treatment of *L. mexicana*

a. *Up to three lesions not requiring immediate therapy, not cosmetically disfiguring and option acceptable to patient:*
   No antileishmanial medication, simple wound care, mostly self-limiting
b. *More than three lesions with diameter ≤30 mm—local treatment:*
   1. Cryotherapy/local infiltration with antimonials two to ten times at 2–8 day intervals (**D**)
   2. 15% paromomycin/12% methylbenzethonium chloride ointment bid for 20 days (**C**) (Arana et al. 2001; Weinrauch et al. 1993)
c. *More than three lesions with diameter >30 mm, delicate location and/or refractory to topical treatment—systemic treatment:*
   1. Ketoconazole (600 mg/day for 28 days) (**B**) (Navin et al. 1992)
   2. Miltefosine (50 mg tid for 28 days) (**B**) (Soto et al. 2004)
   3. Pentavalent antimonials (Sb 20 mg/kg/d for 20 days) (**D**)

Published data on systemic treatment of *L. mexicana* CL are scarce and have involved small patient groups only. Ketoconazole produced superior cure rates at 13 weeks compared to placebo (8/9; 89% vs. 9/16; 56%) and to pentavalent antimonials (8/9; 89% vs. 5/7; 71%) (Navin et al. 1992). Miltefosine had only limited efficacy (9/14; 64%) (Soto et al. 2004), and fluconazole was not tested.

9.2.9 Treatment of Other NWCL Species: *L. naiffi*, *L. lainsoni*, *L. amazonensis*, and *L. venezuelensis*

L. naiffi: It was first described 1989. The few cases described in the literature usually associate *L. naiffi* with low rates of virulence, and the disease evolves with a benign clinical course and a good response to treatment. No association with mucosal leishmaniasis has been observed (Fagundes-Silva et al. 2015). In Surinam, patients with five small lesions in total were successfully treated with pentamidine (van Thiel et al. 2010b), and three small lesions in two patients disappeared without treatment (van der Snoek et al. 2009). However, a case series of *L. naiffi* CL patients in the Brazilian Amazon reported treatment failures in 25% (1/4) of cases treated with antimonials and in 25% (1/4) of cases treated with pentamidine (Fagundes-Silva et al. 2015). Even if there are no data on local treatment, considering the usual benign evolution of the disease and the toxicity of systemic treatment, local treatment with cryotherapy and intralesional antimonials or thermotherapy may be considered the first treatment option.

L. amazonensis: Genetically speaking, this subspecies is closely related to *L. mexicana*, which suggests that a similar treatment approach may be used. However, there are no data to support this.

L. venezuelensis and *L. lainsoni*: Treatment data is not available. Cases were mostly treated as cases of *L. braziliensis*.

9.3 Mucocutaneous or Mucosal Leishmaniasis

Systemic treatment is mandatory in ML cases; the spread and localization makes local treatment impractical or ineffective.

9.3.1 Old World Mucosal Leishmaniasis

- Miltefosine (50 mg tid for 28 days) **(D)** (Mosimann et al. 2016; Neumayr et al. 2012; Stoeckle et al. 2013; Richter et al. 2011)
- Pentavalent antimonials (Sb 20 mg/kg/d for (20–28 days) **(D)** (Faucher et al. 2011; Kharfi et al. 2003)
- Liposomal amphotericin B (3 mg/kg/d on days 1–5 and 10, 17, 24, eventually 31, 38 total dose: 21–40 mg/kg) **(D)** (Richter et al. 2011; Faucher et al. 2011)

There have not been any controlled studies on treating Old World ML and the treatment options listed above were only successfully used and reported in case reports (Neumayr et al. 2012; Richter et al. 2011; Faucher et al. 2011). There are no comparative studies between these treatment options, and preference is guided by practical considerations, such as drug availability and costs.

9.3.2 New World Mucosal Leishmaniasis

1. Pentavalent antimonials (Sb 20 mg/kg/d for 28–30 days) **(A)** (Amato et al. 2008, 2009) + addition of pentoxifylline (400 mg tid for 30 days) **(A)** (Machado et al. 2007; Bafica et al. 2003; Lessa et al. 2001)
2. Liposomal amphotericin B (3 mg/kg/d on days 1–5 and 10, 17, 24, eventually 31, 38: total dose 21–40 mg/kg) (Amato et al. 2008, 2009; Cunha et al. 2015; Rocio et al. 2014) **(C)**
3. Miltefosine (50 mg tid for 28 days) **(B)** (Soto et al. 2007, 2009)

Pentavalent antimonials are still the gold standard of treatment (Amato et al. 2008; Franke et al. 1994), with an overall cure rate of 88% (Amato et al. 2008). Increasing the dosage beyond 20 mg Sb/kg/d for 30 days did not improve the already high cure rate of 91%. However, recurrence rates were high for all dosages used (22 to 25%) (Amato et al. 2009).

Destructive mucosal lesions contain few parasites, while TNF-α levels are high. This suggests that an unmodulated immune response with increased production of pro-inflammatory cytokines (IL 10) is responsible for the tissue damage. Pentoxifylline downregulates TNF-α and inhibits leukocyte migration and adhesion. Combining antimonials (20 mg/kg Sb/d for 30 days) with pentoxifylline (400 mg tid for 30 days) cured 9/10 (Lessa et al. 2001) and 2/2 (Bafica et al. 2003) patients with refractory mucosal leishmaniasis. In a small randomized controlled study, 11/11 (100%) ML patients treated with the above combination were cured, whereas 5/12 (42%) patients treated with antimonials only required a second course of antimonials. Time lapse to cure was 83 days in the antimonials/pentoxifylline treatment group and 145 days in the "antimonials only" group. No relapses were seen in either group at the follow-up visit 2 years later. Pentoxifylline is well tolerated, with only mild adverse effects (gastrointestinal symptoms and arthralgia) (Machado et al. 2007).

Amphotericin B deoxycholate (2–3 mg/kg/d for 20 days) is effective in NWML (Amato et al. 2008). Treatment of ML with liposomal amphotericin B (mean total dose 33–35 mg/kg) cured 49/53 (92%) patients in three studies in Brazil (Cunha et al. 2015; Rocio et al. 2014; Amato et al. 2011). The newer formulations of amphotericin B (colloid dispersion, liposomal) had better cure rates (12/12; 100%) than amphotericin B desoxycholate (5/8; 63%) and higher rates of treatment completion (12/13: 92% vs. 8/17, 53%) (Amato et al. 2009).

In NWML (mainly caused by *L. braziliensis*), miltefosine cured 83% of patients with mild disease (i.e. nasal mucosa) and 58% of patients with more extensive disease (involving the pharynx, larynx and palate) (Soto et al. 2007). Prolonging treatment from 4 to 6 weeks did not substantially increase cure rates (from 71 to 75%) (Soto et al. 2009). A systemic review comparing meglumine antimoniate with miltefosine without respecting the species could not indicate a difference between these treatments, but did not include the combination meglumine antimoniate and pentoxyfilline (Reveiz et al. 2013).

9.3.3 Reported Differences Between ML Due to New World and Old World Species

Five reviews with 43 patients (Neumayr et al. 2012; Richter et al. 2011; Kharfi et al. 2003; Aliaga et al. 2003; Garcia de Marcos et al. 2007) with Mediterranean mucosal leishmaniasis (*L. infantum/donovani*), mostly reported as case reports, indicate some differences between NWML and OWML:

1. The nasal cavity was affected in over 90% of NWML cases but only in 15% of Mediterranean ML cases.
2. Patients with ML acquired in the Mediterranean region had a better prognosis than those who acquired ML in Latin America. 17/17 (100%) patients with OWML treated with meglumine antimoniate (Sb 20 mg/kg/d for 20–28 days) were healed, but one of them had a relapse 1 year later.

3. About half of the patients with ML due to Old World species had some kind of immune suppression.
4. In NWML, destructive lesions with only few parasites and high levels of TNF have been reported. In Mediterranean ML, a high parasite burden was found in the lesions (Aliaga et al. 2003).
5. Host factors might also play a role: more destructive NWML lesions were observed in African descendants than in Latinos. However, a paucity of parasites and a pronounced inflammatory response were observed in lesions from both racial groups (Walton and Valverde 1979).

9.4 Disseminated CL, Diffuse CL and Post-Kala-Azar Dermal Leishmaniasis (PKDL)

Diffuse CL is characterized by a high number of mostly non-ulcerated skin lesions (nodules, patches), a high parasite burden within the lesions and a chronic clinical evolution. Treatment usually results in only transitory improvement, even if repeated and prolonged treatment cycles are applied (Zerpa et al. 2007).

Disseminated CL is characterized by a high number of polymorphic lesions (including ulcerations) with few parasites and frequent involvement of the mucosa. Classically, pentavalent antimonials were used, but in >80% of cases, two to three treatment cycles of 30 days antimonials are required for cure (Vernal et al. 2016). Liposomal amphotericin B has a better cure rate (70%), but cumulative dosages of 30 mg/kg are frequently needed (Machado et al. 2015).

PKDL is a cutaneous sequela with maculopapular or nodular dermatitis developing in some patients suffering from VL. PKDL occasionally presents concomitantly with VL but more commonly occurs after cure of VL. A link between treatment with pentavalent antimonials and the occurrence of PKDL is suspected. In recent years the efficacy of miltefosine treatment in VL significantly declined in India and prolonged treatment (up to 16 weeks) is nowadays necessary to achieve cure. However, prolonging treatment is associated with an increasing rate of gastrointestinal side effects leading to cessation of treatment in a substantial number of patients (Ramesh et al. 2015; Ghosh et al. 2015). Interestingly, only 0.3% of 8311 VL patients treated with liposomal amphotericin B complained of PKDL (Burza et al. 2014), and liposomal amphotericin B is therefore suggested as first-line treatment of PKDL.

9.5 Treatment of Special Groups

9.5.1 Children

In general, the guidelines above also apply to children (Dorlo et al. 2012b). A common problem in children is the presentation of *L. infantum* CL with facial lesions. One is reluctant to do infiltrations on the faces of children under the age of 7. Small nodular lesions may be left alone or treated with cryotherapy only, and multiple or large lesions can be treated with oral fluconazole or miltefosine (2.5–3 mg/kg/d for 28 days).

9.5.2 Pregnancy

Cutaneous leishmaniasis is not known to affect the foetus. Since none of the systemic treatments are known to be safe during pregnancy, systemic treatment should be withheld until after delivery; topical treatment may be applied before (Morgan et al. 2007). However, whether intralesional injections of antimony or topical paromomycin are completely safe during pregnancy is not known. Simple wound care or physical methods like cryotherapy or thermotherapy are preferred, despite the low level of evidence for efficacy. The lesions of pregnant women with *L. braziliensis* CL are larger than in nonpregnant women and have a cauliflower appearance rather than the typical well-demarcated ulcer with raised border (Morgan et al. 2007). In rare situations when lesion localization, size, impact and persistence, despite local therapy, require systemic therapy, liposomal amphotericin B probably has the best benefit-risk ratio, and neither adverse effects on the foetus nor stillbirths were observed when it was used for VL during the first and second trimester (Fontenele e Silva et al. 2013).

9.5.3 Patients Receiving Immunosuppressive Treatment/Coinfection with HIV

Immunosuppressed patients may have a similar presentation than immunocompetent patients, but parasite dissemination, clinical polymorphism and atypical or more severe forms were observed (van Griensven et al. 2014). In most patients treated with a TNF-α antagonist, methotrexate or corticosteroids, the clinical presentation is apparently similar to that of healthy persons. However, long incubation periods of several years, multiple lesions, ML, disseminated CL or the combination with VL have been reported in these groups of patients. The lesions usually respond well to antileishmanial treatment. If possible, immunosuppressive treatment should be discontinued until after the skin lesion has healed and then restarted under close observation (Zanger et al. 2011b; Neumayr et al. 2013).

HIV-positive patients with CL should be carefully assessed for coexisting VL. Localized CL in HIV-infected individuals tends to be associated with minimal immunosuppression and is clinically identical to CL in HIV-negative CL patients but has a higher rate of recurrence after treatment. However, relevant immunosuppression due to HIV facilitates dissemination and may lead to disseminated CL and to VL (Alvar et al. 2008).

9.6 Outlook

Since these treatment recommendations are based on data from patients in endemic regions, they may not apply to travellers (Mosimann et al. 2013) who have different exposure rates and immunity towards *Leishmania* parasites. An international multicentre study, addressing the issue of treating tegumentary leishmaniasis in travellers, is currently ongoing (Leishman working group). This study aims on collecting and linking genomic data of the causative *Leishmania* species to clinical presentation and treatment outcome in travellers with CL and ML. This data is expected to support the currently limited evidence of species-specific treatment in travellers and the compilation of treatment recommendations for this group of patients.

Appendix: Properties of drugs used for both visceral leishmaniasis and tegumentary forms of leishmaniasis

Drug	Administration	Dosage and duration	Half-life	Mode of action	Metabolic and elimination pathways
Systemic pentavalent antimonials	IM/IV	CL old world: 20 mg/kg of Sb^5 base equivalent: 10–20 days	~2 h and 33–76 h[a]	Not fully understood	Renal excretion
		CL new world: 20 mg/kg of Sb5 base equivalent: 20 days		Interference with energy production of *Leishmania* amastigotes and with parasitic DNA topoisomerase I	
		ML: 20 mg/kg of Sb^5 base equivalent 28 days			
		VL immunocompetent: 20 mg Sb^v/kg/day for 28 days			
		VL immunosuppressed: same			
		VL prophylaxis in HIV+: 850 mg Sb^{va} for adults every 3–4 weeks			
Pentamidine	IV (IM)	CL: 4 mg/kg base: 3–4× within 7 days	~ 9–13 h and ~28 d[a]	Interference with the synthesis of *Leishmania* DNA acting on the kinetoblast and on the mitochondrial membrane	Small extent renal excretion (4–17% in 24 h)
		ML: not indicated			
		VL immunocompetent 4 mg/kg every other day or three times per week for ~15–30 doses			
		VL immunosuppressed: same			
		VL prophylaxis in HIV+: 4 mg/kg/day every 2–4 weeks			

(continued)

(continued)

Drug	Administration	Dosage and duration	Half-life	Mode of action	Metabolic and elimination pathways
Miltefosine	Oral	CL: 3 × 50 mg (children:2–2.5 mg/day) for 28 days (Blum et al. 2014)	7 d and 31 d[a]	Not fully understood	Metabolic pathway mediated by phospholipases
		ML3 × 50 mg (children:2–2.5 mg/day) for 28 days (Blum et al. 2014)		Apoptosis and disturbance of lipid dependant cell signalling pathways	Excretion with urine <0.2%
		VL immunocompetent 30–44 kg, 50 mg bid for 28 days; if ≥45 kg, 50 mg tid for 28 days		Antitumor effects	
		VL immunosuppressed: same			
		VL prophylaxis: no data			
Ketoconazole	Oral	CL: 600 mg/day for 28 days	2 h and 8 h[a]	Inhibition of the cytochrome P 450 mediated 1 4α-demethylation of lanosterol, blocking ergosterol synthesis, and causing accumulation of 14α—methyl sterols	Hepatic; mainly biliary excretion
		Not indicated for ML and VL			
Fluconazole	Oral	CL: 400 mg/day for 6 weeks	30 h		Hepatic; mainly renal excretion
		Not indicated for ML and VL			
Liposomal Amphotericin B	IV	CL: 3 mg/kg/day: day 1–5,10; 18 mg/kg total dose equivalent	Terminal: 152 h	Polyenic antibiotic, targets ergosterol in the surface membrane of promastigotes and amastigotes, leading to increased permeability and the influx of ions	No extensive metabolism; very little renal and biliary excretion
		ML: 3–4 mg/kg/day: day 1–5,10; 17, 24, 31, 38: 18 30 (max 40 mg) mg/kg total dose equivalent			
		VL immunocompetent 3 mg/kg/day on days 1–5, 14, and 21 (total dose 21 mg/kg)			
		VL immunosuppressed 4 mg/kg/day on days 1–5, 10, 17, 24, 31, and 38 (total dose 40 mg/kg)			
		VL prophylaxis: amphotericin B lipid complex (iv) 3–5 mg/kg/day every 3 weeks			

(continued)

Drug	Administration	Dosage and duration	Half-life	Mode of action	Metabolic and elimination pathways
15% paromomycin/ 12% methylbenzethonium chloride	Local	CL old world twice daily during 10–20 days		Broad spectrum aminoglycoside antibiotic	–
		CL new world twice daily during 20–30 days			
Intralesional antimonials and cryotherapy (Figs 9.1 and 9.2)	Local	Two to ten times at 2–8 day intervals		Not fully understood	
				Interference with energy production of Leishmania amastigotes and with parasitic DNA topoisomerase I	

[a]Biphasic elimination, typically relating to the distribution and elimination phase of the drug

References

Aliaga L, Cobo F, Mediavilla JD, Bravo J, Osuna A, Amador JM, et al. Localized mucosal leishmaniasis due to Leishmania (Leishmania) infantum: clinical and microbiologic findings in 31 patients. Medicine (Baltimore). 2003;82(3):147–58.

Almeida R, D'Oliveira A Jr, Machado P, Bacellar O, Ko AI, de Jesus AR, et al. Randomized, double-blind study of stibogluconate plus human granulocyte macrophage colony-stimulating factor versus stibogluconate alone in the treatment of cutaneous Leishmaniasis. J Infect Dis. 1999;180(5):1735–7.

Alrajhi AA, Ibrahim EA, De Vol EB, Khairat M, Faris RM, Maguire JH. Fluconazole for the treatment of cutaneous leishmaniasis caused by Leishmania major. N Engl J Med. 2002;346(12):891–5.

Alvar J, Aparicio P, Aseffa A, Den Boer M, Canavate C, Dedet JP, et al. The relationship between leishmaniasis and AIDS: the second 10 years. Clin Microbiol Rev. 2008;21(2):334.

Amato VS, Tuon FF, Bacha HA, Neto VA, Nicodemo AC. Mucosal leishmaniasis—current scenario and prospects for treatment. Acta Trop. 2008;105(1):1–9.

Amato VS, Tuon FF, Imamura R, Abegao dC, Duarte MI, Neto VA. Mucosal leishmaniasis: description of case management approaches and analysis of risk factors for treatment failure in a cohort of 140 patients in Brazil. J Eur Acad Dermatol Venereol. 2009;23(9):1026–34.

Amato VS, Tuon FF, Camargo RA, Souza RM, Santos CR, Nicodemo AC. Short Report: can we use a lower dose of liposomal amphotericin B for the treatment of mucosal American leishmaniasis? Am J Trop Med Hyg. 2011;85(5):818–9.

Andersen EM, Cruz-Saldarriaga M, Llanos-Cuentas A, Luz-Cjuno M, Echevarria J, Miranda-Verastegui C, et al. Comparison of meglumine antimoniate and pentamidine for Peruvian cutaneous leishmaniasis. Am J Trop Med Hyg. 2005;72(2):133–7.

Antezana G, Zeballos R, Mendoza C, Lyevre P, Valda L, Cardenas F, et al. Electrocardiographic alterations during treatment of mucocutaneous leishmaniasis with meglumine antimoniate and allopurinol. Trans R Soc Trop Med Hyg. 1992;86(1):31–3.

Arana BA, Mendoza CE, Rizzo NR, Kroeger A. Randomized, controlled, double-blind trial of topical treatment of cutaneous leishmaniasis with paromomycin plus methylbenzethonium chloride ointment in Guatemala. Am J Trop Med Hyg. 2001;65(5):466–70.

Arevalo J, Ramirez L, Adaui V, Zimic M, Tulliano G, Miranda-Verastegui C, et al. Influence of Leishmania (Viannia) species on the response to antimonial treatment in patients with American tegumentary leishmaniasis. J Infect Dis. 2007;195(12):1846–51.

Armijos RX, Weigel MM, Calvopina M, Mancheno M, Rodriguez R. Comparison of the effectiveness of two topical paromomycin treatments versus meglumine antimoniate for new world cutaneous leishmaniasis. Acta Trop. 2004;91(2):153–60.

Aronson NE, Wortmann GW, Johnson SC, Jackson JE, Gasser RA Jr, Magill AJ, et al. Safety and efficacy of intravenous sodium stibogluconate in the treatment of leishmaniasis: recent U.S. military experience. Clin Infect Dis. 1998;27(6):1457–64.

Aronson NE, Wortmann GW, Byrne WR, Howard RS, Bernstein WB, Marovich MA, et al. A randomized controlled trial of local heat therapy versus intravenous sodium stibogluconate for the treatment of cutaneous Leishmania major infection. PLoS Negl Trop Dis. 2010;4(3):e628.

Asilian A, Davami M. Comparison between the efficacy of photodynamic therapy and topical paromomycin in the treatment of old world cutaneous leishmaniasis: a placebo-controlled, randomized clinical trial. Clin Exp Dermatol. 2006;31(5):634–7. https://doi.org/10.1111/j.1365-2230.2006.02182.x.

Asilian A, Jalayer T, Nilforooshzadeh M, Ghassemi RL, Peto R, Wayling S, et al. Treatment of cutaneous leishmaniasis with aminosidine (paromomycin) ointment: double-blind, randomized trial in the Islamic Republic of Iran. Bull World Health Organ. 2003;81(5):353–9.

Asilian A, Sadeghinia A, Faghihi G, Momeni A. Comparative study of the efficacy of combined cryotherapy and intralesional meglumine antimoniate (Glucantime) vs. cryotherapy and intralesional meglumine antimoniate (Glucantime) alone for the treatment of cutaneous leishmaniasis. Int J Dermatol. 2004;43(4):281–3.

Bafica A, Oliveira F, Freitas LA, Nascimento EG, Barral A. American cutaneous leishmaniasis unresponsive to antimonial drugs: successful treatment using combination of N-methylglucamine antimoniate plus pentoxifylline. Int J Dermatol. 2003;42(3):203–7.

Bailey MS, Lockwood DN. Cutaneous leishmaniasis. Clin Dermatol. 2007;25(2):203–11.

Ballou WR, McClain JB, Gordon DM, Shanks GD, Andujar J, Berman JD, et al. Safety and efficacy of high-dose sodium stibogluconate therapy of American cutaneous leishmaniasis. Lancet. 1987;2(8549):13–6.

Belazzoug S, Neal RA. Failure of Meglumine Antimoniate to cure cutaneous lesions due to Leishmania-major in Algeria. Trans R Soc Trop Med Hyg. 1986;80(4):670–1.

Ben Salah A, Buffet PA, Morizot G, Ben Massoud N, Zaatour A, Ben Alaya N, et al. WR279,396, a third generation aminoglycoside ointment for the treatment of Leishmania major cutaneous leishmaniasis: a phase 2, randomized, double blind, placebo controlled study. PLoS Negl Trop Dis. 2009;3(5):e432.

Ben Salah A, Ben Messaoud N, Guedri E, Zaatour A, Ben Alaya N, Bettaieb J, et al. Topical paromomycin with or without gentamicin for cutaneous leishmaniasis. N Engl J Med. 2013;368(6):524–32.

Blum JA, Hatz CF. Treatment of cutaneous leishmaniasis in travelers 2009. J Travel Med. 2009;16(2):123–31.

Blum J, Desjeux P, Schwartz E, Beck B, Hatz C. Treatment of cutaneous leishmaniasis among travellers. J Antimicrob Chemother. 2004;53:158–66.

Blum J, Lockwood DNJ, Visser L, Harms G, Bailey MS, Caumes E, et al. Local or systemic treatment for new world cutaneous leishmaniasis? Re-evaluating the evidence for the risk of mucosal leishmaniasis. Int Health. 2012;4(3):153–63.

Blum J, Buffet P, Visser L, Harms G, Bailey MS, Caumes E, et al. LeishMan recommendations for treatment of cutaneous and mucosal leishmaniasis in travelers, 2014. J Travel Med. 2014;21(2):116–29.

Boecken G, Weitzel T, Sunderkotter C, Fischer M, Stebut-Borschitz E, Bogdan C, et al. Diagnosis and therapy of cutaneous and musculocutaneous Leishmaniasis in Germany. J Dtsch Dermatol Ges. 2009;7(Suppl 7):s1–38.

Brummitt CF, Porter JA, Herwaldt BL. Reversible peripheral neuropathy associated with sodium stibogluconate therapy for American cutaneous leishmaniasis. Clin Infect Dis. 1996;22(5):878–9.

Bryceson AD, Murphy A, Moody AH. Treatment of 'Old World' cutaneous leishmaniasis with aminosidine ointment: results of an open study in London. Trans R Soc Trop Med Hyg. 1994;88(2):226–8.

Buffet P, Deray G, Grogl M, Martinez F, Jacobs C. Tolerance and pharmacokinetics of antimony in a patient with renal-failure. Nephrol Dial Transplantat. 1995;10(8):1477.

Bumb RA, Prasad N, Khandelwal K, Aara N, Mehta RD, Ghiya BC, et al. Long-term efficacy of single-dose radiofrequency-induced heat therapy vs. intralesional antimonials for cutaneous leishmaniasis in India. Br J Dermatol. 2013;168(5):1114–9.

Burza S, Sinha PK, Mahajan R, Sanz MG, Lima MA, Mitra G, et al. Post Kala-Azar dermal leishmaniasis following treatment with 20 mg/kg liposomal amphotericin B (Ambisome) for primary visceral leishmaniasis in Bihar, India. PLoS Negl Trop Dis. 2014;8(1):e2611. https://doi.org/10.1371/journal.pntd.0002611.

Cardona-Arias JA, Velez ID, Lopez-Carvajal L. Efficacy of thermotherapy to treat cutaneous leishmaniasis: a meta-analysis of controlled clinical trials. PLoS One. 2015;10(5):e0122569. https://doi.org/10.1371/journal.pone.0122569.

Chrusciak-Talhari A, Dietze R, Chrusciak TC, da Silva RM, Gadelha Yamashita EP, de Oliveira PG, et al. Randomized controlled clinical trial to access efficacy and safety of miltefosine in the treatment of cutaneous leishmaniasis caused by Leishmania (Viannia) guyanensis in Manaus, Brazil. Am J Trop Med Hyg. 2011;84(2):255–60.

Cunha MA, Leao AC, de Cassia Soler R, Lindoso JA. Efficacy and safety of liposomal amphotericin B for the treatment of mucosal leishmaniasis from the new world: a retrospective study. Am J Trop Med Hyg. 2015;93(6):1214–8. https://doi.org/10.4269/ajtmh.15-0033.

Delobel P, Pradinaud R. Rhabdomyolysis associated with pentamidine isethionate therapy for American cutaneous leishmaniasis. J Antimicrob Chemother. 2003;51(5):1319–20.

Dorlo TP, van Thiel PP, Huitema AD, Keizer RJ, de Vries HJ, Beijnen JH, et al. Pharmacokinetics of miltefosine in old world cutaneous leishmaniasis patients. Antimicrob Agents Chemother. 2008;52(8):2855–60.

Dorlo TPC, Balasegaram M, Lima MA, de Vries PJ, Beijnen JH, Huitema ADR. Translational pharmacokinetic modelling and simulation for the assessment of duration of contraceptive use after treatment with miltefosine. J Antimicrob Chemother. 2012a;67(8):1996–2004.

Dorlo TPC, Huitema ADR, Beijnen JH, de Vries PJ. Optimal dosing of Miltefosine in children and adults with visceral leishmaniasis. Antimicrob Agents Chemother. 2012b;56(7):3864–72.

el On J, Livshin R, Paz ZE, Weinrauch L. Topical treatment of cutaneous leishmaniasis. J Investig Dermatol. 1985;291(6504):1280–1.

el On J, Halevy S, Grunwald MH, Weinrauch L. Topical treatment of old world cutaneous leishmaniasis caused by Leishmania major: a double-blind control study. J Am Acad Dermatol. 1992;27(2 Pt 1):227–31.

Eldarouti MA, Alrubaie SM. Cutaneous leishmaniasis—treatment with combined cryotherapy and Intralesional Stibogluconate injection. Int J Dermatol. 1990;29(1):56–9.

Emad M, Hayati F, Fallahzadeh MK, Namazi MR. Superior efficacy of oral fluconazole 400 mg daily versus oral fluconazole 200 mg daily in the treatment of cutaneous leishmania major infection: a randomized clinical trial. J Am Acad Dermatol. 2011;64(3):606–8.

Enk CD, Nasereddin A, Alper R, Dan-Goor M, Jaffe CL, Wulf HC. Cutaneous leishmaniasis responds to daylight-activated photodynamic therapy: proof of concept for a novel self-administered therapeutic modality. Br J Dermatol. 2015;172(5):1364–70. https://doi.org/10.1111/bjd.13490.

Esfandiarpour I, Alavi A. Evaluating the efficacy of allopurinol and meglumine antimoniate (Glucantime) in the treatment of cutaneous leishmaniasis. Int J Dermatol. 2002;41(8):521–4.

Esfandiarpour I, Farajzadeh S, Rahnama Z, Fathabadi EA, Heshmatkhah A. Adverse effects of intralesional meglumine antimoniate and its influence on clinical laboratory parameters in the treatment of cutaneous leishmaniasis. Int J Dermatol. 2012;51(10):1221–5. https://doi.org/10.1111/j.1365-4632.2012.05460.x.

Faber WR, Wonders J, Jensema AJ, Chocholova E, Kager PA. Cutaneous leishmaniasis with lymphadenopathy due to Leishmania donovani. Clin Exp Dermatol. 2009;34(5):E196–E8.

Fagundes-Silva GA, Romero GA, Cupolillo E, Yamashita EP, Gomes-Silva A, Guerra JA, et al. Leishmania (Viannia) naiffi: rare enough to be neglected? Mem Inst Oswaldo Cruz. 2015;110(6):797–800. https://doi.org/10.1590/0074-02760150128.

Faris RM, Jarallah JS, Khoja TA, al Yamani MJ. Intralesional treatment of cutaneous leishmaniasis with sodium stibogluconate antimony. Int J Dermatol. 1993;32(8):610–2.

Faucher B, Pomares C, Fourcade S, Benyamine A, Marty P, Pratlong L, et al. Mucosal Leishmania infantum leishmaniasis: specific pattern in a multicentre survey and historical cases. J Infect. 2011;63(1):76–82.

Firdous R, Yasinzai M, Ranja K. Efficacy of glucantime in the treatment of old world cutaneous leishmaniasis. Int J Dermatol. 2009;48(7):758–62.

Firooz A, Khamesipour A, Ghoorchi MH, Nassiri-Kashani M, Eskandari SE, Khatami A, et al. Imiquimod in combination with meglumine antimoniate for cutaneous leishmaniasis: a randomized assessor-blind controlled trial. Arch Dermatol. 2006;142(12):1575–9.

Fontenele e Silva JS, Galvao TF, Pereira MG, Silva MT. Treatment of American tegumentary leishmaniasis in special populations: a summary of evidence. Rev Soc Bras Med Trop. 2013;46(6):669–77. https://doi.org/10.1590/0037-8682-0104-2013.

Franke ED, Llanos-Cuentas A, Echevarria J, Cruz ME, Campos P, Tovar AA, et al. Efficacy of 28-day and 40-day regimens of sodium stibogluconate (Pentostam) in the treatment of mucosal leishmaniasis. Am J Trop Med Hyg. 1994;51(1):77–82.

Garcia de Marcos JA, Dean FA, Alamillos GF, Ruiz Masera JJ, Cortes RB, Vidal JA, et al. Localized leishmaniasis of the oral mucosa. A report of three cases. Med Oral Patol Oral Cir Bucal. 2007;12(4):E281–E6.

Gasser RA Jr, Magill AJ, Oster CN, Franke ED, Grogl M, Berman JD. Pancreatitis induced by pentavalent antimonial agents during treatment of leishmaniasis. Clin Infect Dis. 1994;18(1):83–90.

Ghosh S, Das NK, Mukherjee S, Mukhopadhyay D, Barbhuiya JN, Hazra A, et al. Inadequacy of 12-week Miltefosine treatment for Indian post-Kala-Azar dermal leishmaniasis. Am J Trop Med Hyg. 2015;93(4):767–9. https://doi.org/10.4269/ajtmh.14-0721.

Harms G, Schonian G, Feldmeier H. Leishmaniasis in Germany. Emerg Infect Dis. 2003;9(7):872–5.

Harms G, Scherbaum H, Reiter-Owona I, Stich A, Richter J. Treatment of imported new world cutaneous leishmaniasis in Germany. Int J Dermatol. 2011;50(11):1336–42.

Hellier I, Dereure O, Tournillac I, Pratlong F, Guillot B, Dedet JP, et al. Treatment of old world cutaneous leishmaniasis by pentamidine isethionate. An open study of 11 patients. Dermatology. 2000;200(2):120–3.

Hepburn NC. Thrombocytopenia complicating sodium stibogluconate therapy for cutaneous leishmaniasis. Trans R Soc Trop Med Hyg. 1993;87(6):691.

Hepburn NC, Siddique I, Howie AF, Beckett GJ, Hayes PC. Hepatotoxicity of sodium stibogluconate in leishmaniasis. Lancet. 1993;342(8865):238–9.

Herwaldt BL. Leishmaniasis. Lancet. 1999;354(9185):1191–9.

Herwaldt BL, Berman JD. Recommendations for treating leishmaniasis with sodium stibogluconate (Pentostam) and review of pertinent clinical studies. Am J Trop Med Hyg. 1992;46(3):296–306.

Jamil KM, Haque R, Rahman R, Faiz MA, Bhuiyan AT, Kumar A, et al. Effectiveness study of Paromomycin IM injection (PMIM) for the treatment of visceral leishmaniasis (VL) in Bangladesh. PLoS Negl Trop Dis. 2015;9(10):e0004118. https://doi.org/10.1371/journal.pntd.0004118.

Jebran AF, Schleicher U, Steiner R, Wentker P, Mahfuz F, Stahl HC, et al. Rapid healing of cutaneous leishmaniasis by high-frequency electrocauterization and hydrogel wound care with or without DAC N-055: a randomized controlled phase IIa trial in Kabul. PLoS Negl Trop Dis. 2014;8(2):e2694. https://doi.org/10.1371/journal.pntd.0002694.

Jones TC, Johnson WD Jr, Barretto AC, Lago E, Badaro R, Cerf B, et al. Epidemiology of American cutaneous leishmaniasis due to Leishmania braziliensis braziliensis. J Infect Dis. 1987;156(1):73–83.

Junaid AJ. Treatment of cutaneous leishmaniasis with infrared heat. Int J Dermatol. 1986;25(7):470–2.

Kharfi M, Fazaa B, Chaker E, Kamoun MR. Mucosal localization of leishmaniasis in Tunisia: 5 cases. Ann Dermatol Venereol. 2003;130(1 Pt 1):27–30.

Khatami A, Firooz A, Gorouhi F, Dowlati Y. Treatment of acute old world cutaneous leishmaniasis: a systematic review of the randomized controlled trials. J Am Acad Dermatol. 2007;57(2):335–29.

Killingley B, Lamb LE, Davidson RN. Miltefosine to treat cutaneous leishmaniasis caused by Leishmania tropica. Ann Trop Med Parasitol. 2009;103(2):171–5.

Krause G, Kroeger A. Topical treatment of American cutaneous leishmaniasis with paramomycin and methylbenzethonium chloride: a clinical study under field conditions in Ecuador. Trans R Soc Trop Med Hyg. 1994;88(1):92–4.

Lai AFE, Vrede MA, Soetosenojo RM, Lai AFR. Pentamidine, the drug of choice for the treatment of cutaneous leishmaniasis in Surinam. Int J Dermatol. 2002;41(11):796–800.

Lawn SD, Armstrong M, Chilton D, Whitty CJ. Electrocardiographic and biochemical adverse effects of sodium stibogluconate during treatment of cutaneous and mucosal leishmaniasis among returned travellers. Trans R Soc Trop Med Hyg. 2006;100(3):264–9.

Lessa HA, Machado P, Lima F, Cruz AA, Bacellar O, Guerreiro J, et al. Successful treatment of refractory mucosal leishmaniasis with pentoxifylline plus antimony. Am J Trop Med Hyg. 2001;65(2):87–9.

Lightburn E, Morand JJ, Meynard JB, Kraemer P, Chaudier B, Pages F, et al. Management of American cutaneous leishmaniasis. Outcome apropos of 326 cases treated with high-dose pentamidine isethionate. Med Trop (Mars). 2003;63(1):35–44.

Llanos-Cuentas A, Tulliano G, Araujo-Castillo R, Miranda-Verastegui C, Santamaria-Castrellon G, Ramirez L, et al. Clinical and parasite species risk factors for pentavalent antimonial treatment failure in cutaneous leishmaniasis in Peru. Clin Infect Dis. 2008;46(2):223–31.

Lopez L, Robayo M, Vargas M, Velez ID. Thermotherapy. An alternative for the treatment of American cutaneous leishmaniasis. Trials. 2012;13:58.

Lopez L, Cruz C, Godoy G, Robledo SM, Velez ID. Thermotherapy effective and safer than miltefosine in the treatment of cutaneous leishmaniasis in Colombia. Rev Inst Med Trop Sao Paulo. 2013;55(3):197–204. https://doi.org/10.1590/S0036-46652013000300011.

Machado PR, Lessa H, Lessa M, Guimaraes LH, Bang H, Ho JL, et al. Oral pentoxifylline combined with pentavalent antimony: a randomized trial for mucosal leishmaniasis. Clin Infect Dis. 2007;44(6):788–93.

Machado PR, Ampuero J, Guimaraes LH, Villasboas L, Rocha AT, Schriefer A, et al. Miltefosine in the treatment of cutaneous leishmaniasis caused by Leishmania braziliensis in Brazil: a randomized and controlled trial. PLoS Negl Trop Dis. 2010;4(12):e912.

Machado PR, Rosa ME, Guimaraes LH, Prates FV, Queiroz A, Schriefer A, et al. Treatment of disseminated leishmaniasis with liposomal Amphotericin B. Clin Infect Dis. 2015;61(6):945–9. https://doi.org/10.1093/cid/civ416.

Marsden PD. Mucosal leishmaniasis ("espundia" Escomel, 1911). Trans R Soc Trop Med Hyg. 1986;80(6):859–76.

Mendonca MG, de Brito MEF, Rodrigues EHG, Bandeira V, Jardim ML, Abath FGC. Persistence of Leishmania parasites in scars after clinical cure of American cutaneous leishmaniasis: is there a sterile cure? J Infect Dis. 2004;189(6):1018–23.

Mohammadzadeh M, Behnaz F, Golshan Z. Efficacy of glucantime for treatment of cutaneous leishmaniasis in Central Iran. J Infect Public Health. 2013;6(2):120–4. https://doi.org/10.1016/j.jiph.2012.11.003.

Mohebali M, Fotouhi A, Hooshmand B, Zarei Z, Akhoundi B, Rahnema A, et al. Comparison of miltefosine and meglumine antimoniate for the treatment of zoonotic cutaneous leishmaniasis (ZCL) by a randomized clinical trial in Iran. Acta Trop. 2007;103(1):33–40.

Momeni AZ, Reiszadae MR, Aminjavaheri M. Treatment of cutaneous leishmaniasis with a combination of allopurinol and low-dose meglumine antimoniate. Int J Dermatol. 2002;41(7):441–3.

Morgan DJ, Guimaraes LH, Machado PR, D'Oliveira A Jr, Almeida RP, Lago EL, et al. Cutaneous leishmaniasis during pregnancy: exuberant lesions and potential fetal complications. Clin Infect Dis. 2007;45(4):478–82.

Morizot G, Delgiudice P, Caumes E, Laffitte E, Marty P, Dupuy A, et al. Healing of old world cutaneous leishmaniasis in travelers treated with fluconazole: drug effect or spontaneous evolution? Am J Trop Med Hyg. 2007;76(1):48–52.

Morizot G, Kendjo E, Mouri O, Thellier M, Perignon A, Foulet F, et al. Travelers with cutaneous leishmaniasis cured without systemic therapy. Clin Infect Dis. 2013;57(3):370–80.

Mosimann V, Neumayr A, Hatz C, Blum JA. Cutaneous leishmaniasis in Switzerland—first experience with species specific treatment. Infection. 2013;41(6):1177–82.

Mosimann V, Blazek C, Grob H, Chaney M, Neumayr A, Blum J. Miltefosine for mucosal and complicated cutaneous old world leishmaniasis: a case series and review of the literature. Open Forum Infect Dis. 2016;3(1):ofw008. https://doi.org/10.1093/ofid/ofw008.

Munir A, Janjua SA, Hussain I. Clinical efficacy of intramuscular meglumine antimoniate alone and in combination with intralesional meglumine antimoniate in the treatment of old world cutaneous leishmaniasis. Acta Dermatovenerol Croat. 2008;16(2):60–4.

Nacher M, Carme B, Sainte MD, Couppie P, Clyti E, Guibert P, et al. Influence of clinical presentation on the efficacy of a short course of pentamidine in the treatment of cutaneous leishmaniasis in French Guiana. Ann Trop Med Parasitol. 2001;95(4):331–6.

Nassiri-Kashani M, Firooz A, Khamesipour A, Mojtahed F, Nilforoushzadeh M, Hejazi H, et al. A randomized, double-blind, placebo-controlled clinical trial of itraconazole in the treatment of cutaneous leishmaniasis. J Eur Acad Dermatol Venereol. 2005;19(1):80–3.

Navin TR, Arana BA, Arana FE, Berman JD, Chajon JF. Placebo-controlled clinical trial of sodium stibogluconate (Pentostam) versus ketoconazole for treating cutaneous leishmaniasis in Guatemala. J Infect Dis. 1992;165(3):528–34.

Negera E, Gadisa E, Hussein J, Engers H, Kuru T, Gedamu L, et al. Treatment response of cutaneous leishmaniasis due to Leishmania aethiopica to cryotherapy and generic sodium stibogluconate from patients in Silti, Ethiopia. Trans R Soc Trop Med Hyg. 2012;106(8):496–503.

Neub A, Krahl D, Stich A, Amon U. Cutaneous infection with Leishmania infantum in an infant treated successfully with miltefosine. J Dtsch Dermatol Ges. 2008;6(12):1061–4.

Neumayr ALC, Walter C, Stoeckle M, Braendle N, Glatz K, Blum JA. Successful treatment of imported mucosal Leishmania infantum leishmaniasis with Miltefosine after severe hypokalemia under Meglumine Antimoniate treatment. J Travel Med. 2012;19(2):124–6.

Neumayr A, Buffet P, Visser L, Lockwood DN, Beck B, Schneider T, et al. Clinical aspects and management of cutaneous leishmaniasis in rheumatoid patients treated with TNF-alpha antagonists. TMAID. 2013;11(6):412–20.

Neves LO, Talhari AC, Gadelha EP, Silva Junior RM, Guerra JA, Ferreira LC, et al. A randomized clinical trial comparing meglumine antimoniate, pentamidine and amphotericin B for the treatment of cutaneous leishmaniasis by Leishmania guyanensis. An Bras Dermatol. 2011;86(6):1092–101.

Oliveira-Neto MP, Schubach A, Mattos M, da Costa SC, Pirmez C. Intralesional therapy of American cutaneous leishmaniasis with pentavalent antimony in Rio de Janeiro, Brazil—an area of Leishmania (V.) braziliensis transmission. Int J Dermatol. 1997;36(6):463–8.

Poeppl W, Walochnik J, Pustelnik T, Auer H, Mooseder G. Short Report: Cutaneous leishmaniasis after travel to Cyprus and successful treatment with Miltefosine. Am J Trop Med Hyg. 2011;84(4):562–5.

Rahman SB, ul Bari A, Mumtaz N. Miltefosine in cutaneous leishmaniasis. J Coll Physicians Surg Pak. 2007;17(3):132–5.

Ramesh V, Singh R, Avishek K, Verma A, Deep DK, Verma S, et al. Decline in clinical efficacy of oral Miltefosine in treatment of post Kala-azar dermal leishmaniasis (PKDL) in India. PLoS Negl Trop Dis. 2015;9(10):e0004093. https://doi.org/10.1371/journal.pntd.0004093.

Reithinger R, Mohsen M, Wahid M, Bismullah M, Quinnell RJ, Davies CR, et al. Efficacy of thermotherapy to treat cutaneous leishmaniasis caused by Leishmania tropica in Kabul, Afghanistan: a randomized, controlled trial. Clin Infect Dis. 2005;40(8):1148–55.

Reveiz L, Maia-Elkhoury AN, Nicholls RS, Romero GA, Yadon ZE. Interventions for American cutaneous and mucocutaneous leishmaniasis: a systematic review update. PLoS One. 2013;8(4):e61843. https://doi.org/10.1371/journal.pone.0061843.

Ribeiro AL, Drummond JB, Volpini AC, Andrade AC, Passos VM. Electrocardiographic changes during low-dose, short-term therapy of cutaneous leishmaniasis with the pentavalent antimonial meglumine. Braz J Med Biol Res. 1999;32(3):297–301.

Richter J, Hanus I, Haussinger D, Loscher T, Harms G. Mucosal Leishmania infantum infection. Parasitol Res. 2011;109(3):959–62.

Rocio C, Amato VS, Camargo RA, Tuon FF, Nicodemo AC. Liposomal formulation of amphotericin B for the treatment of mucosal leishmaniasis in HIV-negative patients. Trans R Soc Trop Med Hyg. 2014;108(3):176–8. https://doi.org/10.1093/trstmh/tru011.

Rodrigues ML, Costa RS, Souza CS, Foss NT, Roselino AM. Nephrotoxicity attributed to meglumine antimoniate (Glucantime) in the treatment of generalized cutaneous leishmaniasis. Rev Inst Med Trop Sao Paulo. 1999;41(1):33–7.

Romero GA, Guerra MV, Paes MG, Macedo VO. Comparison of cutaneous leishmaniasis due to Leishmania (Viannia) braziliensis and L. (V.) guyanensis in Brazil: therapeutic response to meglumine antimoniate. Am J Trop Med Hyg. 2001;65(5):456–65.

Roussel M, Nacher M, Fremont G, Rotureau B, Clyti E, Sainte-Marie D, et al. Comparison between one and two injections of pentamidine isethionate, at 7 mg/kg in each injection, in the treatment of cutaneous leishmaniasis in French Guiana. Ann Trop Med Parasitol. 2006;100(4):307–14.

Rubiano LC, Miranda MC, Arenas SM, Montero LM, Rodriguez-Barraquer I, Garcerant D, et al. Noninferiority of miltefosine versus meglumine antimoniate for cutaneous leishmaniasis in children. J Infect Dis. 2012;205(4):684–92.

Sadeghian G, Nilforoushzadeh MA. Effect of combination therapy with systemic glucantime and pentoxifylline in the treatment of cutaneous leishmaniasis. Int J Dermatol. 2006;45(7):819–21.

Sadeghian G, Nilfroushzadeh MA, Iraji F. Efficacy of local heat therapy by radiofrequency in the treatment of cutaneous leishmaniasis, compared with intralesional injection of meglumine antimoniate. Clin Exp Dermatol. 2007;32(4):371–4.

Saenz RE, Paz H, Berman JD. Efficacy of ketoconazole against Leishmania braziliensis panamensis cutaneous leishmaniasis. Am J Med. 1990;89(2):147–55.

Safi N, Davis GD, Nadir M, Hamid H, Robert LL, Case AJ. Evaluation of thermotherapy for the treatment of cutaneous leishmaniasis in Kabul, Afghanistan: a randomized controlled trial. Mil Med. 2012;177(3):345–51.

Salmanpour R, Handjani F, Nouhpisheh MK. Comparative study of the efficacy of oral ketoconazole with intra-lesional meglumine antimoniate (Glucantime) for the treatment of cutaneous leishmaniasis. J Dermatol Treat. 2001;12(3):159–62.

Salmanpour R, Razmavar MR, Abtahi N. Comparison of intralesional meglumine antimoniate, cryotherapy and their combination in the treatment of cutaneous leishmaniasis. Int J Dermatol. 2006;45(9):1115–6.

Schubach A, Haddad F, Oliveira-Neto MP, Degrave W, Pirmez C, Grimaldi G Jr, et al. Detection of Leishmania DNA by polymerase chain reaction in scars of treated human patients. J Infect Dis. 1998;178(3):911–4.

Seaton RA, Morrison J, Man I, Watson J, Nathwani D. Out-patient parenteral antimicrobial therapy--a viable option for the management of cutaneous leishmaniasis. Q J Med. 1999;92(11):659–67.

Shah SA, Memon AA, Auwj-e-Shamim, Baqi S, Witzig R. Low-cost thermotherapy for cutaneous leishmaniasis in Sindh, Pakistan. J Pak Med Assoc. 2014;64(12):1398–404.

Solomon M, Baum S, Barzilai A, Scope A, Trau H, Schwartz E. Liposomal amphotericin B in comparison to sodium stibogluconate for cutaneous infection due to Leishmania braziliensis. J Am Acad Dermatol. 2007;56(4):612–6.

Solomon M, Pavlotsky F, Leshem E, Ephros M, Trau H, Schwartz E. Liposomal amphotericin B treatment of cutaneous leishmaniasis due to Leishmania tropica. J Eur Acad Dermatol Venereol. 2011;25(8):973–7.

Solomon M, Pavlotzky F, Barzilai A, Schwartz E. Liposomal amphotericin B in comparison to sodium stibogluconate for Leishmania braziliensis cutaneous leishmaniasis in travelers. J Am Acad Dermatol. 2013;68(2):284–9.

Sosa N, Capitan Z, Nieto J, Nieto M, Calzada J, Paz H, et al. Randomized, double-blinded, phase 2 trial of WR 279,396 (paromomycin and gentamicin) for cutaneous leishmaniasis in Panama. Am J Trop Med Hyg. 2013;89(3):557–63. https://doi.org/10.4269/ajtmh.12-0736.

Soto J, Berman J. Treatment of new world cutaneous leishmaniasis with miltefosine. Trans R Soc Trop Med Hyg. 2006;100(Suppl 1):S34–40.

Soto J, Buffet P, Grogl M, Berman J. Successful treatment of Colombian cutaneous leishmaniasis with four injections of pentamidine. Am J Trop Med Hyg. 1994;50(1):107–11.

Soto J, Fuya P, Herrera R, Berman J. Topical paromomycin/methylbenzethonium chloride plus parenteral meglumine antimoniate as treatment for American cutaneous leishmaniasis: controlled study. Clin Infect Dis. 1998;26(1):56–8.

Soto J, Toledo J, Gutierrez P, Nicholls RS, Padilla J, Engel J, et al. Treatment of American cutaneous leishmaniasis with miltefosine, an oral agent. Clin Infect Dis. 2001;33(7):E57–61.

Soto JM, Toledo JT, Gutierrez P, Arboleda M, Nicholls RS, Padilla JR, et al. Treatment of cutaneous leishmaniasis with a topical antileishmanial drug (WR279396): phase 2 pilot study. Am J Trop Med Hyg. 2002;66(2):147–51.

Soto J, Arana A, Toledo J, Rizzo N, Vega JC, Diaz A, et al. Miltefosine for new world cutaneous leishmaniasis. Clin Infect Dis. 2004;38:1266–72.

Soto J, Toledo J, Valda L, Balderrama M, Rea I, Parra R, et al. Treatment of Bolivian mucosal leishmaniasis with miltefosine. Clin Infect Dis. 2007;44(3):350–6.

Soto J, Rea J, Balderrama M, Toledo J, Soto P, Valda L, et al. Efficacy of miltefosine for Bolivian cutaneous leishmaniasis. Am J Trop Med Hyg. 2008;78(2):210–1.

Soto J, Rea J, Valderrama M, Toledo J, Valda L, Ardiles J, et al. Efficacy of extended (six weeks) treatment with miltefosine for mucosal leishmaniasis in Bolivia. Am J Trop Med Hyg. 2009;81(3):387–9.

Soto J, Rojas E, Guzman M, Verduguez A, Nena W, Maldonado M, et al. Intralesional antimony for single lesions of bolivian cutaneous leishmaniasis. Clin Infect Dis. 2013;56(9):1255–60. https://doi.org/10.1093/cid/cit049.

Soto J, Paz D, Rivero D, Soto P, Quispe J, Toledo J, et al. Intralesional Pentamidine: a novel therapy for single lesions of Bolivian cutaneous leishmaniasis. Am J Trop Med Hyg. 2016;94(4):852–6. https://doi.org/10.4269/ajtmh.15-0640.

Soto-Mancipe J, Grogl M, Berman JD. Evaluation of pentamidine for the treatment of cutaneous leishmaniasis in Colombia. Clin Infect Dis. 1993;16(3):417–25.

Sousa AQ, Frutuoso MS, Moraes EA, Pearson RD, Pompeu MML. High-dose oral fluconazole therapy effective for cutaneous leishmaniasis due to Leishmania (Vianna) braziliensis. Clin Infect Dis. 2011;53(7):693–5.

Stoeckle M, Holbro A, Arnold A, Neumayr A, Weisser M, Blum J. Treatment of mucosal leishmaniasis (*L. infantum*) with miltefosine in a patient with good syndrome. Acta Trop. 2013;128(1):168–70.

Sundar S, Jha TK, Thakur CP, Bhattacharya SK, Rai M. Oral miltefosine for the treatment of Indian visceral leishmaniasis. Trans R Soc Trop Med Hyg. 2006;100(Suppl 1):S26–33.

Sundar S, Jha TK, Thakur CP, Sinha PK, Bhattacharya SK. Injectable paromomycin for visceral leishmaniasis in India. N Engl J Med. 2007;356(25):2571–81. https://doi.org/10.1056/NEJMoa066536.

Tallab TM, Bahamdam KA, Mirdad S, Johargi H, Mourad MM, Ibrahim K, et al. Cutaneous leishmaniasis: schedules for intralesional treatment with sodium stibogluconate. Int J Dermatol. 1996;35(8):594–7.

Tappe D, Muller A, Stich A. Resolution of cutaneous old world and new world leishmaniasis after oral miltefosine treatment. Am J Trop Med Hyg. 2010;82(1):1–3.

Valencia BM, Miller D, Witzig RS, Boggild AK, Llanos-Cuentas A. Novel low-cost thermotherapy for cutaneous leishmaniasis in Peru. PLoS Negl Trop Dis. 2013;7(5):e2196. https://doi.org/10.1371/journal.pntd.0002196.

van der Meide WF, Sabajo LOA, Jensema AJ, Peekel I, Faber WR, Schallig HDFH, et al. Evaluation of treatment with pentamidine for cutaneous leishmaniasis in Suriname. Int J Dermatol. 2009;48(1):52–8.

van der Snoek EM, Lammers AM, Kortbeek LM, Roelfsema JH, Bart A, Jaspers CAJJ. Spontaneous cure of American cutaneous leishmaniasis due to Leishmania naiffi in two Dutch infantry soldiers. Clin Exp Dermatol. 2009;34(8):E889–E91.

van Griensven J, Carrillo E, Lopez-Velez R, Lynen L, Moreno J. Leishmaniasis in immunosuppressed individuals. Clin Microbiol Infect. 2014;20(4):286–99. https://doi.org/10.1111/1469-0691.12556.

van Griensven J, Gadisa E, Aseffa A, Hailu A, Beshah AM, Diro E. Treatment of cutaneous Leishmaniasis caused by Leishmania aethiopica: a systematic review. PLoS Negl Trop Dis. 2016;10(3):e0004495. https://doi.org/10.1371/journal.pntd.0004495.

van Thiel PP, Leenstra T, Kager PA, de Vries HJ, van Vugt M, van der Meide WF, et al. Miltefosine treatment of Leishmania major infection: an observational study involving Dutch military personnel returning from northern Afghanistan. Clin Infect Dis. 2010a;50(1):80–3.

van Thiel PPAM, van Gool T, Kager PA, Report BAC. First cases of cutaneous leishmaniasis caused by Leishmania (Viannia) naiffi infection in Surinam. Am J Trop Med Hyg. 2010b;82(4):588–90.

Velez I, Lopez L, Sanchez X, Mestra L, Rojas C, Rodriguez E. Efficacy of miltefosine for the treatment of American cutaneous leishmaniasis. Am J Trop Med Hyg. 2010;83(2):351–6.

Vernal S, De Paula NA, Gomes CM, Roselino AM. Disseminated Leishmaniasis by Leishmania viannia subgenus: a series of 18 cases in southeastern Brazil. Open Forum Infect Dis. 2016;3(1):ofv184. https://doi.org/10.1093/ofid/ofv184.

Walton BC, Valverde L. Racial-differences in Espundia. Ann Trop Med Parasitol. 1979;73(1):23–9.

Weinrauch L, Livshin R, Even-Paz Z, el On J. Efficacy of ketoconazole in cutaneous leishmaniasis. Arch Dermatol Res. 1983;275(5):353–4.

Weinrauch L, Cawich F, Craig P, Sosa JX, el On J. Topical treatment of new world cutaneous leishmaniasis in Belize: a clinical study. J Am Acad Dermatol. 1993;29(3):443–6.

Wise ES, Armstrong MS, Watson J, Lockwood DN. Monitoring toxicity associated with parenteral sodium Stibogluconate in the day-case management of returned travellers with new world cutaneous leishmaniasis. PLoS Negl Trop Dis. 2012;6(6):e1688.

Wortmann GW, Aronson NE, Byrd JC, Grever MR, Oster CN. Herpes zoster and lymphopenia associated with sodium stibogluconate therapy for cutaneous leishmaniasis. Clin Infect Dis. 1998;27(3):509–12.

Wortmann G, Miller RS, Oster C, Jackson J, Aronson N. A randomized, double-blind study of the efficacy of a 10- or 20-day course of sodium stibogluconate for treatment of cutaneous leishmaniasis in United States military personnel. Clin Infect Dis. 2002;35(3):261–7.

Wortmann G, Zapor M, Ressner R, Fraser S, Hartzell J, Pierson J, et al. Liposomal amphotericin B for treatment of cutaneous leishmaniasis. Am J Trop Med Hyg. 2010;83(5):1028–33.

Zanger P, Kotter I, Raible A, Gelanew T, Schonian G, Kremsner PG. Case report: successful treatment of cutaneous leishmaniasis caused by Leishmania aethiopica with liposomal Amphotericin B in an immunocompromised traveler returning from Eritrea. Am J Trop Med Hyg. 2011a;84(5):692–4.

Zanger P, Kotter I, Kremsner P, Gabrysch S. Tumor necrosis factor alpha antagonist drugs and leishmaniasis in Europe. Tropical Med Int Health. 2011b;16:172.

Zerehsaz F, Salmanpour R, Handjani F, Ardehali S, Panjehshahin MR, Tabei SZ, et al. A double-blind randomized clinical trial of a topical herbal extract (Z-HE) vs. systemic meglumine antimoniate for the treatment of cutaneous leishmaniasis in Iran. Int J Dermatol. 1999;38(8):610–2.

Zerpa O, Ulrich M, Blanco B, Polegre M, Avila A, Matos N, et al. Diffuse cutaneous leishmaniasis responds to miltefosine but then relapses. Br J Dermatol. 2007;156(6):1328–35.

Control and Public Health Aspects 10

Marleen Boelaert, Sakib Burza, and Gustavo Romero

10.1 Introduction

As amply shown in the previous chapters, there is no single leishmaniasis disease, and so there is no single magic bullet to control it. Visceral leishmaniasis (VL) and cutaneous leishmaniasis (CL) are transmitted by different sand fly species, each with specific vector behavior, and therefore control methods also need to be specific. Some species of *Leishmania* are maintained in animal reservoir hosts, but not all. The World Health Organization (WHO) distinguishes 12 distinct disease systems across the globe and recommends specific control strategies for each (WHO 2010). Adequate control strategies therefore have to be tailored to the specific features of the ecosystem that sustains the leishmanial parasite, taking into account its specific human, vector, reservoir, and environmental characteristics. Control of the leishmaniasis requires an interdisciplinary "One Health" approach. We describe first the burden of leishmaniasis in public health terms, discuss the options for control in more detail, and describe the main control approaches that are in place today.

M. Boelaert (✉)
Unit of Epidemiology and Control of Neglected Tropical Diseases, Department of Public Health, Institute of Tropical Medicine, Antwerp, Belgium
e-mail: mboelaert@itg.be

S. Burza
Médecins Sans Frontières- Operational Centre Barcelona, Barcelona, Spain
e-mail: Sakib.BURZA@barcelona.msf.org

G. Romero
Center for Tropical Medicine, Faculty of Medicine, University of Brasilia, Federal District, Brazil
e-mail: gromero@unb.br

F. Bruschi, L. Gradoni (eds.), *The Leishmaniases: Old Neglected Tropical Diseases*, https://doi.org/10.1007/978-3-319-72386-0_10

10.2 Burden of Disease

10.2.1 Case Numbers

In 2012 WHO estimated there were between 200,000 and 400,000 new cases of visceral leishmaniasis (VL) per year and 0.7–1.3 million of cutaneous leishmaniasis (CL). These figures were estimates based on the cases reported by endemic countries to WHO but corrected for a certain degree of underreporting (Alvar et al. 2012).

WHO regularly reviews the global burden estimates. Recently, the WHO collated national data on VL and CL incidence in the 25 highest endemic countries (WHO 2016). VL high-burden countries reported a total of 30,758 VL cases and CL high-burden countries a total of 153,027 CL cases. These numbers include both primary and relapse cases. More than 90% of the global new VL cases were reported from seven countries: Brazil, Ethiopia, India, Kenya, Somalia, South Sudan, and Sudan. However, there is often a large gap between the number of reported cases and the actual number of VL or CL in the community. Usually the true numbers are largely underreported, for many reasons. In endemic areas, VL patients typically have poor access to the health system, so will never reach a health facility where their case can be included in the records and notified. Private practitioners also manage leishmaniasis cases, but will not notify the cases to the national surveillance system. For CL, underreporting is even more problematic, as patients in rural areas with typically self-healing ulcers may see no point in consulting a health worker.

WHO recently conducted a Delphi survey to estimate the amount of underreporting, which suggested rates of 1.2 in India, 1.2–1.8 in Bangladesh, Ethiopia, Nepal and South Sudan, and between 2 and –4 in Somalia, Spain, and Sudan. In Latin America, a substantial effort was made to develop a reporting system that gives adequate information on the actual numbers of CL and VL cases in the continent in an attempt to minimize underreporting (PAHO 2016).

But to make the issue even more complicated, underreporting rates tend to vary over time, depending on the intensity of control and the accessibility of care. When in 1992 the total number of reported VL cases in the world was 82,000 (Ashford et al. 1992), WHO extrapolated this figure at the time to an annual estimated case load of 500,000 based on a 1:5 ratio of declared to undeclared cases in community surveys in India (Desjeux 1992). Today the 1:5 underreporting ratio for VL seems to have substantially declined in India to 1:1.2, through a combination of improvement of community facilities and access to diagnostics, mandatory reporting of the disease, and the general intense efforts of government and NGOs to make VL medicines more freely available in the elimination initiative.

When studied in population-based longitudinal studies, the observed incidence rates of VL (also known as kala-azar) in endemic communities vary between 2/1000 person-years in Kenya (Schaefer et al. 1995) and 14/1000 person-years in Ethiopia (Ali and Ashford 1994a, b). In the Indian subcontinent, incidence rates lie within similar ranges. These rates are comparable or higher than the incidence rates of tuberculosis. There are almost no data available on incidence rates of CL.

10.2.2 Socioeconomic Factors

Leishmaniasis is part of a group known as the neglected tropical diseases (NTD) (Molyneux et al. 2005). The term "neglected" originally referred to the domain of drug research and development, as the pharmaceutical industry has few incentives to invest in these diseases. Today leishmaniasis still is one of those diseases for which we lack effective, affordable, and easy to use drugs. Until its emergence as a coinfection of HIV/AIDS, VL was not perceived as a direct threat to industrial countries (Dujardin et al. 2008). Only in recent years has leishmaniasis attained more global attention. This was triggered initially by the 2005 VL elimination initiative in Southeast Asia (see below), and later additional momentum was provided by the London Declaration on NTDs in 2012 (2012).

It is still fair to say that the leishmaniases often remain a hidden problem, with many patients living in remote areas with poor access to health services. Moreover, reported figures that are aggregated at country or regional level do not reflect the real impact of VL in affected communities, because VL has a focal distribution. In its endemic form, VL can profoundly affect the community because it is an ultimately fatal disease if left untreated and may result in significant impoverishment of households (Meheus et al. 2006). In addition, VL can have a disastrous impact when it strikes a nonimmune population, for instance, during mass population displacements due to conflict.

The leishmaniases are poverty-related diseases. However, most of the socioeconomic research has been limited to VL. In the Indian subcontinent, the transmission of VL is related to housing conditions as the vector *Phlebotomus argentipes* breeds in cracks of mud-plastered houses and moist soils. In 2000, Thakur described the socioeconomic conditions of 938 VL patients from Bihar in India (Thakur 2000). Seventy-five percent of them were poor (defined as daily income below US $1), and most of them lived in mud-walled or grass-covered houses. Boelaert et al. (2009) described how in the same state, which is one of the economically most deprived in India, villages affected by VL were literally the "poorest of the poor," with 80% of the families belonging to the poorest two quintiles of the state's wealth distribution.

VL is not only a disease of the poor; it is also a source of poverty. In every family where a VL case occurs, many days of productive life are lost due to this severely debilitating disease, and this indirect cost may represent up to 60% of the total household cost (Meheus et al. 2006). Other studies conducted in India (Sarnoff et al. 2010) (Sundar et al. 2010), Nepal (Rijal et al. 2006; Adhikari and Maskay 2003; Adhikari et al. 2009), and Bangladesh (Anoopa Sharma et al. 2006) concur that VL puts a very heavy economic burden on the household of a case. The related expenditure is often catastrophic, requiring the sale of assets and livestock or taking loans.

In summary, VL is one of the most neglected parasitic diseases in the world. Although at the global level, the numbers of VL might be lower in comparison to those of HIV/AIDS, malaria, and tuberculosis, the social and economic implications of VL are profound in the affected communities. VL is clearly a poverty-related disease, affecting indeed the poorest of the poor but also jeopardizing the economic development of affected areas.

10.3 Evidence on Control Strategies

Control strategies for this vector-born disease can target the human host, the vector, or the animal reservoir in case of zoonotic disease. Vaccination of humans is unfortunately not yet an option, though research in this area is advancing.

The fact that self-healing CL with *Leishmania major* usually leads to benign lesions as well as protection from reinfection has been known since ancient times. People in Asia and the Middle East inoculated exudates from active sores into the buttocks of children to spare them the development of disfiguring CL scars on the face. The first true vaccines containing live *L. major* promastigotes exploited the same principle. They were used in the Central Asian republics of the former USSR and Israel (Duthie et al. 2012). The practice of "leishmanization," as the procedure was known, is now discontinued as it is considered unsafe. Vaccines containing killed promastigotes mixed with BCG as adjuvant failed to protect against leishmaniasis (Noazin et al. 2009). Second-generation vaccines, which consist of attenuated parasites, recombinant molecules, or parasite DNA, mixed together in a cocktail vaccine, and recombinant organisms carrying leishmanial genes are in development.

As there is currently no vaccine on the market to protect humans from *Leishmania* infection, there are only a few ways to protect individuals: avoiding exposure to natural zoonotic foci and protecting oneself against sand fly bites with repellents and other devices.

10.3.1 Case Detection and Management

The first and foremost reason to promote early case detection and treatment is that treatment outcome is worse in patients with advanced VL or CL. In Sudan, adult VL patients with severe anemia, malnutrition, and a long duration of disease have a high risk of dying during treatment (Collin et al. 2004). In Brazil a meta-analysis by Belo and co-workers identified jaundice, thrombocytopenia, hemorrhage, HIV coinfection, diarrhea, age <5 and age >50 years, severe neutropenia, dyspnea, and bacterial infections as the strongest predictors of poor prognosis and death (Belo et al. 2014). Moreover, in the case of anthroponotic VL or CL transmission, there is almost no other option to control the disease given the absence of a vaccine and the limited effectiveness of vector control in many places. Early diagnosis and treatment is thus important for both clinical and public health reasons. Mathematical modeling suggests that asymptomatic carriers of *Leishmania* parasites may contribute to transmission (Stauch et al. 2011), but currently there are no treatments available safe enough to justify use in otherwise healthy persons.

A major breakthrough for the control programs was achieved when in 1997 a rapid diagnostic test (RDT) was found highly sensitive and specific for the diagnosis of VL (Sundar et al. 1998). This, combined with the oral drug miltefosine, meant that the detection and treatment of VL that was previously in the hands of specialist clinicians could now be decentralized and implemented in primary care centers.

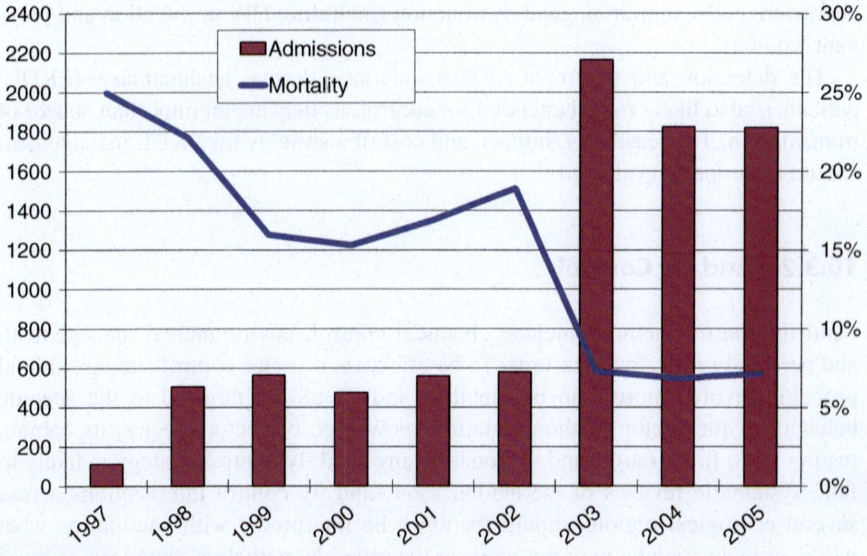

Fig. 10.1 VL case numbers and case fatality in MSF program in Ethiopia. (Courtesy of K. Ritmeijer, MSF)

Figure 10.1 demonstrates the potential impact of such RDT on access to care—in 2003, the Médecins Sans Frontières (MSF) program in Ethiopia was able to reach three times more patients after the introduction of the RDT.

The wide geographical heterogeneity of the leishmaniases extends to variable diagnostic accuracy of RDTs (Cunningham et al. 2012) and variable drug efficacy. Therefore, even in diagnosis and treatment, there is no one size fits all strategy. The VL elimination initiative that was launched in 2005 by the governments of India, Nepal, and Bangladesh opted for a case finding and treatment strategy based on the rK39 ICT as diagnostic tool and miltefosine as the first-line drug. Aside from the obvious benefits of an oral treatment, the decision to switch was driven by increasing failure rates of the antimonials in the region.

More recently, the governments of India, Nepal, and Bangladesh have adopted the use of a single infusion of liposomal amphotericin B (10 mg/kg body weight) as first-line treatment. This has been implemented at both primary and secondary settings, with over 10,000 patients treated as of the end of 2016. This has been made possible by a WHO donation mechanism for select low-resource settings. Newer lower-dose combinations of existing drugs also remain available.

However, in African VL, monotherapy with liposomal amphotericin B preparations was not shown effective (WHO 2010). As such, in such regions where antimonials remain effective, they continue to be widely used, but their toxicity profile requires reconsideration of this policy and the development of improved therapeutic options.

In Brazil, a 7-day L-Amb regimen is available for VL patients <1 and >50 years old, severe disease, presence of renal, cardiac, or hepatic failure, comorbidities characterized by immunological dysfunction (including HIV coinfection and pregnant women).

The detection and treatment of post kala-azar dermal leishmaniasis (PKDL) patients is also likely to be beneficial for control, as they are an important source of transmission. The feasibility, impact, and cost of a strategy for PKDL management should be properly evaluated.

10.3.2 Sand Fly Control

Sand fly control methods include chemical control, environmental management, and personal protection measures. To be effective, a vector control strategy should probably involve more than one method and should be targeted to the specific behavior of the vector. Without detailed knowledge of vector species, its habitat, resting sites, flight range, and seasonality, any sand fly control strategy is likely to fail. Systematic reviews of the evidence on sand fly control interventions across several ecological regions should therefore be interpreted with caution, as what works in one econiche may not work in the other. Nonetheless, the merit of these reviews is that they tend to emphasize rigor of study design and give more weight to well-designed experimental than to observational descriptive studies when it comes to grading the evidence.

A recent Cochrane review (Gonzalez et al. 2015) concluded that insecticides to reduce sand fly density may be effective at reducing the incidence of CL, but that there was insufficient evidence from trials to know whether it is better to spray the internal walls of houses or to treat bednets, curtains, bedsheets, or clothing. One four-arm RCT from Afghanistan compared indoor residual spraying (IRS), insecticide-treated bednets (ITNs), and insecticide-treated bedsheets (ITS), with no intervention. Over the 15-month follow-up, the three intervention areas each had a lower incidence of CL than the control area (IRS: risk ratio (RR) 0.61, 95% confidence interval (CI) 0.38–0.97; ITNs: RR 0.32, 95% CI 0.18–0.56; ITS: RR 0.34, 95% CI 0.20–0.57). There was no statistical difference between the three interventions. One additional trial of ITNs from Iran was underpowered to show a difference. We describe each of these methods in some more detail.

10.3.2.1 Insecticide Spraying

For IRS to be effective, the sand fly vectors should be mainly endophilic and an important part of transmission should be indoor. IRS is inappropriate where the vector is not peridomestic, as is the situation in Sudan where transmission of VL takes place mainly outside the villages in *Acacia* sp. and *Balanites aegyptiaca* woodlands and where activities such as wood cutting and shepherding expose people to sand flies outdoors.

Several classes of insecticides are used in IRS including organochlorines (e.g., DDT), organophosphates (e.g., malathion), carbamates (e.g., propoxur), and synthetic pyrethroids (e.g., deltamethrin, permethrin, and lambda-cyhalothrin) (WHO 2010). The choice of insecticide for sand fly control has to be strictly regulated at the national level. Countries need to embed this sand fly control strategy in an integrated vector management policy including an insecticide rotation plan.

Insecticide spraying during the malaria eradication campaigns of the 1950s is often credited for a simultaneous reduction of *L. donovani* transmission in the Indian subcontinent, but in other parts of the world, this effect was less clear, presumably due to differences in vector ecology and in disease transmission characteristics. Also, within months of the malaria eradication campaign ending, the first cases of kala-azar reappeared in India and culminated in a very large VL epidemic in Bihar at the end of the 1970s (Joshi et al. 2003; Kishore et al. 2006).

Direct evidence of the effect of house spraying on *Leishmania* transmission is scanty as most of the trials only looked at the effect on vector density and did not study clinical outcomes. As reviewed by Ostyn et al. (2008), house spraying reduced the density of endophilic species such as *Lutzomyia verrucarum*, *Lu. peruensis* (Davies et al. 2000), *Lu. longipalpis* and *Lu. ovallesi* (Feliciangeli et al. 2003a, b), and *Lu. intermedia* (Falcao et al. 1991) in the New World and *P. papatasi* (Benzerroug et al. 1992) in the Old World. In contrast, house spraying failed to reduce the density of exophilic sand flies such as *L. nuneztovari* in Bolivia (Le Pont et al. 1989). A more recent cluster randomized trial showed that indoor residual spraying was effective to prevent CL in Morocco (Faraj et al. 2016).

The effectiveness of these spraying programs is not the only issue of concern; other problems are their side effects in health and environment and their sustainability. Several factors such as cost of the insecticides, poor quality of implementation, logistical constraints, the low acceptance by the community and low community participation, and the emergence of resistance compromise the longer-term effectiveness and sustainability of this intervention. The VL elimination initiative in the Indian subcontinent has IRS as one of its main pillars, Nepal and Bangladesh using synthetic pyrethroids, and India, until recently, DDT. Resistance to DDT was reported from India in 2008 (Ostyn et al. 2008) and again highlighted more recently by Coleman et al. (2015), Kumar et al. (2015). This emerging resistance has motivated a switch to synthetic pyrethroid spraying in India in 2016.

10.3.2.2 Insecticide-Treated Materials

Insecticide treated bednets combine the individual protection provided by a net (barrier effect) with the insect-killing effect of the insecticide. As the insecticide has also a deterrent and repellent effect, mesh size does not matter so much. The bednets act as "baited traps" with the odor of the sleeper as bait. As reviewed by Ostyn et al. (2008), insecticide-impregnated bednets (ITN) are considered very effective for reducing man-vector contact and intra- and peridomiciliary transmission of vector-borne diseases. Most ITNs are impregnated with synthetic pyrethroids. Compared with house spraying, ITNs have a number of advantages: (a) their effectiveness does not depend on the endo- or exophilic behavior of the vectors; (b) less insecticide is used; and (c) the family has control over its use and will depend less on the actual performance of a disease control program.

At least six studies have tested the effect of treated bednets on the incidence of CL under field conditions (Tayeh et al. 1997; Nadim et al. 2000; Reyburn et al. 2000; Alten et al. 2003; Jalouk et al. 2007; Gunay et al. 2014).

Three of them included entomological evaluations of deltamethrin (25 mg/m^2)-treated bednets by comparison of indoor sand fly abundance using sticky traps. All studies failed to find any statistical difference, possibly because of small subsamples (10 houses per arm) or because sand fly abundance before the intervention was different in treatment and control groups. Nevertheless, all six trials noted a reduction in CL incidence in the intervention group using ITNs. A household study in Kabul compared ITN with two other interventions (house spraying and impregnated bed sheets) and showed a reduction in CL incidence from 7.2% in the control group to 2.4% in the houses with ITNs (Reyburn et al. 2000).

There is limited evidence that insecticide-treated bednets (ITNs) also provide protection against VL. Case-control studies conducted in Bangladesh and Nepal showed that sleeping under a (non-impregnated) bednet during the warm months was a protective factor against VL (Bern et al. 2000, 2005). However a large randomized controlled trial in Nepal and India testing the efficacy of long-lasting ITNs to prevent *L. donovani* infection and VL failed to show a protective effect in villages where a large proportion of households used untreated nets (Picado et al. 2010a, b).

Depending on sleeping traditions of the population and biting habits of the local vector, other insecticide-impregnated materials have been evaluated (e.g., curtains, blankets) (Reyburn et al. 2000). There is no strong evidence that using clothes treated with insecticide (e.g., soldiers' uniforms) protects individuals against sand fly bites and reduces the risk of leishmaniasis. Repellents that are applied as topical solution to the skin may be an alternative for personal protection against sand fly bites. The use of repellents as a long-term public health intervention in endemic regions may be difficult, as they require repeated applications (i.e., protection lasts for a maximum of 10 h). Nevertheless, insect repellents may be the only option in regions where *Leishmania* transmission occurs outdoors. The efficacy of repellents combined with other vector control tools (i.e., ITNs) should be further explored (Gonzalez et al. 2015).

10.3.3 Reservoir Control

In zoonotic VL, the control of the animal reservoir remains a challenge. Dogs are the main reservoir of *L. infantum* in zoonotic VL in the Mediterranean Basin and in South America. CL and MCL in South America are typically transmitted in a sylvatic or peridomestic cycle, which makes reservoir control even more complex.

10.3.3.1 Test-and-Treat or Test-and-Cull Dogs

Despite some evidence from trials in Brazil showing a decreased incidence of seroconversion during an 18-month follow-up in children following a test-and-cull campaign in dogs, the efficiency and acceptability of this strategy is increasingly debated and has failed to control VL in Brazil (Costa 2011). Culled pet dogs are quickly replaced by fully susceptible young pets and these maintain *Leishmania* transmission (Nunes et al. 2008). Test-and-treat strategies have been recommended without convincing evidence on their impact. Treating infected dogs may not be an effective control measure as relapses are frequent, and dogs can become infectious again within weeks after treatment. Furthermore, treating dogs could be a relevant factor for inducing parasite resistance to drugs used for treating human cases in a scenario of extreme scarcity of new antileishmanial drugs (Maia et al. 2013).

10.3.3.2 Dog Collars

Deltamethrin-treated dog collars protected against *Leishmania* infection in dogs (Maroli et al. 2001). When treated collars were applied to all dogs in endemic villages in Iran, the risk of infection in children was also reduced (Gavgani et al. 2002). Unfortunately the dog collars need to be replaced every 6 months and dogs tend to lose them rapidly. Ongoing community-based trials in Brazil and in Iran will shed light on the effectiveness of this intervention when used on a large scale to protect people from VL.

10.3.3.3 Dog Vaccines

Ultimately, the best strategy would be to vaccinate the dogs if the effectiveness in reducing transmission to humans is demonstrated. Four vaccines were registered for use in dogs: two in Brazil and two in Europe. Their impact on the transmission to humans in zoonotic VL has not yet been firmly established.

10.3.3.4 Other

If the reservoirs are peridomestic species such as the fat sand jird, reservoir of *L. major* in North Africa, their elimination using rodenticides or destroying the rodent burrows may be an effective control measure. However, when the *Leishmania* cycle is maintained by sylvatic reservoirs (i.e., the climbing rat for *L. mexicana*), there are no effective methods to control them. Controlling zoonotic disease seems even more complex in the New World where almost all the *Leishmania* species have sylvatic hosts. Even clearing the forest may not be adequate, as shown in eastern Brazil where *L. braziliensis* adapted to the deforestation (Jirmanus et al. 2012).

Novel approaches are currently being explored, as evidence accumulates that sand fly larvae feed on rodent feces (Ingenloff et al. 2013). Therefore, trials are underway with insecticide baited traps (Mascari et al. 2012).

Furthermore, environmental management may prove an effective way of combating rodent reservoirs.

10.3.4 Epidemiological Surveillance and Outbreak Response

The importance of adequate national and international surveillance systems for leishmaniasis cannot be stressed enough. The main parameter to track the epidemiological trend is the number of reported leishmanial cases. As discussed above, this indicator is subject to a fluctuating underreporting ratio and trends are not easy to interpret. Standard case definitions for reporting are required, quality control on reporting, prompt analysis, and feedback mechanisms, allowing for adequate outbreak response. WHO recommends that in endemic malarious areas, "*visceral leishmaniasis should be suspected when fever lasts for more than two weeks and no response has been achieved with anti-malaria drugs (assuming drug resistant malaria has also been considered*)." Source: WHO Recommended Surveillance Standards. Second edition WHO/CDS/CSR/ISR/99.2 Available on http://www.who.int/csr/resources/publications/surveillance/whocdscsrisr992.pdf?ua=1 (Accessed on July 15 2017).

Frequently used case definitions of suspect and confirmed VL in endemic areas are shown in the text box below.

In endemic areas

- A **suspect case of visceral leishmaniasis** is somebody with a history of fever of more than 2 weeks and enlarged spleen not responding to antimalarial treatment.
- A **confirmed case of visceral leishmaniasis** is a suspect case with serological and/or parasitological confirmation.

A surveillance system is unlikely to capture 100% of cases, even in countries where VL is notifiable. Studies conducted in the Indian subcontinent showed that poverty, illiteracy, and long distances to health centers were common barriers for patients to seek early professional medical care. Until recently, when VL was clinically suspected, patients had to be referred to a hospital for diagnostic confirmation and treatment, an effort which many of the patients could not afford. Under these conditions, the disease was inevitably underreported; such as was the case, for example, in Bihar state in India mentioned above. The actual VL incidence documented in community surveys in 2003 was eight and in 2006 five times higher than the official case notification. Because of the intense elimination efforts in this region, access to care improves, and underreporting ratios change. More recent estimates in 2015 (unpublished) indicate that only 20–30% of VL cases in the community are not recorded in the official data. For a correct interpretation of trends in reported cases, it is therefore essential to keep track of the extent of underreporting, possible differences in reporting units/methods over time and between surveillance sites. Underreporting ratios may change over time, mainly depending on availability to and access to diagnosis and treatment.

When a surveillance system functions well, the information it generates should lead to meaningful action. This action can be of different types: surveillance data can help the health authorities to evaluate the real extent of the problem and the main population at risk; they can help improve the clinical management and follow-up of patients and implement active case detection where and when necessary. Surveillance data can also help to identify and solve the technical and operational difficulties faced by the disease control program and facilitate the evaluation of the impact of the intervention.

10.4 Leishmaniasis Control by Region

10.4.1 VL Elimination in Asia

There are two historic examples of successful leishmaniasis control programs in Asia. In China, VL was one of the major parasitic diseases before the creation of the People's Republic in 1949. From the areas north of the Yangtze river, over 530,000 cases were reported in 1951 (Guan and Shen 1991). Between 1950 and 1958, a nationwide control campaign (mass treatment of patients, killing of infected dogs, and use of insecticides) brought the disease under control in the plains where the anthroponotic form had reigned. Transmission could however not be stopped in the mountainous and desert region where sporadic cases of zoonotic transmission continue to be reported until today (Lun et al. 2015). India controlled a large VL epidemic in the 1970s in Bihar through a combination of active case detection and treatment in the community and residual spraying in the houses (Thakur et al. 1994). Unfortunately, the epidemic resurged in the mid-1980s. In 2005 the governments of Bangladesh, Nepal, and India signed a memorandum of understanding to eliminate VL "as a public health problem" by 2015, a target that can be understood as "intense control," as the aim was to reduce annual incidence to below 1 per 10,000 at (sub) district level. The deadline was later postponed to 2017. As explained above, untreated VL and PKDL patients are considered the main reservoir of parasites in this region, hence the emphasis on adequate detection and treatment of all clinical VL and PKDL cases. Insecticide residual spraying (IRS) of houses and animal shelters is the second pillar in the Indian subcontinent where the vector (*P. argentipes*) is restricted to the intra- and peri-domiciliary area.

A substantial decline in VL cases has been observed in recent years. In 2016 only 244 cases were reported from Nepal, 270 from Bangladesh, and 6221 from India (see Fig. 10.2). Nepal has reached the elimination threshold for more than three consecutive years now, while Bangladesh reached it in 2016. In India, some more effort will be required still, as at least 8% of the endemic blocks are still above

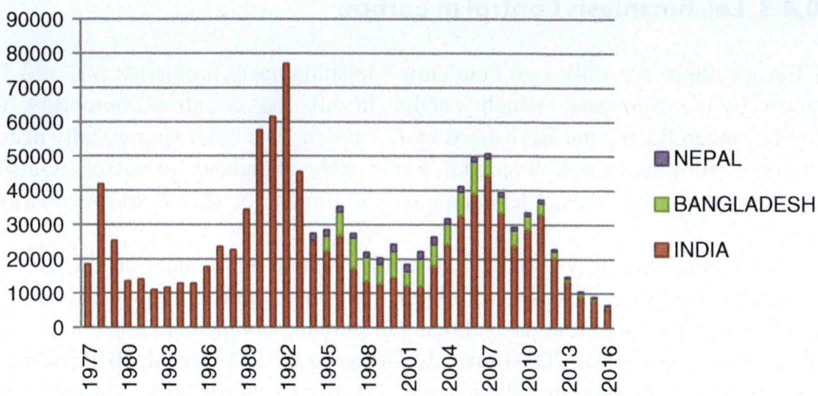

Fig. 10.2 Number of VL cases in India, Nepal, and Bangladesh (1977–2016)

threshold. Additionally, in part due to increasing community awareness and active case detection, the number of reported PKDL cases in India has substantially increased: from 421 cases in 2014 to 1487 in 2016. A further challenge is the emergence of HIV-VL coinfection in the adult population in Bihar that has been signaled at sentinel sites (Burza et al. 2014). Unless adequately addressed, both groups of patients will remain reservoirs of the *Leishmania* parasite and, without appropriate treatment, potential incubators of resistant parasites moving into the post-elimination period.

The major question for the next few years will be how to sustain this achievement and whether a more ambitious objective of zero transmission is technically and operationally feasible. Several knowledge gaps persist that should be addressed if one wants to make a case for regional elimination of transmission: (1) how big is the role of PKDL cases and asymptomatic carriers in maintaining transmission to humans (Singh et al. 2016; Cameron et al. 2016), and (2) is there an animal reservoir (Bhattarai et al. 2010)?

10.4.2 VL Control in East Africa

East Africa is the region that reports most VL cases after the Indian subcontinent, with 15,533 cases reported in 2014 (WHO 2016). Sudan, South Sudan, and Ethiopia bear the highest burden. Civil unrest, forced migration, and famine have led to large epidemics in the past. Between 1984 and 1994, a VL epidemic caused the death of an estimated 100,000 people in the Western Upper Nile province of war-torn Southern-Sudan (Seaman et al. 1996).

VL is thought to be primarily anthroponotic, despite identification of some potential animal reservoirs. So far, the main VL control strategy is through provision of early diagnosis and care. Programs give special attention to PKDL and to HIV-VL coinfection.

10.4.3 Leishmaniasis Control in Europe

In Europe there are only two "endemic" leishmaniasis problems: VL and CL caused by *L. infantum*, which occurs in all the countries bordering the Mediterranean Basin, and CL caused by *L. tropica*. The latter sporadically occurs in Greece, but poses a risk of spreading due to the abundance of vectors. Outside this, many imported cases of leishmaniasis continue to be seen in European travel clinics.

L. infantum, causing VL and CL, is both a human and animal health issue. In humans it causes sporadic cases in children, immune-compromised individuals, and transplant recipients, but larger outbreaks in immune-competent people (e.g., over 500 cases in Madrid since 2009) have also been reported (Arce et al. 2013; Herrador et al. 2015). As observed in Italy, it seems the disease is expanding toward central Europe, probably due to climate change (Maroli et al. 2008).

Canine leishmaniasis is a major veterinary problem with a seroprevalence of 25% in domestic dogs. In Europe, no particular measures are taken at veterinary public health level but insecticide-treated collars and spot-on insecticides are commonly used to protect individual dogs against sand fly bites in endemic areas. For personal protection against bites of peridomestic sand flies, insecticide-treated nets and topically applied insecticides are recommended by public health authorities.

10.4.4 VL Control in Latin America

The control of zoonotic VL due to *L. infantum* in Latin America is one of the most challenging chapters of VL control and has over the years led to disappointing results (Romero and Boelaert 2010). Brazil accounts for more than 90% of the VL disease burden in the continent. While the officially reported data in Brazil show a stable trend of VL at country level for the past decade, a more in-depth analysis demonstrates that this masks a shift in the spatial distribution of disease. VL cases occur in new areas and municipalities while its incidence decreases in others, leading to a misleading picture of apparent stagnation of case numbers at the aggregate level but also posing a major challenge for the national control program.

Since 1953, when VL control was initiated in Brazil using a combination of DDT spraying, treatment of human cases, and selective elimination of seropositive dogs, control programs have—with some rare exceptions—failed to contain the geographical dispersion of the disease. From being primarily confined to rural areas, the disease spread to the peri-urban areas of medium- and large-size cities (Costa 2011). Until the 1990s, VL was considered a public health problem only in Brazil, but more recently, the rapid expansion to Paraguay and Argentina and the increasing number of cases observed in Central America and Colombia point to the failure of traditional control strategies. This worrisome evolution has triggered the pursuit of innovative and more efficient control approaches (PAHO 2016; WHO 2016).

Today, point-of-care diagnostic test for early detection of human cases and access to free treatment are available for most of the affected population in Brazil and Colombia, but this is not always the case in Paraguay and Central American countries. Despite reasonably wide access to diagnosis and treatment, the case fatality rate of VL remains very high in Brazil, with some facility-based data reporting mortality as high as 10%. This phenomenon has been partially attributed to HIV coinfection, but other factors such as late presentation, undernutrition, and specific communities with poor access to adequate diagnosis and treatment are likely to contribute to these poor outcomes and remain areas on which public health efforts should be focused.

As dogs are the main reservoir of VL in Latin America, most of the control efforts involve culling of seropositive dogs, and spraying of pyrethroid has been applied less intensively. However, none of the currently applied control interventions are based on strong scientific evidence, and there are no consistent data about the impact of the current control strategies on human and canine incidence rates

(Gonzalez et al. 2015). Mathematical modeling of VL control in the Americas concluded that culling seropositive dogs would be less effective than vector control, presuming that the insecticide effect would be sustained (Dye 1996). Such a premise is hard to achieve after banning DDT because of the short residual effect of pyrethroids (Feliciangeli et al. 2003a).

Recently, mathematical modeling based on real-world data of the (imperfect) routine dog culling program demonstrated that this intervention leads to different outcomes, depending on canine infection rates and intervention coverage. This raises reasonable doubts about the impact of the current control measures (Costa 2011). Moreover, the VL control strategy in Brazil has come under fierce criticism by the target population because of the poor acceptability of dog culling and the discomfort caused by indoor insecticide spraying. Health officials responsible for conducting the control activities in the field recognized the high failure rate and the weak coverage related to its inherent complexity and high costs of the control interventions (Zuben and Donalisio 2016). Such examples of low acceptability of IRS are not limited to Brazil; poor community uptake has been seen across all contexts where IRS is an important control strategy.

As there is no optimal solution to control, the efforts for vaccine development and alternative approaches to vector control are critical, but again, so far any consistent efficacy/effectiveness data about the effect of current vaccines or insecticide-impregnated dog collars on *L. infantum* transmission to humans has yet to be seen.

Meanwhile, caring for the approximately 4000 VL human cases annually reported from Latin America requires the development of a strong strategy focused on improving access to and quality of care in order to drastically reduce the case fatality rate, including the strengthening of clinical research with special emphasis on drug development, implementing effective point-of-care diagnosis, and developing and improving tools for early detection of cases with poor prognosis (Belo et al. 2014; Romero 2016).

10.4.5 CL Control in North Africa and Middle East

In North Africa and the Middle East both zoonotic CL, caused predominantly by *L. major*, and anthroponotic CL, caused by *L. tropica*, are widespread (Hotez et al. 2012). The largest number of *L. major* cases occurs in arid areas of Iran, Saudi Arabia, Morocco, Tunisia, Syria, Libya, and Iraq, transmitted by the sand fly *P. papatasi* or a closely related species (Hotez et al. 2012) and maintained in a rodent reservoir, such as jirds, fat sand rats, or great gerbils. The burrows of these rodents are the breeding sites: the larvae thrive in the moist soil and the adults use the burrows as their daytime resting sites. According to the literature, the most effective way to control *L. major* is to combine reservoir and vector control, but there is scanty evidence on the effectiveness of this approach (Reithinger et al. 2007). Guidelines for leishmaniasis control published by WHO in 1988 suggested placing poisoned baits in burrow entrances to control the foci where *Meriones* species are the reservoir host (World Health Organization 1990).

10.4.6 CL Control in Central Asia

In Central Asian regions where sand flies *P. papatasi* and *P. duboscqi* breed in rodent burrows, the destruction of these burrows, combined with both reservoir and vector elimination, is suggested as control method (Saf'janova 1971). In the former Soviet Union the destruction of the rodent burrows was used to control *P. papatasi* between the 1940s and 1980s (Lane 1991).

10.4.7 Cutaneous and Mucocutaneous Leishmaniasis Control in Latin America

In Latin America the multiplicity of sand fly species, *Leishmania* species, vertebrate reservoirs, and ecological niches where CL and the rare mucocutaneous form of the disease (MCL) occur have been a challenge for planning and executing control interventions. The origin of MCL is not well understood, precluding the development of effective measures to prevent this severe form of the disease. Adequate treatment of CL with systemically applied drugs has been proposed as a way to prevent MCL, but until now, there is no consistent evidence to support that hypothesis (Blum et al. 2012). In most countries, the prevention measures are limited to health education messages for reducing exposure to sand fly bites. No current intervention has been proven effective against any of these clinical forms, and case detection and treatment of cases continue to be the only control strategy. Indoor residual spraying with insecticides could be useful in some specific econiches (Davies et al. 2000).

References

Adhikari SR, Maskay NM. The economic burden of Kala-azar in households of the Danusha and Mahottari districts of Nepal. Acta Trop. 2003;88:1–2.

Adhikari SR, Maskay NM, Sharma BP. Paying for hospital-based care of Kala-azar in Nepal: assessing catastrophic, impoverishment and economic consequences. Health Policy Plan. 2009;24:129–39.

Ali A, Ashford RW. Visceral leishmaniasis in Ethiopia. III. The magnitude and annual incidence of infection, as measured by serology in an endemic area. Ann Trop Med Parasitol. 1994a;88:43–7.

Ali A, Ashford RW. Visceral leishmaniasis in Ethiopia. IV. Prevalence, incidence and relation of infection to disease in an endemic area. Ann Trop Med Parasitol. 1994b;88:289–93.

Alten B, Caglar SS, Kaynas S, Simsek FM. Evaluation of protective efficacy of K-OTAB impregnated bednets for cutaneous leishmaniasis control in Southeast Anatolia-Turkey. J Vector Ecol. 2003;28:53–64.

Alvar J, Velez ID, Bern C, Herrero M, Desjeux P, Cano J, Jannin J, Den BM. Leishmaniasis worldwide and global estimates of its incidence. PLoS One. 2012;7:e35671.

Anoopa Sharma D, Bern C, Varghese B, Chowdhury R, Haque R, Ali M, Amann J, Ahluwalia IB, Wagatsuma Y, Breiman RF, Maguire JH, Mcfarland DA. The economic impact of visceral leishmaniasis on households in Bangladesh. Trop Med Int Health. 2006;11:757–64.

Arce A, Estirado A, Ordobas M, Sevilla S, Garcia N, Moratilla L, de la Fuente S, Martinez AM, Perez AM, Aranguez E, Iriso A, Sevillano O, Bernal J, Vilas F. Re-emergence of leishmaniasis in Spain: community outbreak in Madrid, Spain, 2009 to 2012. Euro Surveill. 2013;18:20546.

Ashford DA, Desjeux P, Deraadt P. Estimation of population at risk of infection and number of cases of leishmaniasis. Parasitol Today. 1992;8:104–5.

Belo VS, Struchiner CJ, Barbosa DS, Nascimento BW, Horta MA, da Silva ES, Werneck GL. Risk factors for adverse prognosis and death in American visceral leishmaniasis: a meta-analysis. PLoS Negl Trop Dis. 2014;8:e2982.

Benzerroug EH, Benhabylles N, Izri MA, Belahcene EK. Les pulverisations intra- et peri-domiciliaires de DDT dans la lutte contre la leishmaniose cutanee zoonotique en Algerie. Ann Soc Belg Med Trop. 1992;72:5–12.

Bern C, Joshi AB, Jha SN, Das ML, Hightower AW, Thakur GD, Bista MB. Factors associated with visceral leishmaniasis in Nepal: bed-net use is strongly protective. Am J Trop Med Hyg. 2000;63:184–8.

Bern C, Hightower AW, Chowdhury R, Ali M, Amann J, Wagatsuma Y, Haque R, Kurkjian K, Vaz LE, Begum M, Akter T, Cetre-Sossah CB, Ahluwalia IB, Dotson E, Secor WE, Breiman RF, Maguire JH. Risk factors for Kala-Azar in Bangladesh. Emerg Infect Dis. 2005;11:655–62.

Bhattarai NR, Van der Auwera G, Rijal S, Picado A, Speybroeck N, Khanal B, De Doncker S, Das ML, Ostyn B, Davies C, Coosemans M, Berkvens D, Boelaert M, Dujardin JC. Domestic animals and epidemiology of visceral leishmaniasis, Nepal. Emerg Infect Dis. 2010;16:231–7.

Blum J, Lockwood DN, Visser L, Harms G, Bailey MS, Caumes E, Clerinx J, van Thiel PP, Morizot G, Hatz C, Buffet P. Local or systemic treatment for new world cutaneous leishmaniasis? Re-evaluating the evidence for the risk of mucosal leishmaniasis. Int Health. 2012;4:153–63.

Boelaert M, Meheus F, Sanchez A, Singh SP, Vanlerberghe V, Picado A, Meessen B, Sundar S. The poorest of the poor: a poverty appraisal of households affected by visceral leishmaniasis in Bihar, India. Trop Med Int Health. 2009;14:639–44.

Burza S, Mahajan R, Sanz MG, Sunyoto T, Kumar R, Mitra G, Lima MA. HIV and visceral leishmaniasis coinfection in Bihar, India: an underrecognized and underdiagnosed threat against elimination. Clin Infect Dis. 2014;59:552–5.

Cameron MM, Acosta-Serrano A, Bern C, Boelaert M, den Boer M, Burza S, Chapman LA, Chaskopoulou A, Coleman M, Courtenay O, Croft S, Das P, Dilger E, Foster G, Garlapati R, Haines L, Harris A, Hemingway J, Hollingsworth TD, Jervis S, Medley G, Miles M, Paine M, Picado A, Poche R, Ready P, Rogers M, Rowland M, Sundar S, DE Vlas SJ, Weetman D. Understanding the transmission dynamics of Leishmania donovani to provide robust evidence for interventions to eliminate visceral leishmaniasis in Bihar, India. Parasit Vectors. 2016;9:25.

Coleman M, Foster GM, Deb R, Pratap Singh R, Ismail HM, Shivam P, Ghosh AK, Dunkley S, Kumar V, Coleman M, Hemingway J, Paine MJ, Das P. DDT-based indoor residual spraying suboptimal for visceral leishmaniasis elimination in India. Proc Natl Acad Sci U S A. 2015;112:8573–8.

Collin S, Davidson R, Ritmeijer K, Keus K, Melaku Y, Kipngetich S, Davies C. Conflict and kala-azar: determinants of adverse outcomes of kala-azar among patients in southern Sudan. Clin Infect Dis. 2004;38:612–9.

Costa CH. How effective is dog culling in controlling zoonotic visceral leishmaniasis? A critical evaluation of the science, politics and ethics behind this public health policy. Rev Soc Bras Med Trop. 2011;44:232–42.

Cunningham J, Hasker E, Das P, El Safi S, Goto H, Mondal D, Mbuchi M, Mukhtar M, Rabello A, Rijal A, Sundar S, Wasunna M, Adams E, Menten J, Peeling R, Boelaert M. A global comparative evaluation of commercial immunochromatographic rapid diagnostic tests for visceral leishmaniasis. Clin Infect Dis. 2012;55:1312–9.

Davies CR, Llanos-Cuentas EA, Campos P, Monge J, Leon E, Canales J. Spraying houses in the Peruvian Andes with lambda-cyhalothrin protects residents against cutaneous leishmaniasis. Trans R Soc Trop Med Hyg. 2000;94:631–6.

Desjeux P. Human leishmaniasis: epidemiology and public health aspects. World Health Stat Q. 1992;45:267–75.

Dujardin JC, Campino L, Canavate MC, Dedet JP, Gradoni L, Soteriadou K, Mazeris A, Ozbel Y, Boelaert M. Spread of vector-borne diseases and neglect of leishmaniasis, Europe. Emerg Infect Dis. 2008;14:1013–8.

Duthie MS, Raman VS, Piazza FM, Reed SG. The development and clinical evaluation of second-generation leishmaniasis vaccines. Vaccine. 2012;30:134–41.

Dye C. The logic of visceral leishmaniasis control. Am J Trop Med Hyg. 1996;55:125–30.

Falcao AL, Falcao AR, Pinto CT, Gontijo CM, Falqueto A. Effect of deltamethrin spraying on the sandfly populations in a focus of American cutaneous leishmaniasis. Mem Inst Oswaldo Cruz. 1991;86:399–404.

Faraj C, Yukich J, Adlaoui EB, Wahabi R, Mnzava AP, Kaddaf M, El Idrissi AL, Ameur B, Kleinschmidt I. Effectiveness and cost of insecticide-treated bed nets and indoor residual spraying for the control of cutaneous leishmaniasis: a cluster-randomized control trial in Morocco. Am J Trop Med Hyg. 2016;94:679–85.

Feliciangeli MD, Mazzarri MB, Blas SS, Zerpa O. Control trial of Lutzomyia longipalpis s.l. in the Island of Margarita, Venezuela. Trop Med Int Health. 2003a;8:1131–6.

Feliciangeli MD, Mazzarri MB, Campbell-Lendrum DH, Maroli M, Maingon R. Cutaneous leishmaniasis vector control perspectives using lambdacyhalothrin residual house spraying in El Ingenio, Miranda State, Venezuela. Trans R Soc Trop Med Hyg. 2003b;97:641–6.

Gavgani AS, Hodjati MH, Mohite H, Davies CR. Effect of insecticide-impregnated dog collars on incidence of zoonotic visceral leishmaniasis in Iranian children: a matched-cluster randomised trial. Lancet. 2002;360:374–9.

Gonzalez U, Pinart M, Sinclair D, Firooz A, Enk C, Velez ID, Esterhuizen TM, Tristan M, Alvar J. Vector and reservoir control for preventing leishmaniasis. Cochrane Database Syst Rev. 2015;5(8):CD008736.

Guan LR, Shen WX. Recent advances in visceral leishmaniasis in China. Southeast Asian J Trop Med Public Health. 1991;22:291–8.

Gunay F, Karakus M, Oguz G, Dogan M, Karakaya Y, Ergan G, Kaynas S, Kasap OE, Ozbel Y, Alten B. Evaluation of the efficacy of Olyset® plus in a village-based cohort study in the Cukurova plain, Turkey, in an area of hyperendemic cutaneous leishmaniasis. J Vector Ecol. 2014;39:395–405.

Herrador Z, Gherasim A, Jimenez BC, Granados M, San Martin JV, Aparicio P. Epidemiological changes in leishmaniasis in Spain according to hospitalization-based records, 1997–2011: raising awareness towards leishmaniasis in non-HIV patients. PLoS Negl Trop Dis. 2015;9:e0003594.

Hotez PJ, Savioli L, Fenwick A. Neglected tropical diseases of the Middle East and North Africa: review of their prevalence, distribution, and opportunities for control. PLoS Negl Trop Dis. 2012;6:e1475.

Ingenloff K, Garlapati R, Poche D, Singh MI, Remmers JL, Poche RM. Feed-through insecticides for the control of the sand fly Phlebotomus argentipes. Med Vet Entomol. 2013;27:10–8.

Jalouk L, Al Ahmed M, Gradoni L, Maroli M. Insecticide-treated bednets to prevent anthroponotic cutaneous leishmaniasis in Aleppo governorate, Syria: results from two trials. Trans R Soc Trop Med Hyg. 2007;101:360–7.

Jirmanus L, Glesby MJ, Guimaraes LH, Lago E, Rosa ME, Machado PR, Carvalho EM. Epidemiological and clinical changes in American tegumentary leishmaniasis in an area of Leishmania (Viannia) braziliensis transmission over a 20-year period. Am J Trop Med Hyg. 2012;86:426–33.

Joshi AB, Bhatt LR, Regmi S, Ashford RW. An assessment of the effectiveness of insecticide spray in the control of *Visceral leishmaniasis* in Nepal. J Nepal Health Res Counc. 2003;1:1–6.

Kishore K, Kumar V, Kesari S, Dinesh DS, Kumar AJ, Das P, Bhattacharya SK. Vector control in leishmaniasis. Indian J Med Res. 2006;123:467–72.

Kumar V, Shankar L, Kesari S, Bhunia GS, Dinesh DS, Mandal R, Das P. Insecticide susceptibility of Phlebotomus argentipes & assessment of vector control in two districts of West Bengal, India. Indian J Med Res. 2015;142:211–5.

Lane RP. The contribution of sandfly control to leishmaniasis control. Ann Soc Belg Med Trop. 1991;71(Suppl 1):65–74.

Le Pont F, Padilla JM, Desjeux P, Richard A, Mouchet J. [Impact of the spraying of deltamethrin in a focus of leishmaniasis in Bolivia] Impact de pulverisations de deltamethrine dans un foyer de leishmaniose de Bolivie. Ann Soc Belg Med Trop. 1989;69:223–32.

Lun ZR, Wu MS, Chen YF, Wang JY, Zhou XN, Liao LF, Chen JP, Chow LM, Chang KP. Visceral leishmaniasis in China: an endemic disease under control. Clin Microbiol Rev. 2015;28:987–1004.

Maia C, Nunes M, Marques M, Henriques S, Rolao N, Campino L. In vitro drug susceptibility of Leishmania infantum isolated from humans and dogs. Exp Parasitol. 2013;135:36–41.

Maroli M, Mizzoni V, Siragusa C, D'orazi A, Gradoni L. Evidence for an impact on the incidence of canine leishmaniasis by the mass use of deltamethrin-impregnated dog collars in southern Italy. Med Vet Entomol. 2001;15:358–63.

Maroli M, Rossi L, Baldelli R, Capelli G, Ferroglio E, Genchi C, Gramiccia M, Mortarino M, Pietrobelli M, Gradoni L. The northward spread of leishmaniasis in Italy: evidence from retrospective and ongoing studies on the canine reservoir and phlebotomine vectors. Tropical Med Int Health. 2008;13:256–64.

Mascari TM, Stout RW, Foil LD. Laboratory evaluation of oral treatment of rodents with systemic insecticides for control of bloodfeeding sand flies (Diptera: Psychodidae). Vector Borne Zoonotic Dis. 2012;12:699–704.

Meheus F, Boelaert M, Baltussen R, Sundar S. Costs of patient management of visceral leishmaniasis in Muzaffarpur, Bihar, India. Trop Med Int Health. 2006;11:1715–24.

Molyneux DH, Hotez PJ, Fenwick A. "Rapid-impact interventions": how a policy of integrated control for Africa's neglected tropical diseases could benefit the poor. PLoS Med. 2005;2:e336.

Nadim A, Motabar B, Houshmand B, Keyghobadi K, Aflatonian M. Evaluation of pyrethroid impregnated bednets for control of anthroponotic cutanious leishmaniasis in BAM (Islamic Republic of Iran). Geneva: World Health Organization; 2000.

Noazin S, Khamesipour A, Moulton LH, Tanner M, Nasseri K, Modabber F, Sharifi I, Khalil EA, Bernal ID, Antunes CM, Smith PG. Efficacy of killed whole-parasite vaccines in the prevention of leishmaniasis: a meta-analysis. Vaccine. 2009;27:4747–53.

Nunes CM, Lima VM, Paula HB, Perri SH, Andrade AM, Dias FE, Burattini MN. Dog culling and replacement in an area endemic for visceral leishmaniasis in Brazil. Vet Parasitol. 2008;153:19–23.

Ostyn B, Vanlerberghe V, Picado A, Dinesh DS, Sundar S, Chappuis F, Rijal S, Dujardin JC, Coosemans M, Boelaert M, Davies C. Vector control by insecticide-treated nets in the fight against visceral leishmaniasis in the Indian subcontinent, what is the evidence? Trop Med Int Health. 2008;13:1073–85.

PAHO. RePLeish. 3a Reunion de los Programas Nacionales de Leishmaniasis de los Paises prioritarios de las Américas; 2016. http://www.paho.org/hq/index.php?option=com_docman&task=doc_download&Itemid=&gid=34537&lang=pt: Panamerican Health Organization.

Picado A, Das ML, Kumar V, Kesari S, Dinesh D, Roy L, Rijal S, Das P, Rowland M, Sundar S, Coosemans M, Boelaert M, Davies C. Effect of village-wide use of long-lasting insecticidal nets on visceral leishmaniasis vectors in India and Nepal: a cluster randomized trial. PLoS Negl Trop Dis. 2010a;4:e587.

Picado A, Singh SP, Rijal S, Sundar S, Ostyn B, Chappuis F, Uranw S, Gidwani K, Khanal B, Rai M, Paudel IS, Das ML, Kumar R, Srivastava P, Dujardin JC, Vanlerberghe V, Andersen EW, Davies CR, Boelaert M. Longlasting insecticidal nets for prevention of Leishmania donovani infection in India and Nepal: paired cluster randomised trial. BMJ. 2010b;341:c6760.

Reithinger R, Dujardin JC, Louzir H, Pirmez C, Alexander B, Brooker S. Cutaneous leishmaniasis. Lancet Infect Dis. 2007;7:581–96.

Reyburn H, Ashford R, Mohsen M, Hewitt S, Rowland M. A randomized controlled trial of insecticide-treated bednets and chaddars or top sheets, and residual spraying of interior rooms for the prevention of cutaneous leishmaniasis in Kabul, Afghanistan. Trans R Soc Trop Med Hyg. 2000;94:361–6.

Rijal S, Koirala S, Van der Stuyft P, Boelaert M. The economic burden of visceral leishmaniasis for households in Nepal. Trans R Soc Trop Med Hyg. 2006;100:838–41.

Romero GAS. O controle de leishmaniose visceral no Brasil: transformar é preciso. Cad. Saúde Pública. 2016;32(6), eCO010616. Available from https://doi.org/10.1590/0102-311XCO01061.

Romero GA, Boelaert M. Control of visceral leishmaniasis in latin america-a systematic review. PLoS Negl Trop Dis. 2010;4:e584.

Saf'janova VM. Leishmaniasis control. Bull World Health Organ. 1971;44:561–6.

Sarnoff R, Desai J, Desjeux P, Mittal A, Topno R, Siddiqui NA, Pandey A, Sur D, Das P. The economic impact of visceral leishmaniasis on rural households in one endemic district of Bihar, India. Trop Med Int Health. 2010;15(Suppl 2):42–9.

Schaefer KU, Kurtzhals JA, Gachihi GS, Muller AS, Kager PA. A prospective sero-epidemiological study of visceral leishmaniasis in Baringo District, Rift Valley Province, Kenya. Trans R Soc Trop Med Hyg. 1995;89:471–5.

Seaman J, Mercer AJ, Sondorp E. The epidemic of visceral leishmaniasis in western Upper Nile, southern Sudan: course and impact from 1984 to 1994. Int J Epidemiol. 1996;25:862–71.

Singh OP, Hasker E, Boelaert M, Sundar S. Elimination of visceral leishmaniasis on the Indian subcontinent. Lancet Infect Dis. 2016;16:e304–9.

Stauch A, Sarkar RR, Picado A, Ostyn B, Sundar S, Rijal S, Boelaert M, Dujardin JC, Duerr HP. Visceral leishmaniasis in the Indian subcontinent: modelling epidemiology and control. PLoS Negl Trop Dis. 2011;5:e1405.

Sundar S, Reed SG, Singh VP, Kumar PC, Murray HW. Rapid accurate field diagnosis of Indian visceral leishmaniasis. Lancet. 1998;351:563–5.

Sundar S, Arora R, Singh SP, Boelaert M, Varghese B. Household cost-of-illness of visceral leishmaniasis in Bihar, India. Trop Med Int Health. 2010;15(Suppl 2):50–4.

Tayeh A, Jalouk L, Al-Kahaimi AM. A cutaneous leishmaniasis control trial using pyrethroid-impregnated bednets in villages near Aleppo, Syria; 1997. p 42.

Thakur CP. Socio-economics of visceral leishmaniasis in Bihar (India). Trans R Soc Trop Med Hyg. 2000;94:156–7.

Thakur CP, Francis S, Therese S, Victoria S, Puspa S. A kala-azar control programme for remote tribal communities. World Health Forum. 1994;15:245–7.

WHO. Control of the leishmaniases: report of a meeting of the WHO Expert Committee on the control of leishmaniases. Geneva: World Health Organization, 22–26 March 2010. WHO Technical Report Series; no. 949; 2010.

WHO. Wkly Epidemiol Rec. 2016;22:285–96.

World Health Organization. Control of leishmaniasis. Report of a WHO Expert Committee. 2012. London Declaration on Neglected Tropical Diseases; 1990. [Online]. [Accessed].

Zuben AP, Donalisio MR. Difficulties in implementing the guidelines of the Brazilian Visceral Leishmaniasis Control Program in large cities. Cad Saude Publica. 2016;32. https://doi.org/10.1590/0102-311X00087415.

The manufacturer's authorised representative in the EU is Springer Nature Customer Service Centre GmbH, Europaplatz 3, 69115 Heidelberg, Germany. If you have any concerns regarding our products, please contact ProductSafety@springernature.com

Printed and bound by CPI Group (UK) Ltd, Croydon, CR0 4YY
01/10/2026
02225866-0002